J. Manning

MICROBE HUNTERS —THEN AND NOW

EDITED BY

HILARY KOPROWSKI, M.D.

Professor, Department of Microbiology and Immunology
Director, Center for Neurovirology
Thomas Jefferson University
Philadelphia, Pennsylvania

MICHAEL B.A. OLDSTONE, M.D.

Professor, Department of Neuropharmacology
Head, Division of Virology
The Scripps Research Institute
La Jolla, California

MEDI-ED PRESS
BLOOMINGTON, ILLINOIS

Library of Congress Cataloging-in-Publication Data
Microbe hunters, then and now / edited by Hilary Koprowski,
  Michael B.A. Oldstone
      454 p.   cm.
      Includes bibliographical references and index.
      ISBN 0-936741-11-2
      1. Medical microbiology--History. I. Koprowski, Hilary.
  II. Oldstone, Michael B.A.
      [DNLM: 1. Microbiology--history. 2. Research Personnel. 3. Virus
  Diseases--prevention & control. QW 11.1 M626 1996]
  QR46.M5365 1996
  616'.01'09--dc20
  DNLM/DLC
  for Library of Congress   96-28898
                                                                            CIP

MEDI-ED PRESS
Constitution Place, Suite A
716 East Empire Street
Bloomington, Illinois 61701
1-800-500-8205

MICROBE HUNTERS —THEN AND NOW

Contents

PLANT VIRUSES

BACTERIA

PARASITES

OTHER ISSUES

Biographical Data on the Editors

HILARY KOPROWSKI, M.D.

Hilary Koprowski is currently Professor of the Department of Microbiology and Immunology, Director of the Center for Neurovirology and Director of the Biotechnology Foundation Laboratories at Thomas Jefferson University in Philadelphia..

He is a member of the American Academy of Arts and Sciences, the National Academy of Sciences, the New York Academy of Sciences and twenty-eight other learned institutions. He has published more than 800 scientific papers and is recipient of more than eighteen awards including the Philadelphia Award and Scott Award.

Dr. Koprowski was born in Poland and received an MD degree from Warsaw University in 1939. He worked in Italy and Brazil before coming to the United States in 1944, where he started his work at Lederle Laboratories. As Associate Director, he initially worked on viral encephalitis. He soon became interested in research on rabies and vaccines against poliomyelitis. In 1950, specific antibodies from his polio vaccine were developed in a child and this was repeated with 20 more children who also developed immunity. The vaccine was successfully used in the firstmass vaccination in Africa and later in Poland and Croatia. In 1957, he assumed the directorship of the Wistar Institute, which he built into a world-class biomedical research organization.

Among his many other major achievements were development of monoclonal antibodies for the detection of cancer cells and the first cure of pancreatic and colon cancers. He is actively engaged in research work on multiple sclerosis and continues his work in the field of neurotropic viruses.

MICHAEL B.A. OLDSTONE, M.D.

Michael B.A. Oldstone is a member (professor) at the Scripps Research Institute, La Jolla, California. He is Head of the Division of Virology in the Department of Neuropharmacology and directs a research laboratory studying viral pathogenesis. He is currently a Scientific Counselor for the National Institute of Allergy and Infectious Disease, a member of the World

Health Organization Scientific Advisory Group of Experts (for infectious diseases), and Editor of Virology.

Dr. Oldstone received an MD degree from the University of Maryland School of Medicine in Baltimore in 1961 and participated in the MD/PhD Program at Johns Hopkins McCullom Pratt Institute of Biochemistry, Department of Microbiology. After completing two residencies, one in medicine and a second in neurology, he was appointed to a postdoctoral research fellowship at the Laboratory of Frank Dixon, Department of Experimental Pathology, at Scripps Clinic and Research Foundation at La Jolla, California, where he remained on staff in the Department of Experimental Pathology and later in the Department of Immunopathology.

Dr. Oldstone is the recipient of many honors and awards, including the Rous-Whipple Award for Research Excellence in Investigative Pathology, American Association of Pathologists, FASEB; the Kotz Award and Lecturer, Visiting Professor of Neuroscience, NINDS; the National Institutes of Health Visiting Scholar Award (NIDR); the Abraham Flexner Award and Lectureships for Contributions in Biomedical Research, Flexner Foundation, Vanderbilt University; and The Cotzias Award and Lecture for Research Excellence in Nervous System Disease, American Academy of Neurology.

He is a leader in the field of viral pathogenesis. Dr. Oldstone has published numerous scientific papers and articles in the field of virology and immunology and has directed the training of more than 50 fellows and students.

Contributors

EDITORS

HILARY KOPROWSKI, M.D.

Professor, Department of Microbiology and Immunology, Director, Center for Neurovirology, Thomas Jefferson University, 1020 Locust Street, Jefferson Alumni Hall, Room M-85, Philadelphia, Pennsylvania 19107, USA. Telephone: (215) 503-4761; Telefax: (215) 923-6786.

MICHAEL B. A. OLDSTONE, M.D.

Department of Neuropharmacology, Division of Virology, The Scripps Research Institute, 10550 N. Torrey Pines Road, La Jolla, California 92037, USA. Telephone: (619) 554-8054; Telefax: (619) 554-9981.

CONTRIBUTING AUTHORS

ALAN G. BARBOUR, M.D.

Departments of Microbiology and Molecular Genetics and Medicine, University of California, Irvine, College of Medicine, Irvine, California 92717 USA.

ROGER N. BEACHY

Division of Plant Biology, Department of Cell Biology, The Scripps Research Institute, 10666 North Torrey Pines Road, La Jolla, California 92037, USA. Telephone: (619) 554-2550; Telefax: (619) 554-6188.

NICO A. BOS

Department of Histology and Cell Biology, University of Groningen, Groningen, The Netherlands.

ETHEL R. CEBRA

Department of Biology, University of Pennsylvania, Leidy Labs, Kaplan Wing, Philadelphia, Pennsylvania 19104 USA. Telephone: (215) 898-5599; Telefax: (215) 898-9786.

JOHN J. CEBRA

Department of Biology, University of Pennsylvania, Leidy Labs, Kaplan Wing, Philadelphia, Pennsylvania 19104, USA. Telephone: (215) 898-5599; Telefax: (215) 898-9786.

BERNHARD DIETZSCHOLD, D.V.M.

Center for Neurovirology, Department of Microbiology and Immunology, Thomas Jefferson University, Jefferson Alumni Hall, Room 455, 1020 Locust Street, Philadelphia, Pennsylvania 19107 USA. Telephone: (215) 503-4695; Telefax: (215) 923-7145.

FRANK FENNER

The John Curtin School of Medical Research, Australian National University, GPO Box 334, Canberra, Australian Capital Territory 2601, Australia. Telephone: (61) 6 249 2526; Telefax: (61) 6 247 4823; E mail: fenner @jcsmr.anu.edu.au.

JOHN H. FITCHEN

Division of Plant Biology, Department of Cell Biology, The Scripps Research Institute, 10666 North Torrey Pines Road, La Jolla, California 92037 USA. Telephone: (619) 554-2550; Telefax: (619) 554-6188.

TERYL K. FREY, PH.D.

Department of Biology, Georgia State University, University Plaza, Atlanta, Georgia 30303, USA. Telephone: (404) 651-1938; Telefax: (404) 651-2509; E mail: tfrey@gsu.edu.

ZHEN FANG FU, D.V.M., PH.D.

Center for Neurovirology, Department of Microbiology and Immunology, Thomas Jefferson University, Jefferson Alumni Hall, Room 455, 1020 Locust Street, Philadelphia, Pennsylvania 19107, USA. Telephone: (215) 503-4695; Telefax: (215) 923-7145.

ROBERT C. GALLO, M.D.

University of Maryland, Institute for Human Virology, Medical Biotechnology Center, University of Maryland at Baltimore, 725 W. Lombard St., 2nd Floor, Baltimore, Maryland 21201, USA.

ANNE A. GERSHON

Department of Pediatrics, Columbia University College of Physicians & Surgeons, Black Building 4-427, 650 W. 168th Street, New York, New York 10032, USA. Telephone: (212) 305-1556; Telefax: (212) 305-2284.

D.A. HENDERSON, M.D., M.P.H.

University Distinguished Service Professor, The Johns Hopkins University, Baltimore, Maryland 21205, USA. Telephone: (410) 955-9489; Telefax: (410) 889-6514.

MAURICE R. HILLEMAN, PH.D., D.SC.

Director, Merck Institute of Therapeutic Research, Merck Research Laboratories, UM4-1, West Point, Pennsylvania 19486, USA. Telephone: (215) 652-8913; Telefax: (215) 652-2154.

MARTIN M. KAPLAN

Consultant, WHO, and Director of Geneva Pugwash Conferences on Science and World Affairs, 69, rue de Lausanne, 1202 Geneva, Switzerland. Telephone: 41-22 906 16 51; Telefax: 41 22 731 0194; E mail: pugwash@hei.unige.ch.

SAMUEL L. KATZ, M.D.

Wilburt Cornell Davison Professor of Pediatrics, Division of Infectious Diseases, Department of Pediatrics, Duke University School of Medicine, Box 2925, Durham, North Carolina 27710, USA. Telephone: (919) 684-3734; Telefax: (919) 681-8934; E mail: katz0004@mc.duke.edu

EDWIN D. KILBOURNE, M.D.

Professor, Department of Microbiology & Immunology, New York Medical College, Basic Science Building, Room 315, Valhalla, New York 10595, USA. Telephone: (914) 993-4193; Telefax: (914) 993-4176.

SAUL KRUGMAN, M.D. (DECEASED)

Professor of Pediatrics, New York University Medical Center, New York, New York.

PHILIP LARUSSA

Department of Pediatrics, Columbia University College of Physicians & Surgeons, Black Building 4-427, 650 W. 168th Street, New York, New York 10032, USA. Telephone: (212) 305-1556; Telefax: (212) 305-2284.

B.W.J. MAHY, PH.D., SC.D.

Director, Division of Viral and Rickettsial Diseases, National Center for Infectious Diseases, Centers for Disease Control and Prevention, Building 6, Room 110 MSA30, 1600 Clifton Road, Atlanta, Georgia 30333, USA. Telephone: (404) 639-3574; Telefax: (404) 639-3163.

MARIANNE MANCHESTER

Department of Neuropharmacology, Division of Virology, The Scripps Research Institute, 10550 N. Torrey Pines Road, La Jolla, California 92037, USA. Telephone: (619) 554-8054; Telefax: (619) 554-9981.

FRANCOIS X. MESLIN

Chief, Veterinary Public Health, WHO Av. Appia, 1211 Geneva 27, Switzerland. Telephone: 41-022 701 25 75; Telefax: 41 022 731 48 95; E mail: meslin@who.ch..

JOHN B. MOLONEY

6814 Greyswood Road, Bethesda, Maryland 20817, USA. Telephone: (301) 365-1503; Telefax: (309) 365-1503.

THOMAS P. MONATH, M.D.

Vice President, Research & Medical Affairs, OraVax Inc., 230 Albany Street, Cambridge, Massachusetts 02139, USA. Telephone: (617) 494-1339; Telefax: (617) 494-8872.

MONIQUE MOREAU, PH.D.

Pasteur Mérieux Sérums & Vaccins, Research Development, Bacteriology, 1541, avenue Marcel Mérieux, 69280 Marcy l'Etoile Cédex, France. Telephone: 33-(78) 87 33 91; Telefax: 33-(78) 87-39-58.

AKIO NOMOTO, PH.D.

Professor, Department of Microbiology, Institute of Medical Science, The University of Tokyo, 4-6-1 Shirokanedai, Minato-ku, Tokyo 108, Japan.

ENZO PAOLETTI

Virogenetics Corporation, 465 Jordan Road, Troy, New York 12180 USA.

C.J. PETERS

Division of Viral and Rickettsial Diseases, National Center for Infectious Diseases, Centers for Disease Control and Prevention, Building 6, Room 110 MSA30, 1600 Clifton Road, Atlanta, Georgia 30333 USA. Telephone: (404) 639-3574; Telefax: (404) 639-3163.

STANLEY B. PRUSINER

Departments of Neurology and Biochemistry and Biophysics, University of California, San Francisco, California. Address correspondence to: Department of Neurology, HSE-781, University of California, San Francisco, California 94143-0518, USA. Telephone: (415) 476-4482; Telefax: (415) 476-8386.

FRIEDRICH SCHEIFLINGER

Staff Scientist, IMMUNO AG, Uferstrasse 15, A-2304 Orth/Donau, Austria. Telephone: 43 2212-2701; Telefax: 43 2212-2716.

KHUSHROO E. SHROFF

Department of Biology, University of Pennsylvania, Leidy Labs, Kaplan Wing, Philadelphia, Pennsylvania 19104 USA. Telephone: (215) 898-5599; Telefax: (215) 898-9786.

THOMAS M. SHINNICK, PH.D.

Chief, Immunology and Molecular Pathogenesis Section, Tuberculosis/Mycobacteriology Branch, Division of AIDS, STD, and TB Laboratory Research, Centers for Disease Control and Prevention, 1600 Clifton Road, NE, Atlanta, Georgia 30333, USA. Telephone: (404) 639-3601; Telefax: (404) 639-1287; E mail: tms1@ciddas1.em.cdc.gov.

JOHN SKEHEL, F.R.S.

Director, National Institute for Medical Research, The Ridgeway, Mill Hill, London NW7 1AA, United Kingdom. Telephone: 44 (0) 181 959-3666; Telefax: 44 (0) 181 906-4477; E mail: m-brenna@nimr.mrc.ac.uk.

WILLIAM W. STEAD, M.D., M.A.C.P.

Director, Tuberculosis Program, Arkansas Department of Health, and Professor of Medicine, University of Arkansas College of Medicine. Address correspondence to: Tuberculosis Program, Arkansas Department of Health, 4815 W. Markham Street, Mail Slot 45, Little Rock, Arkansas 72205-3867, USA. Telephone (501) 661-2415; Telefax: (501) 661-2759.

SHARON STEINBERG

Department of Pediatrics, Columbia University College of Physicians & Surgeons, Black Building 4-427, 650 W. 168th Street, New York, New York 10032, USA. Telephone: (212) 305-1556; Telefax: (212) 305-2284.

ELLEN G. STRAUSS

Senior Research Associate, Division of Biology, 156-29, California Institute of Technology, Pasadena, California 91125, USA. Telephone: (818) 395-4903; Telefax: (818) 449-0756; E mail: straussj@starbase1.caltech.edu.

JAMES H. STRAUSS

Professor, Division of Biology, 156-29, California Institute of Technology, Pasadena, California 91125, USA. Telephone: (818) 395-4903; Telefax: (818) 449-0756; E mail: straussj@starbase1.caltech.edu.

HOWARD STREICHER, M.D.

Laboratory of Tumor Cell Biology, National Cancer Institute, National Institutes of Health, Building 37, Room 6B04, Bethesda, Maryland 20892-4255 USA. Telephone: (301) 496-6007; Telefax: (301) 496-8394.

JOHN A. TINE, PH.D.

Research Scientist, Virogenetics Corporation, 465 Jordan Road, Troy, New York 12180, USA.

TON VAN HELVOORT, PH.D.

P.O.Box 514, 6190 BA Beek, The Netherlands. E mail: TvanHelvoort @compuserve.com.

EBERHARD WECKER, MD

Emeritus Professor, Institut fur Virologie und Immunbiologie, University of Würzburg, Versbacher Strasse 7, 97078 Würzburg, Germany. Telephone: 49 (931) 75535.

THOMAS H. WELLER, M.D.

Richard Pearson Strong Professor of Tropical Public Health, Emeritus, Harvard School of Public Health, and Consultant, Viral and Parasitic Diseases: International Health, 56 Winding River Road, Needham, Massachusetts 02192, USA. Telephone: (617) 235-3905; Telefax: (617) 235-3905.

T. ULF WESTBLOM, M.D.

Professor of Internal Medicine, Texas A&M University, College of Medicine, Chief, Medical Service, Department of Veterans Affairs Central Texas Healthcare System, 1901 South 1st Street, Temple, Texas 76504, USA. Telephone: (817) 771-4546; Telefax: (817) 899-4026.

Foreword

Microbe Hunters!

Since 1927, I have witnessed the development of vaccines for a wide range of diseases, from vesicular fever to poliomyelitis. But it was because of rabies and Hilary Koprowski that the Marcel Merieux Foundation was born at Pensieres. On a historical note, Marcel Merieux was with Louis Pasteur in 1894, together with Emile Roux, Albert Calmette, Elie Metchnikoff, and Alexander Yersin, pioneers in microbe hunting.

Present-day researchers continue the fight to control or eradicate disease. During our meeting in May 1995, the following appeal was formulated:

> The participants at a meeting in Les Pensieres, a World Health Organization Collaborating Center, on *Microbe Hunters—Then and Now,* express their concern on the reappearance of the Ebola virus, the recurrence of diphtheria, the urgency to investigate the new vaccine against malaria and to accelerate the means of struggle against AIDS. They ask Dr. Charles Merieux to intercede personally on these matters with the President of the French Republic before the Congress of Cannes when France will transfer the presidency of the European Union to Spain. It behooves the country of Pasteur to mobilize the rich countries to help the others—especially Africa—with a priority on health in the framework of a new Euro-Mediterranean and African policy for the five coming years.

I wish to assure you that this appeal was heard, as evidenced during The Sabin Foundation's tribute to Jonas Salk on December 7, 1995, when Hilary Koprowski and I announced the mobilization of efforts in fighting AIDS.

At a meeting on prions in Val de Grace on March 20, 1996, our friend Gajdusek recalled that, with Hilary Koprowski, we were among the first to organize a conference devoted to slow viruses. But the latest meeting addressed the now famous bovine spongiform encephalopathy [mad cow disease].

Microbe hunters will always be there and the scientific world needs their experience to remind us that microbe hunting is a rational endeavor and to prevent the hysteria surrounding mad cow disease from spreading to, for example, AIDS research.

Tradition is not the enemy but rather the framework of courage.

Charles Merieux
Marcel Merieux Foundation
Pensieres, France

Prologue

Two hundred and fifty years ago, Antony Leeuwenhoek,
who was a matter-of-fact man, looked through a magic eye,
saw microbes, and so began this history.

PAUL DE KRUIF—*Microbe Hunters*

Seventy years ago, Paul de Kruif published the classic book *Microbe Hunters*, which portrayed the historic figures primarily responsible for opening the fields of microbiology and immunology. Beginning with Leeuwenhoek who described "the nimble creatures" under the microscope in the sixteenth century, the book chronicles individual heroes and their major contributions to the time of Erlich, the inventor of the "silver bullet," i.e., chemotherapy, and founder of modern immunology in the late nineteenth century. Neither the heroes of the book nor its millions of readers around the globe would have imagined the rapid progress that has been made in the fields these scientists opened to the world.

Who would have imagined the complete elimination of that ancient disease, smallpox, which killed nearly three hundred million people in the twentieth century alone, three times more than all the wars in this century? Poliomyelitis, an epidemic disease in the nineteenth century and even more frequent in the twentieth century, is also near eradication, contrary to expectation. These triumphs of vaccination are also well illustrated in the control of yellow fever, measles, and rubella. However, contrasting examples exist, including the lack of control and the emergence of different patterns in ancient infections such as tuberculosis, as well as the emergence of new epidemiologic patterns in the oldest recorded infectious disease of man, rabies, despite the availability of an effective vaccine.

Viral vaccines are beginning to have an impact on human cancers, 20 to 30% of which are thought to be associated with viral infections. A primary example is the development of hepatitis B virus vaccine as a means for the prevention of liver cancer. In contrast, no successful vaccine has yet been developed against retroviruses. A varicella vaccine now holds promise in preventing the painful recurrence of the herpes zoster-induced disease, shingles.

As ancient plagues are brought under control, new ones are emerging. Hemorrhagic fevers made a formidable appearance in all continents during the second half of the twentieth century, and the names Ebola, Hanta, and Lassa instill fear. Malaria affects millions of people worldwide and causes

19

more than two million deaths annually, although there are high hopes that a vaccine under development will prevent this disease.

In this book, we present the history of and progress in the development of vaccines against important human infections. Our intent is also to acquaint the reader with the more recent discoveries and advances in the field of microbiology, local mucosal immunity, and plant virology. The need to provide vaccines to populations of the world at low cost and efficiently has led to novel strategies such as the production of edible vaccines in plants and to an understanding of mucosal immunity. The development of such approaches is discussed, along with the possibility of vaccination against ever-spreading Lyme disease as well as the evidence of an infectious origin for peptic ulcer and the changing therapeutic approaches to treat this disorder. Lastly, the mysteries of chronic degenerative and often genetic diseases of the nervous system are presented in the context of prions.

Like de Kruif's selection of several prominent microbe hunters as his heroes, we too were forced to portray only some of the major figures in microbe hunting. However, our task of singling out any one individual as being responsible for achievements in a given field was considerably more difficult, given the enormous progress made in microbiology in the past 70 years. We have therefore assembled pairs of eminent contributors, one of whom presents the historical aspect of a microbial disease and the other, a selected subject dealing with the current status of the same infection. For infections of recent history and new approaches to their analysis, we have selected only one person to deal with the subject. Although the contributors were selected based on their eminence in their field of research, we fully appreciate the important contributions of other scientists throughout the world and offer our sincere apologies for the sin of omission.

We are indebted to the Marcel Merieux Foundation and especially Charles Merieux who provided support in bringing together the contributors of this book for discussion of their work. We also gratefully acknowledge the Rockefeller Foundation Bellagio Study and Conference Center, Villa Serbelloni, for providing an environment for us to assemble this volume.

Hilary Koprowski   Michael B. A. Oldstone
Philadelphia, Pennsylvania   La Jolla, California

ANIMAL VIRUSES

*Of course, it is sure as the sun following the dawn of tomorrow
that the high deeds of the microbe hunters have not come to an end;
there will be others to fashion magic bullets. And they will be waggish men
and original...for it is not from a mere combination of incessant work
and magnificent laboratories that such marvelous cures are to be got.*

PAUL DE KRUIF—*Microbe Hunters*

Smallpox

The number of deaths from smallpox in the twentieth century was three times higher than all deaths from armed conflicts. Yet today, smallpox does not exist as a disease anywhere in the world. The program that successfully eradicated smallpox led to similar campaigns against other contagious diseases, and notable success was achieved with elimination of paralytic poliomyelitis from the entire Western hemisphere.

The history of smallpox is quilted in the story of human migrations and wars, dramatically favoring one population or army over another. The sixteenth century was a time of exploration, and smallpox was spread across oceans by mariners as well as along land routes by armies and caravans. European explorers and the colonists who followed came to the continents of America, Australia, and South Africa and brought smallpox as part of their luggage. The arrival of smallpox played a major role in the Spanish conquest of Mexico and Peru, the Portuguese colonization of Brazil, the settlement of North America by the English and French, and the settlements in Australia. By the turn of the eighteenth century, the disease had become endemic in the major cities of Europe and the British Isles. Nearly one-tenth of all mankind had been killed, crippled or disfigured by smallpox. The disease continued through most of the twentieth century killing over 300 million people.

The remarkable tale of the origin and spread of smallpox is retold here by Frank Fenner, followed by D.A. Henderson's account of the battle to eradicate smallpox.

1

History of Smallpox

Frank Fenner

Most of this account is devoted to summarizing the history of smallpox and the discovery of methods to prevent it, with only a brief mention of the discovery of the microbe that causes the disease.

The Disease

CLINICAL DESCRIPTION

The clinical features of most cases of smallpox were so characteristic that outbreaks of what Edward Jenner called "the most dreadful scourge of the human species" were recorded reasonably accurately, even in ancient times. Apart from suggestions that smallpox in Europe may have been rather mild between the tenth century, when it first occurred there, and the sixteenth century (Carmichael and Silverstein 1987), the only form of smallpox that was recognized until the late nineteenth century was what we now call variola major. Strains of this virus differed in virulence, with those in the Indian subcontinent being more virulent than African strains. Mild strains of the kind described as variola minor, with a case-fatality rate of only 1%, were not recognized until the very end of the nineteenth century, in South Africa and the United States. Thus, most of this historical survey focuses on variola major.

As seen in India during the first half of the twentieth century, about 90% of cases of smallpox among unvaccinated persons were of the "ordinary" type, in which the case-fatality rate varied from 10 to 60%, depending on the strain of virus, the age of the patient and the severity of the rash. The rash was usually very distinctive, with many discrete skin lesions that were pustular when fully developed (Figure 1A) and were more numerous on the face and extremities than on the trunk. There were two rarer and usually fatal forms: the "flat" and the "hemorrhagic" types of smallpox, which accounted for the other 10% of cases. Persons who had recently been successfully vaccinated

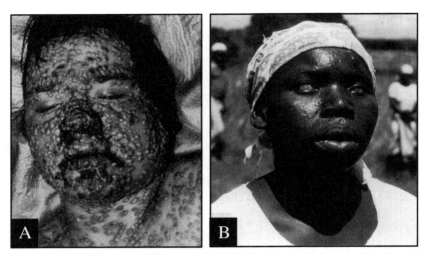

FIGURE 1. CLINICAL FEATURES OF SMALLPOX. A. Severe facial rash of "ordinary" type smallpox, illustrating the distinctive nature of the pustules of smallpox. B. Pockmarks and blindness in a Zairean woman who had recovered from smallpox. (B, from Fenner et al 1988).

were usually completely protected, but if the interval after vaccination was more than five years, they might suffer from "modified" smallpox. Without revaccination, protection gradually waned to such an extent that infections could be fatal.

Not only was smallpox an acute disease with very obvious and distinctive skin lesions, but infection with variola major left its traces in many survivors in the form of distinctive pockmarks, usually most numerous on the face. Sometimes heavily pockmarked persons were also blinded, although this was a rather rare complication (Figure 1B).

CONTAGION AND EPIDEMIOLOGY

Smallpox was a specifically human disease, i.e., there was no animal reservoir, and subclinical infections in unvaccinated persons were almost unknown. The incubation period was 10 to 14 days, during which the infected person was quite well and, importantly in terms of disease spread, able to travel. The onset was sudden, with fever, headache, and malaise, and two to three days later, rash.

Patients became infectious after the rash had appeared, when the lesions of the enanthem, i.e., the lesions on the buccal mucous membranes, broke down and virus was excreted in the saliva and pharyngeal secretions. Infectivity declined rapidly as the lesions of the enanthem healed and those of the exanthem scabbed. Although there was a large amount of virus in the skin scabs, these were not an important source of person-to-person infection, although they contributed to the contamination of fomites, such as blankets.

Recovery was followed by lifelong immunity and no recurrences. These features meant that smallpox could be maintained as an endemic disease only in populations of about 200,000 persons; thus, small isolated populations, such as Eskimos, Amazonian Indians or the inhabitants of oceanic islands, never constituted a continuing focus of infection, although if smallpox was introduced, they suffered devastating outbreaks.

Infection occurred by close face-to-face contact, such as is common between members of the same household. Rarely, smallpox was spread by fomites and, even more rarely, by an airborne route through coughing, early in the course of disease, which could cause problems in crowded buses or trains and in hospitals.

SMALLPOX IN THE ANCIENT WORLD

Although there are many hints of the occurrence of smallpox before and during the first millennium of the Christian era, there is no documentation as old or as reliable as the early descriptions of rabies (see Chapter 3). Extensive skin lesions that may have been due to smallpox were described in three Egyptian mummies, the most famous being that of Ramses V, who died as a young man in 1157 BC. Some modern historians have interpreted writings about various ancient "plagues," such as those of the Hittites in 1346 BC, of Syracuse in 595 BC, and of Athens in 490 BC, as being due to smallpox, but it is difficult to find reliable descriptions of the disease before about the fourth century AD.

Generalizing from the scanty data available, it seems probable that smallpox was endemic in the basin of the River Ganges from as early as 500 BC, although there is no description of such a disease in the ancient writings of the *Atharvaveda*. There are reasonable records from China (Figure 2), where the disease is said to have entered from the west in AD 48; it persisted in the dense agricultural populations in the valleys of the great rivers. There are reliable descriptions of its movement to the Korean Peninsula in 583 AD, and it occurred for the first time in Japan in 585. Ko Hung, a famous Chinese physician and alchemist, differentiated smallpox and measles in a book completed in 340 AD. In the western part of the Eurasian landmass, an early major spread of smallpox occurred during the great Islamic expansion across North Africa and into the Iberian Peninsula in the eighth and ninth centuries. In 622 AD, Ahrun, a Christian priest living in Alexandria 30 years before the Arab conquest of Egypt, wrote clear descriptions differentiating smallpox and measles, as did the great Arab physician Al-Razi in 910 AD.

One feature common to all the earliest writers whose descriptions are now widely accepted—Ko Hung in China, Vaghbata in India and Al-Razi in Asia Minor—is that they describe smallpox as primarily a disease of children, indicating that the disease was endemic in the large populations they described.

By 1000 AD, smallpox was probably endemic in the more densely populated parts of the Eurasian landmass from Japan to Spain and including the

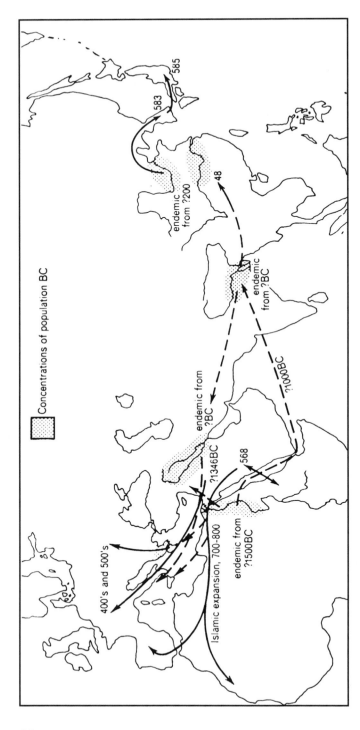

FIGURE 2. THE PROBABLE SPREAD OF SMALLPOX in the Eurasian landmass and the northern coast of Africa before 1000 AD. Many of the early dates (BC) are guesses; the unmarked dates (AD) are more firmly based. (From Fenner et al 1988).

African countries on the southern rim of the Mediterranean Sea. Over the next few centuries, with the movement of people to and from Asia Minor during the Crusades, smallpox became well established in Europe. In Africa, caravans crossing the Sahara to the densely populated kingdoms of West Africa carried smallpox with them, and the disease was repeatedly introduced into the port cities of East Africa by Arab traders.

SPREAD TO OTHER CONTINENTS BY EUROPEAN COLONIZATION

By the sixteenth century, smallpox was becoming steadily more serious in European countries, and statistics of smallpox deaths began to be collected at about this time in Geneva, London, and Sweden. The stage was set for the next explosive spread of the disease with the development in Europe of ocean-going ships and the movement of European explorers and colonists to the newly discovered continents of the Americas, Australia, and South Africa (Figure 3).

Smallpox played a crucial role in the Spanish conquest of Mexico and Peru, Portuguese colonization of Brazil, and the settlement of North America by the English and the French. In Africa, smallpox was first introduced into the populous states north of the Gulf of Guinea with caravans traveling from North Africa and into more southerly parts of western Africa by the Portuguese in 1490. These foci in Africa served as the source of the next great expansion of the disease—into the Americas.

Shipments of slaves from West African ports across the Atlantic Ocean to the Americas resulted in the first outbreak of the disease in the western hemisphere among African slaves on the island of Hispaniola in 1518 (Henige 1986). Here smallpox and other infectious diseases rapidly destroyed the once large indigenous population and spread to neighboring islands of the Caribbean. In 1518–1519, the Cortés expedition began its conquest of Mexico, but his Spanish soldiers carried no smallpox. In 1520 the Governor of Cuba sent a second expedition to bring Cortés under control. This expedition included one African slave who had been infected in Cuba, and he initiated an epidemic among the native American Indians, first in Yucatan and then in central Mexico. It was this epidemic that gave the Conquistadors their easy victory over the Aztecs. From central Mexico the epidemic moved southward, 1524–1527, ahead of Pizarro's invasion of Peru in 1533, and made victory possible there also. Further south, slaves brought smallpox from the Portuguese colonies in Africa to the Portuguese colony in Brazil. Epidemics of smallpox and other infectious diseases played a major role in destroying Indian resistance to the European invasions and so much reduced the Indian populations that a very large slave trade from Africa was established to provide labor for the new colonies.

Colonization of the east coast of North America was begun by France, Great Britain, and The Netherlands almost a century later than the invasion of Central and South America by the Spanish and Portuguese. Once again,

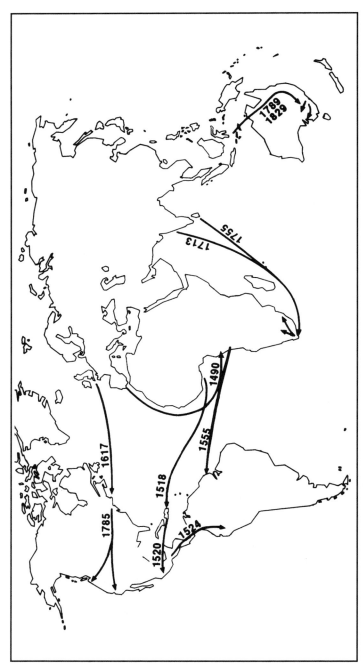

FIGURE 3. THE SPREAD OF SMALLPOX ASSOCIATED WITH THE EUROPEAN COLONIZATION of the Americas, Australia, and southern Africa. This differs from the figure published in Fenner et al (1988) in that newly available information makes it clear that the introduction of smallpox into Hispaniola occurred via slaves from Africa and not from Spain. The introduction of virus that caused the outbreak near Sydney in 1789 (and another epidemic in southern Australia in 1829-1831) resulted from smallpox introductions via Macassarese trepang fishermen landing on the northern coast of Australia, spreading through Australian aboriginal tribes to the southern part of Australia.

smallpox followed, initially from English settlers who arrived in 1617 (and later with African slaves) and set off an epidemic among the Indians that cleared a place for the settlers who came from Plymouth in 1620. Smallpox spread west among the Indian tribes far ahead of European settlement.

The disease was also influential during the wars between Europeans in North America. It played a part in the wars between Great Britain and France in what was to become Canada. At about this time, smallpox was also used for biological warfare, the best recorded occasion being in 1763, when contaminated blankets were given to members of hostile Indian tribes by the British. In the Revolutionary War, the long duration of the siege of Boston was due in part to the existence of smallpox in the city and Washington's fear of exposing his army to the disease. When the British finally left, Washington ordered "one thousand men who had had smallpox" to occupy the city.

Smallpox was introduced into southern Africa by ships docking at Cape Town, initially in 1713, when it was introduced in contaminated bed linen from a ship returning from India, again in 1755 in a ship from Ceylon, and in 1767 in a Danish ship coming from Europe. The first two introductions caused many deaths among the European settlers, but by 1767, they had learned the value of "variolation," i.e., the inoculation of pus from the pustules, and the death rate was much lower. However, each introduction had disastrous results for the aboriginal inhabitants, the Hottentots and Bushmen, with cases after the 1755 outbreak spreading as far as the Kalahari Desert.

Finally, smallpox was recorded in Australia in 1789 in Sydney, a year after the British had established a penal settlement there. It devastated several aboriginal tribes, but only one case was recorded in the tiny European settlement. There is no reliable evidence that it was introduced by the European invaders; it is thought to have been part of an epidemic introduced somewhat earlier into the aboriginal population of northern Australia by trepang fisherman sailing from Macassar in modern Indonesia (Campbell 1996). Several decades later, epidemics in 1829–1830 (Campbell 1983, 1985) and in the 1860s (Campbell 1996) spread from the northern coast of Australia through the sparse aboriginal population to the south of the continent. As in the Americas, the depopulation caused by smallpox simplified the expansion of the British occupation.

THE SPREAD OF VARIOLA MINOR

The mild form of smallpox, variola minor, was not recognized with certainty until the closing years of the nineteenth century, almost simultaneously in the United States and in southern Africa. By then, vaccination was well established as a preventive measure, but variola minor was not taken very seriously by the public or the health authorities, and it spread all over North and then South America (Figure 4), from the United States to several countries in Europe, and also to Australia, where it caused an epidemic in

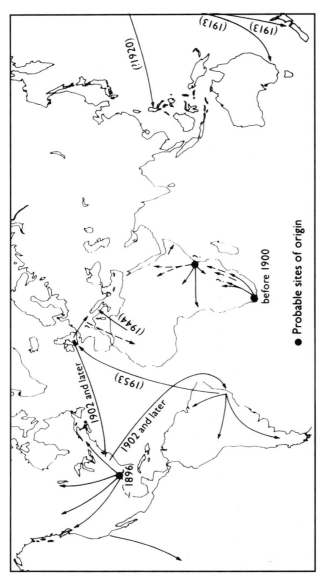

FIGURE 4. THE SPREAD OF VARIOLA MINOR in Africa and variola minor (alastrim) in the Americas and from the United States to Europe, Australia, and New Zealand during the twentieth century. Figures in parentheses indicate dates of importation that did not result in long-continued endemic disease. Variola minor was recognized in southern Africa at the end of the nineteenth century. It occurred, together with variola major, in most countries in southern, eastern and central Africa at least from the 1930s onward. It was probably carried from Europe across the Mediterranean Sea to northern Africa (and back to Italy in 1944). In Africa, there were probably independent foci of origin in eastern and southern Africa. No strains of confirmed southern African origin have been available for PCR testing. One African strain of variola minor virus tested (from Somalia) had a hemagglutinin molecule resembling that of variola major virus; another, from Sierra Leone, resembled the alastrim strains from South America and England (the latter derived from the United States). (From Fenner et al 1988).

1914–1917 that infected over 2,000 people, with only two deaths. Recent nucleotide sequence studies of the hemagglutinin genes of several strains of variola virus show that the African and American strains of variola minor virus were the result of different mutations in local strains of variola major virus (Ropp et al 1995).

RECOGNITION OF THE VIRUS

Jenner and his contemporaries spoke of the "virus" of smallpox and of his vaccine, but they used the word in the general sense of "poison"; viruses in the modern sense of the word were not recognized as causes of diseases of animals until the demonstration by Loeffler and Frosch (1898) that foot-and-mouth disease was caused by a "filterable virus." Later that year, Sanarelli (1898) suggested that a fatal disease of rabbits, myxomatosis, was caused by a virus (in the modern sense); this was the first poxvirus to be recognized. Even before, in 1886 John Buist, a Scottish bacteriologist, reported that he had seen what must have been the virions of a poxvirus in stained smears of smallpox vaccine, although he regarded them as spores (Buist 1887; Gordon 1937). Calmette and Guérin, best known for their development of the bacillus (BCG) for vaccination against tuberculosis, were also involved in smallpox vaccine production in the Pasteur Institute, and in 1901 they described numerous minute refractile particles in vaccine lymph which they suggested might be the "virulent elements." Both Prowazek (1905) and Paschen (1906) developed staining methods to demonstrate what they called the "elementary bodies," and in 1906, Negri showed that the infectivity remained after vaccine lymph had been passed through a filter that held back bacteria, and that pox diseases were indeed caused by "filterable viruses." The actual shape of the virions of poxviruses was first demonstrated by the pioneers of electron microscopy, Ruska and Kausche (1943), and the development of negative staining electron microscopy by Brenner and Horne (1959) provided the means to make a rapid diagnosis of smallpox from scab material, a technique that proved invaluable during the Intensified Smallpox Eradication Programme.

IMMUNIZATION AGAINST SMALLPOX

VARIOLATION

No history of smallpox is complete without mention of the discovery of methods of preventing it. Smallpox is unique among infectious diseases in that it has a long history of effective means of preventive inoculation, what we now call "immunization." The first method used was the inoculation of pus from the pustules, a process that was called "inoculation" and later "variolation." This procedure appears to have been developed independently in India and China in the eleventh or twelfth century and was later seen in Asia Minor. In 1721, variolation was introduced from the Ottoman Empire

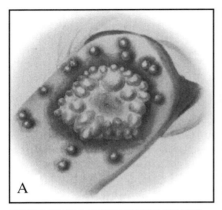 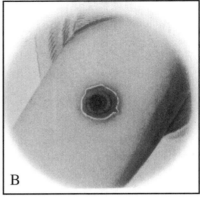

FIGURE 5. HALFTONE PHOTOGRAPH prepared from an engraving by George Kirtland of colored drawing made in 1801 by Captain C. Gold, showing lesions 13 days after inoculation with variola major virus (variolation) (A) and Jenner's vaccine virus (vaccination) (B). They were published by Kirtland in 1806 and independently reproduced (together with other engravings illustrating the lesions at different stages) in the Jenner Centenary number of the *British Medical Journal*, 23 May 1896.

into Bohemia and England by Dr. Johann Adam Riemann and Lady Mary Wortley Montagu, respectively. Although it produced fearsome skin lesions at the inoculation site (Figure 5A) and sometimes a generalized rash and had a case-fatality rate of 1 to 2%, it was deemed much better than natural smallpox, which at this time was an almost universal disease in Europe, with a case-fatality rate of over 40% in babies and about 25% overall.

VACCINATION

As a boy, Edward Jenner (born in 1749) had been variolated and as a physician, he was himself a variolator. But his astute observations in the early 1790s that cowpox appeared to protect milkmaids against smallpox, followed by his experiments in 1796–1798 showing that this was indeed the case, forever altered the history of infectious diseases. Vaccination was introduced (Figure 5B) and wherever it was assiduously practiced, it had a dramatic effect on the incidence of smallpox (Figure 6). In 1881, Pasteur gave the term "vaccination" its general meaning to describe this kind of preventive inoculation to honor Jenner at an international medical congress in London.

Development of modern smallpox vaccine. The development of methods of preparing the vaccine for large-scale use is only briefly summarized here, since the eradication of smallpox is detailed in Chapter 2. Jenner had obtained his vaccine from a girl infected with cowpox from a cow. The vaccine was maintained by arm-to-arm passage in human subjects, although from time to time, the vaccine was lost and further strains were obtained from cows or occasionally horses. Arm-to-arm vaccination was not a safe practice; sometimes the vaccine became contaminated with variola virus and there were many

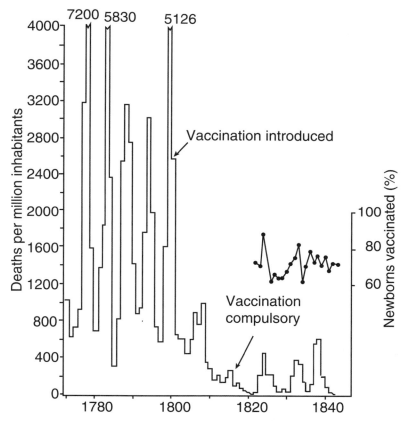

FIGURE 6. SMALLPOX DEATHS IN SWEDEN per million population between 1722 and **1843. Note the dramatic decline after the introduction of vaccination in 1801 and the percentages of newborn children known to have been vaccinated in infancy from 1820 onward. (From Fenner et al 1988).**

cases of the transmission of syphilis and at least one reliable description of a hepatitis epidemic caused by arm-to-arm vaccination. To overcome these risks and to provide a reliable source of virus, Italian authorities in 1840 began producing vaccine in cows for use in human vaccination, and after considerable discussion at a medical congress in Lyons in 1864, the practice gradually spread through Europe, although arm-to-arm vaccination was still popular in Britain until it was banned in 1898. Methods of production and harvesting were gradually improved, and storage in glycerol became standardized. By the 1960s, vaccine was being produced in the skin of calves, sheep, and buffalo, on the chorioallantoic membrane of developing eggs, and in tissue culture, but

✓

growth in calf skin was the most commonly used method throughout the Intensified Smallpox Eradication Programme.

Although vaccinia virus is relatively heat-stable, failure rates with liquid vaccine were very high in tropical countries. This problem was eventually overcome by vacuum-drying the virus as initially done by Camus (1909) in France and later by Otten (1927) in The Netherlands East Indies. In the 1950s, Collier (1955), working in England, adapted freeze-drying techniques for large-scale commercial use, and large amounts of heat-stable vaccine became available. Thus it was finally possible to contemplate the global eradication of smallpox.

REFERENCES

Detailed references are given in Hopkins (1983) and Fenner et al (1988).

Brenner S, Horne RW (1959). A negative staining method for high resolution electron microscopy of viruses. Biochim Biophys Acta 34: 103-110.

Buist JB (1887). Vaccinia and variola: A study of their life-history. London: Churchill.

Calmette A, Guérin C (1901). Récherche sur la vaccin expérimentale. Ann Inst Pasteur 15: 161-168.

Campbell J (1983). Smallpox in aboriginal Australia, 1829-31. Hist Stud 20: 536-556.

Campbell J (1985). Smallpox in aboriginal Australia, the early 1830s. Hist Stud 21: 336-358.

Campbell J (1996). Smallpox in Aboriginal Australia. (in press).

Camus L (1909). Quelque modifications à la préparation et à la conservation du vaccin sec. C R Soc Biol 67: 626-929.

Carmichael AG, Silverstein AM (1987). Smallpox in Europe before the seventeenth century: virulent killer or benign disease? J Hist Med Allied Sci 42: 147-168.

Fenner F, Henderson DA, Arita I, Jezek Z, Ladnyi ID (1988). Smallpox and its eradication. Geneva: World Health Organization.

Gordon M (1937). Virus bodies. John Buist and the elementary bodies of vaccinia. Edinburgh Med J 44: 65-71.

Henige D (1986). When did smallpox reach the New World (and why does it matter)? In: PE Lovejoy, ed. Africans in bondage. Madison: University of Wisconsin Press, 11-26.

Hopkins DR (1983). Princes and peasants. Smallpox in history. Chicago: University of Chicago Press.

Loeffler, Frosch (1898). Berichte der Kommission zur Erforschung der Maul- und Klauenseuche bei dem Institut für Infektionskrankheiten in Berlin. Zentralbl Bakt 23: 371-391.

Negri A (1906). Über Filtration der Vaccinevirus. Z Hyg Infektkr 54: 327-346.

Paschen E (1906). Was wissen wir über den Vakzine-erreger. Münch med Wochschr 53: 2391-2393.

Pasteur L (1881). Vaccination in relation to chicken cholera and splenic fever. Vol 1. London: Trans Int Med Congr, Seventh Session. 370-378.

Prowazek S (1905). Untersuchungen über die Vaccine. Arb Reichsgesundheitsamte 22: 535-556.

Ropp SL, Jin Q, Knight JC, Massung RF, Esposito JJ (1995). PCR strategy for identification and differentiation of smallpox and other orthopoxviruses. J Clin Microbiol 33: 2069-2076.

Ruska H, Kausche GA (1943). Über Form, Grossenverteilung und Struktur eineger Elementarkorper. Zentralbl Bakt 150: 311-318.

Sanarelli G (1898). Das myxomatogene Virus. Beitrag zur Stadium der Krankheitserreger ausserhalb des Sichtbaren. Zentralbl Bakt Abt. 1, Orig. 23: 865-873.

2

Smallpox Eradication

D. A. HENDERSON

In May 1980, it was possible for the first time to affirm that a disease had been eradicated throughout the world (Fenner et al 1988). That the disease was smallpox was singularly appropriate. For more than three centuries, smallpox had been the most universally feared of all diseases, responsible globally for more deaths than any other infection. Even in the twentieth century, one contrasts the estimated 100 million deaths caused by armed conflict with a figure at two to three times greater for deaths from smallpox. We should also recall that it was Jenner's effort to protect against smallpox that is celebrated as the dawn of the science of immunology.

What is little appreciated, however, is that the smallpox eradication program was begun reluctantly and almost not at all. From a contemporary vantage point, it would seem obvious that smallpox of all human diseases was the best possible candidate for eradication. Humans were the only host; a highly effective and inexpensive vaccine was available; and detection of cases was easy, given its unique clinical presentation. Moreover, governments across the world had been conducting vaccination programs and controlling outbreaks for more than a century. Nevertheless, the notion that smallpox, as a practical matter, might be eradicated from a defined geographical area was not even proposed until the 1950s and not endorsed until a decade later.

This seems surprising, but then global policy-making in health often appears to be as much quixotic as rational. Preceding smallpox in a parade of failed eradication attempts was a hookworm campaign begun in 1909, yellow fever in 1913, yaws in the 1950s, and malaria in 1955. Each was launched on a premise as much based on evangelistic conviction as on dispassionate scientific assessment, and each was eventually destined to collapse.

While Jenner and Jefferson had expressed hopes early in the nineteenth century that smallpox might someday be conquered, the first serious proposal to undertake smallpox eradication as a practical matter was advanced in 1953 by Dr. Brock Chisholm, the first Director General of the World Health Organization (WHO). In his fifth and last year as Director General, he proposed that smallpox eradication be undertaken as a global effort, an overarching campaign which would demonstrate the importance of WHO to every member state. In the discussion that ensued at the WHO Assembly, virtually every industrialized country and several developing ones argued that such a program would be too vast and complicated and the matter was unceremoniously dropped. Instead, a malaria eradication program was launched the following year.

The one country that was not a party to these earlier discussions was the USSR, which for many years had not actively participated in WHO. In 1958, it resumed active participation and immediately dropped a figurative bombshell. Its principal delegate, the distinguished academician and Vice Minister of Health Victor Zhdanov proposed a 10-year program for the eradication of smallpox. He argued that the USSR had succeeded in eradicating smallpox over its vast and ethnically heterogeneous society and that there was no reason other countries around the world could not do likewise. WHO estimated the cost of such a program at $100 million. The Assembly voted to accept the program and to provide, from its regular budget, $100,000. Fiscal annihilation was diplomatically more polite than outright rejection. Not surprisingly, little was achieved over the next seven years.

Meanwhile, between 1957 and 1967, WHO and its member countries invested nearly $1500 million in a steadily failing malaria eradication effort. As frustration heightened, the very concept of eradication became increasingly doubted by scientists and public administrators alike. The prevalent mood was well captured by the literate and widely read Dr. René Dubos who, referring to smallpox and malaria eradication in particular, wrote in 1965:

> Social considerations, in fact, make it probably useless
> to discuss the theoretical flaws and technical difficul-
> ties of eradication programs, because more earthly fac-
> tors will certainly bring them soon to a gentle and silent
> death...eradication programs will eventually become a
> curiosity item on library shelves, just as have all social
> utopias.

Meanwhile, Russian delegates to successive World Health Assemblies increasingly protested the lack of progress in the smallpox program. But, the allocation for smallpox in WHO's regular budget remained small, as did voluntary contributions for smallpox activities. The United Nations International Children's Fund (UNICEF), disturbed by the costly and failing malaria experience, refused to assign any funds to a new eradication effort.

Until 1966, the problem appeared to be insolvable. A substantial increase in WHO's regular budget was out of the question because of growing resistance of most industrialized countries, the principal contributors to WHO, to any further increases. The Director General, for his part, was committed to malaria eradication and diverted whatever discretionary funds he could to deal with mounting difficulties in that program.

Under pressure to do more for smallpox eradication, WHO's Director General Marcelino Candau in 1966 took an ingenious tack. He proposed a two-part budget, consisting of the conventional budget and a special budget of $2.4 million specifically for smallpox eradication. It basically said to the Assembly that if they wanted a smallpox eradication program, they would have to vote for a much larger overall budget. If both parts were accepted, the overall increase in the budget would be 16%. As expected, almost every industrialized country vehemently protested the size of the increase and many others expressed doubts about the wisdom of the program. Given the previous unhappy experiences with eradication efforts, there were understandable reservations about the credibility of its public health proponents. After a lengthy contentious debate, the budget was finally accepted by a margin of only two votes, the narrowest vote in the history of the World Health Assembly.

Smallpox eradication was thus launched, albeit with grave reservations about both prospects for success and the wisdom of the investment. One can only wonder what might have eventuated if two countries at the Assembly had voted differently and terminated the smallpox eradication effort, for it was this program that provided both the foundation and courage for the launch of global programs on immunization. From that effort, the Children's Vaccine Initiative arose, as well as the more recent polio eradication campaign.

The basic strategy of smallpox eradication was comparatively straightforward. First, it set as a goal the protection by immunization of 80% of the population through systematic vaccination campaigns. Second, it called for the development of national surveillance programs involving weekly reporting of all cases of smallpox, investigation of cases that occurred, and containment of these cases. Third, it stipulated that a research program be undertaken from the outset to address a range of questions extending from the "bench" to the "bush."

The immunization program exceeded all expectations in almost every country. Better than 90% coverage was commonly achieved and resistance to vaccination was infrequent. Given reasonable organization, vaccinators could each readily average more than 100 vaccinations per day even in areas where travel was largely by horseback or on foot. A quality control program was stipulated. In simple terms, an assessment team of two persons would check a 5 to 10% sample of villages some 7 to 10 days after mass vaccination. This served as an important stimulus to do a competent job, since teams obliged to repeat the effort because of poor results did so without payment of per diem allowances. Not all countries agreed to the quality control component, arguing that it was a waste to assign personnel and transport simply for checking the work of others.

Surveillance proved to be, by far, the most critical element of the strategy and the most difficult to persuade countries to implement. The foundation of a surveillance system is a reporting network of health posts and hospitals that report new cases each week and an epidemiological unit that investigates and contains outbreaks. By knowing where the cases were and in what groups, immunization activities could be redirected to maximum effect and progress in the campaign monitored. However, many government and international staff protested the diversion of staff from the task of mass vaccination and insisted that complex reporting networks could never be established in developing countries. In fact, such networks were regularly established within 12 to 18 months, and the data greatly facilitated the work everywhere. Still, the concept was and remains a difficult and elusive one, as witnessed by the lack of global surveillance for other vaccine-preventable diseases in many countries, now more than 20 years after the inception of the expanded program on immunization.

Finally, it is important to say a word about research. From the beginning of the program, there was a commitment to a wide-ranging research agenda—from improvements in vaccine protection and quality, to field epidemiology in order to better understand the natural disease, to elaborate ecological studies intended to confirm the absence of an animal reservoir of smallpox. The budgeted WHO funds for research were minuscule but sufficient to regularly convene a multidisciplinary scientific group. This group, on its own initiative, addressed many questions defined in the course of the meetings. The impact of the research findings is best summarized by stating that virtually every premise of the program significantly changed from what it had been at the outset—from our understanding of the epidemiology of the disease, to the way vaccine was produced and tested, to the techniques of vaccination, to the actual implementation of the program in the field.

The smallpox eradication campaign, so tentatively launched in January 1967, exceeded all expectations. During a period of just over 10 years, programs were conducted in more than 50 countries with a population of more than 1000 million persons. Smallpox cases numbering 10 to 15 million per year with two to three million deaths dropped to zero in a period of only 10 years and 9 months. As many as 200,000 national staff and more than 700 international staff participated in the effort. Success can be attributed to many factors, including an international organization that united such an effort as well as countless acts of dedication, sacrifice, and imagination by all its participants. But the program fulfilled three additional important principles:

1. A clearly defined ultimate goal with ongoing measurements to evaluate progress. Smallpox eradication clearly meant zero cases of smallpox and the surveillance program provided the data needed to measure success in reaching that objective.

2. Clearly defined procedural objectives with quantitative measurements for monitoring progress on a continuing basis. In brief, systematic immunization programs with specific time-targets and systems for quality control.

3. A broad-ranging research agenda, closely linked to the program, which continually asked the question of whether this program or any of its components could be conducted better, faster, more efficiently, or with greater certainty, through use of other means, other tools, or other approaches.

The one international program that today closely follows the three principles is the program for Guinea worm eradication, a program now making extraordinary progress. The program for poliomyelitis eradication is beginning to fulfill these requisites and is rapidly gaining momentum. Hopefully, there will soon be others.

REFERENCES

Fenner F, Henderson DA, et al (1988). Smallpox and its eradication. Geneva: World Health Organization, Chapters 9, 10, and 31.

Dubos R (1965). Man adapting. New Haven: Yale University Press.

Vignette

Rabies

In 1954, a house painter in Anaheim, California, suffering from rabies, had bitten numerous people in the course of one night and scared many officers of public safety before he was subdued by the town policemen. This terrifying aspect of the disease, known for more than 20 centuries BC, has struck the fantasy of philosophers, naturalists, scientists, poets, and playwrights.

As recently as 1995, the Nobel Prize winner Garcia Marques wrote a novel in which rabies played an important role in deciding the fate of the characters. In the play by Curt Goetz, *Der Hund im Hirn* (*Dog in the Brain*) (In: *Sämtliche Bühnenwerke*, Wilhelm Heyne Verlag, Munich, Germany, 1977), the hero, a university professor, plays on his wife's lover's fear of rabies in order to stop their love affair.

"Lyssa" (for "rabies") was a terrifying word in the Greek vocabulary and the invective "rabid dog" applied to Hector in the *Iliad* indicates that Homer was acquainted with rabies. Indeed, four centuries later, Aristotle provided a clear description of animal rabies, although he excluded the possibility that man could also become infected with rabies. Studies by the sixteenth century savant, Girolamo Fracastoro, and by Louis Pasteur represent two milestones in the history of rabies. Fracastoro provided the clearest description of human rabies and its fatal outcome in his treatise "The Incurable Wound." He also quite accurately described the incubation period, although with the continual change in the infectivity patterns of rabies, incubation times today may differ. Following Galtier's adaptation of rabies to rabbits, Pasteur's monumental work resulted in the production of the first rabies vaccine. Yet unlike smallpox, a disease that has been eradicated throughout the world, and polio, which is in the process of being eradicated, rabies remains present in parts of the world. This contrasting situation of rabies with smallpox and polio rests primarily in the fact that rabies virus can infect all warm-blooded animals, whereas infection with smallpox and polio is limited to humans. Thus, the fight against rabies in countries where wildlife is primarily the source of infection is often an uphill battle, particularly in places where bats are carriers of the virus. The fact that the incubation period of rabies can be as long as six years after a bite does not make control of rabies any easier. However, the weaponry of modern science holds promise in controlling even this, the oldest scourge of mankind.

3

A Brief History of Rabies

Martin M. Kaplan and Francois X. Meslin

In July 1885, Joseph Meister, a nine-year-old boy who had been bitten savagely in the face and elsewhere by a rabid dog two days previously, was brought to Louis Pasteur for treatment. Despite his own fear of failure using his still experimental rabies vaccine, Pasteur ordered the intervention because of the desperate pleas of Joseph's mother. Joseph Meister survived and lived until 1940, working as the gatekeeper of the Pasteur Institute in Paris. We now know that the chances of successful vaccine intervention days after severe exposure are not good, especially where facial wounds are involved. But the event gave Pasteur confidence to continue the pursuit of an effective anti-rabies vaccine.

One year later, luck again was an even greater factor in Pasteur's widely heralded treatment of 19 Russians brought from Smolensk to his laboratory in Paris two weeks after being severely bitten by a rabid wolf; only 3 of the 19 died. Experience with wolf bites in the present century has shown that at least nine deaths could have been expected, even with vaccination. As Pasteur himself said, "Fortune favors him who is prepared."

In the history of medicine, rabies appears to be the oldest disease clearly recognized as communicable. Let us therefore go back to ancient times and proceed to the present in the history of this fascinating malady. We focus here on the search for the "causative microbe," a term introduced by Pasteur, and end with an overview of the present situation regarding this most dreaded of diseases.

* * *

Achilles called Hector a "rabid dog" in the Iliad (before 700 BC), and the following reference appears in the Mesopotamian "Laws of Eshnunna" (before 1800 BC):

> If a dog is mad and the authorities have brought the
> fact to the knowledge of its owner; if he does not keep
> it in and it bites a man and causes his death, then the

> owner shall pay two-thirds of a mina (40 shekels) of
> silver. If it bites a slave and causes his death, he shall
> pay 15 shekels of silver.

Aristotle, though a great philosopher, was rather undependable in some of his observations of nature. One of his notable errors concerned rabies, stating that human beings do not contract rabies from the bite of a mad dog, although all other animals are susceptible.

In ancient times, summer months saw an increase in the number of mad dogs, so that Sirius was called the "dog star." Democritus in 500 BC described rabies in dogs and domestic animals. In the first century AD, Aulus Cornelius Celsus prescribed a cure for rabies by dunking the patient in a pool, and if Celsus' preventive measures had been followed, including immediate excision of the bitten tissue and cauterization of the wound with a hot iron, the victim's life might have been spared.

In 1546, the Italian physician Girolamo Fracastoro wrote a vivid description of rabies infection in humans:

> Its incubation following a bite by a rabid animal is so
> stealthy, slow and gradual that the infection is very
> rarely manifest before the 20th day, in most cases after
> the 30th, and in many cases not until four or six months
> have elapsed. There are cases recorded in which it be-
> came manifest a year after the bite.

Once the disease takes hold,

> ...the patient can neither stand nor lie down, like a mad
> man he flings himself hither and thither, tears his flesh
> with his hands, and feels intolerable thirst. This is the
> most distressing symptom, for he so shrinks from wa-
> ter and all liquids that he would rather die than drink or
> be brought near to water. It is then that they bite other
> persons, foam at the mouth, their eyes look twisted,
> and finally they are exhausted and painfully breathe
> their last.

It is worth citing a series of little-known but prescient observations made in 1812 by the physician James Thacher of Plymouth, Massachusetts. He recommended excision and cauterization of the wounded area and also suggested that the offending animal be held for observation rather than be killed. He further advised that normal dogs receive inoculation with saliva from a rabid animal and that excision of the injected areas be performed periodically to determine how long the poison would persist locally before death occurred—an early forerunner of present rationale for local treatment of wounds (Spink 1978).

Koprowski (1995) notes:

> Rabies in the Americas was first reported by Fray Jose
> Gil Ramirez in Mexico in 1709, but it is possible that it

was the disease already described in the early sixteenth century by the first bishop of the New World, Petrus Martyr-Anglerius, who observed Spanish soldiers suffering from madness after being bitten "by large bats, not smaller than turtle doves, which come from the marshes on the river and attack our men with deadly bites." This may have been the first time that vampire bats were incriminated as carriers of rabies in the Americas.

Steele and Fernandez (1991) mention many historical figures who wrote about the cause and treatment of rabies, ranging from the bizarre to the recognizably accurate. They include Pliny, Ovid, Xenophon, Virgil, Horace, Avicenna, Maimonides, and others. The same authors cite the many epizootics of rabies that plagued Europe starting in the thirteenth century and became widespread in the nineteenth century in North and South America and the Far East.

In the early part of the nineteenth century, German and French workers proved experimentally that saliva from rabid humans and dogs was infective for healthy dogs. Thus was confirmed the centuries-old suspicion that the causative agent of rabies was associated with the saliva of a mad dog.

In 1881 Victor Galtier in Lyon succeeded in transmitting the disease to rabbits, making them the animal of choice for laboratory work that was then taken up by Pasteur and his colleagues. Pasteur noticed that Emile Roux, another colleague, was trying to determine how long the "virus" would persist at 37°C in the spinal cords of rabbits. Pasteur modified the procedure by adding potassium hydroxide as a drying agent and by maintaining the flasks at room temperature. This led to the celebrated Pasteur treatment of 14 to 21 inoculations of increasing doses of living fixed virus, which was widely used until the 1950s. Roux never quite forgave Pasteur for exploiting his (Roux's) original work, but when Pasteur was later attacked for alleged failures of his vaccine, Roux loyally came to his defense.

Less known is the fact that Pasteur wanted to include himself as one of the subjects to demonstrate the safety and effectiveness of his vaccine. In a letter to a friend dated March 28, 1885, he said "I have not yet dared to treat human beings after bites from rabid dogs, but the time is not far off, and I am much inclined to begin by myself—inoculating myself with rabies, and then arresting the consequences—for I am beginning to be very sure of my results." Later, after the successful treatment of Joseph Meister, Pasteur again wanted to take the vaccine following a laboratory accident in which his colleague Grancher accidentally inoculated himself with a syringe containing live virus. According to an account by Adrien Loir, nephew of Madame Pasteur, he was dissuaded from doing so by Loir, Grancher, and Eugene Viala, who, however, vaccinated themselves (Valery-Radot 1971).

Taking Pasteur's work as the principal landmark, it is interesting and rather curious that knowledge of rabies advanced very little for more than half a century. There was inconclusive work on anti-rabies serum by Babes and Lepp in 1889, the discovery by Negri in 1903 of the pathonogmonic bodies in brain cells so useful in diagnosis, the minor modifications to Pasteur's vaccine by addition of chemicals that acted mainly as preservatives (e.g., phenol, glycerin, ether, formalin), and the use of mice by Webster as standard experimental procedure (Bunn 1991; Steele and Fernandez 1991). However, the major questions remained, including the efficacy of rabies vaccines, despite the fact that millions of people had been vaccinated since the vaccine was developed. Indeed, it was the doubt and confusion about the use of anti-rabies vaccines that led to WHO's early involvement in rabies work.

The systematic search for the causal agent of rabies began with Pasteur. He failed in many attempts to isolate an agent from the saliva of mad dogs. He and his assistant Roux gradually became convinced that the "virus" attacked the nervous system because of symptoms and a long and variable period of incubation. Pasteur succeeded in transmitting the disease in laboratory animals by inoculating portions of the medulla, using sterilized instruments (Valery-Radot 1971). By 1887, Babes isolated several organisms from the nervous system of rabid animals which he claimed would cause rabies. In 1906, he confirmed the presence of Negri bodies but concluded that they represented a reaction to the infection and were not parasites, as originally believed by Negri who had thought that they were protozoa. This followed much work carried out by various investigators between the 1880s and the turn of the century (Steele and Fernandez 1991; Vodopija and Clark 1991).

Remlinger in 1903 claimed to have produced rabies in rabbits with Berkefeld V filtrates of the central nervous tissue from rabid dogs and rabbits. Work continued in many laboratories to isolate and characterize the virus. A real advance was made in 1936 when Webster and Clow first grew the virus in tissue culture. In the same year, Galloway and Elfford reported the size (100–150 nm) of the rabies virus by ultrafiltration, and in 1962, Almeida and colleagues reported 400-nm particles by negative contrast microscopy. In 1963, several groups reported the bullet-shaped morphology of the virus for the first time (Johnson 1965; see Steele and Fernandez 1991 for review).

International action against rabies was set in motion at the first International Rabies Conference, held in Paris in 1927, where the Health Organization of the League of Nations was asked to collect and publish statistics on treatment in an attempt to determine the best method of vaccination. A large body of data on over 1.42 million treated persons was assembled, but no real conclusions were drawn (McKendrick 1940). The need for a second International Rabies Conference became clear, but this was never held because of World War II.

In 1945, following the War's end, an Interim Commission was set up to make recommendations for the establishment of the World Health Organization (WHO). In its report to the First World Health Assembly (WHA) held in

June 1948, the Interim Commission recommended that rabies be given early consideration by WHO. The Commission's report stated that the inquiry on vaccination histories had failed to answer the question of vaccine efficacy. The second WHA (1949) authorized the formation of the Expert Committee on Rabies in 1950.

Thus began a series of meetings and collaborative efforts by WHO which continue today. Hilary Koprowski was chairman of the 8th Expert Committee meeting in 1991 and is a co-editor of the fourth edition of the WHO publication "Laboratory Techniques in Rabies" (Meslin, Kaplan and Koprowski 1996).

THE IRAN INCIDENT

There were some experimental indications that rabies antiserum might have value in the post-exposure treatment of humans. An opportunity to test this possibility arose in 1954 when 29 patients were brought to the Pasteur Institute in Teheran from a mountain village where the same rabid wolf had bitten them some 30 hours previously. One of the patients received the usual course of potent vaccine (21 inoculations) prepared at the Iran Institute, while the other group received the vaccine course plus one inoculation of Lederle's high-titer antiserum prepared in horses. The results were clear-cut: survival in the antiserum-treated group was significantly greater than in the group receiving vaccine alone. However, no significant differences were observed in patients receiving one *vs* two inoculations of antiserum. In light of the high incidence of serum sickness due to the equine origin of the serum, use of a single inoculation was eventually recommended.

RABIES VACCINATION TODAY

VACCINES AND OTHER PRODUCTS FOR HUMAN USE

Around 1955, there was a transition from vaccines prepared from animal nerve tissue to embryonated eggs and very soon thereafter to adaptation of rabies virus to cultures of human diploid cells. This vaccine, which remains the reference in comparative studies of immunogenicity, appeared on the French market in 1974 and a little later in North America. The late 1970s and the 1980s saw the development of vaccines prepared on various cellular substrates such as primary explant cells of hamster, dog or fetal calf kidney, fibroblasts of chicken embryo or diploid cells from rhesus monkey fetal lung, and finally cells from continuous cell lines. The production of some of these vaccines, i.e., those of dog and calf origin, was stopped at the end of the 1980s, whereas millions of doses of vaccines prepared on cells or embryonating eggs have been sold throughout the world. However, in most countries where canine rabies is hyperendemic, vaccine produced on imported or locally derived animal nerve tissue and administered to patients is still much more abundant than modern vaccine produced on cells or embryonated eggs. Indeed, vaccines similar to those used in 1885 by Pasteur are found alongside products of the "second-latest" generation prepared on cells of continuous lines cultivated on microcarriers and highly purified. Still, there is a noticeable trend toward the modern vaccines, due perhaps to popular demand, pressure from health authorities, and/or commercial interests.

An additional concern is the inadequate supply of rabies immunoglobulins or serum, whether homologous or heterologous. It is therefore rare for a patient at risk of rabies who requires the systematic use of serum and vaccine to receive the combined therapy that WHO recommends.

The increased availability of modern vaccines is highly dependent on its increased affordability. In an effort to reduce the direct costs (vaccine, injection equipment, nursing staff) and indirect costs (transportation, accommodation, lost days of work, etc.) of vaccine treatment, efforts are underway to reduce the volume of vaccine injected during a course of treatment and/or the number of visits required for treatment, while ensuring effectiveness for all categories of exposure.

Clinical trials have been conducted over the last 15 years to reduce direct costs, especially those related to the purchase of a cell culture vaccine, which is usually imported by developing countries (Nicholson 1996). Reduced treatment regimens were evaluated at the last WHO Expert Committee on Rabies, and regimens were adopted: the "2.1.1." regimen, consisting of two intramuscular doses on day 1 and a single-dose booster on days 7 and 21, and the "2.2.2.0.1.1." regimen, consisting of a dose injected intradermally at two sites on days 0, 3 and 7, and at one site on days 30 and 90. Other regimens involving vaccine injection in eight sites on the first day, four sites on day 7, and one site on days 30 and 90 (804011) offer an additional safety margin. However, none of these regimens is entirely satisfactory and research into new reduced regimens such as a 40201 using a total amount of 0.8 ml of vaccine and requiring only three visits on days 0, 7, and 30 is continuing. The use of highly purified horse immunoglobulins, which are safer than the heterologous products of the previous generation, should provide at least a partial solution to the problem of cost and insufficient supply of human immunoglobulin. Nevertheless, recommendations may be simply academic, since in most cases, immunoglobulins are not available or affordable.

STRATEGIES FOR RABIES CONTROL

Canine rabies still prevails in most countries of Africa, Asia, and Central and Latin America. An estimated 2.4 billion people live in areas where dog rabies exists. In these areas, the dog is responsible for most of the estimated 40,000 human deaths annually due to rabies worldwide, although reliable data are scarce in many countries, and the real impact of the disease on human and animal health cannot be fully assessed. In fact, the worldwide human death rate due to rabies, most often following exposure to dogs, may be 60,000 or 70,000 annually if higher case estimates are used for very populated rabies-infected countries in Africa and Asia.

Dramatic decreases in the number of human cases due to dog rabies have been reported in several countries. In China, the number of cases has declined within four to five years from 5,000 to 1,000 each year. Thailand now reports approximately 100 cases annually, representing a three-fold reduction in case number compared to reports of the late 1970s. In China and Thailand, progress seems to be related to the increased availability of affordable, safe, and efficient post-exposure treatment. In other countries, e.g., Tunisia, Sri Lanka, and Vietnam, large-scale dog vaccination campaigns combined with an increased number of post-exposure treatments in humans are responsible for most of the improvements. In many other countries where only small efforts have been made to prevent rabies in humans (through post-exposure treatment and public education) or in dogs (through immunization or population control), little impact on rabies is observed.

VACCINES FOR ANIMALS

Scaled-up efforts toward the oral immunization of foxes through vaccine-containing bait distributed in the wild has led to a dramatic reduction in the incidence of rabies. This positive trend started in 1989 and the frequency of cases has decreased to less than 20% of the 1989 level in countries that have been conducting oral immunization campaigns since 1992 or before (Table 1).

Oral immunization of foxes started in 1978 in Europe, with small-scale tests dominating until 1985. Within this period, about 1.6 million baits were distributed in Switzerland, Germany, and Italy. Since then, a total of 17 countries, including the Federation of Russia, Estonia, and Belarus, have participated, and the land areas targeted for fox vaccination have increased enormously. Currently, nearly 15 million vaccine baits are distributed annually by airplanes or hunters. Since 1978, 73.7 million baits were produced

TABLE 1. PRINCIPAL TYPES OF RABIES VACCINE for veterinary use

A.   Inactivated (no living virus)

   Nerve tissue—from sheep and suckling mice

   Cell culture—various cell lines

B.   Modified live virus

   Parenteral use—ERA (tissue culture) and Flury (eggs)

   Oral use—attenuated virus strains, SAD and SAD derivatives

   – recombinant of thymidine kinase-deficient vaccinia virus with glycoprotein gene of rabies virus

   – recombinant rabies vaccines with raccoon and avian poxviruses, baculovirus, adenovirus and BCG, salmonellae

Adapted from the 8th WHO Expert Committee on Rabies (1992)

53

overall and spread over an area of 4.9 million kilometers, which is more than 1.58 times the area of the current member states of the European Community (including Sweden, Finland, and Austria).

In North America, the oral immunization technique is being used to control fox rabies in Ontario, Canada, and raccoon rabies in parts of the United States (e.g., New York, New Jersey). A project will be initiated soon to control rabies in coyotes in southern Texas.

VACCINES AND OTHER PRODUCTS—RESEARCH AND DEVELOPMENT

Oral vaccination of dogs is currently under study. Since 1988, WHO has continuously promoted international collaboration and coordinated research in this effort, especially with respect to safety of candidate vaccines and bait and bait delivery systems for dogs and for non-target species. Between 1991 and 1994, WHO established guidelines for determining oral vaccine efficacy in laboratory dogs and for bait development, bait preference trials, and for the evaluation of bait delivery systems in the field. However, a number of safety requirements still remain to be fulfilled.

From 1982 to 1990, WHO also coordinated efforts to characterize Lyssaviruses using monoclonal antibodies, which has led to the application of genetic methods for virus identification. The recent results and the advantages/disadvantages of serological *vs* genetic typing techniques for epidemiological, phylogenetic, and diagnostic purposes and their potential implications for vaccine strain selection were reviewed at a meeting in November 1994.

Studies have shown that mice injected with anti-rabies monoclonal antibodies are protected against a virulent challenge with the virus. These results suggest the potential usefulness of monoclonal antibodies, which can easily be produced in large quantities, in post-exposure treatment of humans (Koprowski 1995). A recombinant vaccine using an avian poxvirus (canarypox) as a carrier for the rabies glycoprotein gene was tested in volunteers for safety and immunogenicity. This vaccine might be of some interest, especially in zones where rabies is highly endemic and might be economically attractive if antigens against the major infant diseases could be incorporated and expressed by the same virus vector. However, such immunization does not appear to obviate the need for at least one booster after contact that is presumed infective.

There is some evidence that an embryonic egg vaccine might induce an earlier antibody response (at day 7) when applied intradermally instead of intramuscularly, possibly due to its content of free rabies nucleoprotein. The role of the rabies nucleoprotein as an enhancer of the immune response requires further study, however.

The production of vaccine for human use from substrates previously reserved for animals (e.g., the BHK-21 cell line) is an option that should be explored, although the risks attached to use of these products after ß-propiolactone inactivation must be assessed further, especially with respect to the biological activity of the residual cellular DNA from those cells. Still, the use of such cells enables production of antigens in larger quantities than is obtained with more conventional substrates.

References

Baer G, ed. (1991). The natural history of rabies. Ed 2. Boston: CRC Press.

Bunn TO (1991). Canine and feline vaccines. In: Baer G. The natural history of rabies. Ed 2. Boston: CRC Press.

Bynum WF (1995). The scientist as anti-hero. Nature 375: 25.

Johnson H (1965). Rabies. In: Horsfall FL, Tammon, I, eds. Viral and rickettsial infections of man. Philadelphia: JP Lippincott, 814-840.

Koprowski H (1995). Visit to an ancient curse. Scientific American, Science and Medicine, May/June, 1995: 48-57.

McKendrick AG (1940). A ninth analytical review of reports from Pasteur Institute on the results of antirabies treatment. Bull Health Org League of Nations 9: 31-78.

Meslin F-X, Kaplan MM (1996). General considerations in the production and use of brain tissue and purified chick embryo rabies vaccines for human use. In: Meslin F-X, Kaplan MM, Koprowski HK, eds. Laboratory techniques in rabies. Geneva: World Health Organization, Chapter 19.

Meslin F-X, Stohr K (1994). Results of the world survey of rabies for 1989. In: Rodriguez J et al, eds. Proc. 3rd Int. Meeting on Advances and Research Towards Rabies Control in the Americas, Mexico, 5-7 October 1992, 1994. Washington, DC: Pan-American Health Organization, 165-170.

Morgeaux S, Tordo N, Gaunter C et al (1993). Beta-treatment impairs the biological activity of residual DNA from BHK-21 cells infected with rabies virus. Vaccine 11: 89-90.

Nicholson KG (1996). Cell culture vaccines for human use: General considerations. In: Meslin F-X, Kaplan MM, Koprowski HK, eds. Laboratory techniques in rabies. Ed 4. Geneva: World Health Organization, Chapter 22.

Petricciani JC (1993). Ongoing tragedy of rabies. Lancet 342: 1067.

Rupprecht CE, Hanlon CA, Dietzschold B, et al (1993). Utility of monoclonal antibodies in the post exposure treatment of rabies. In: Proceedings of the Symposium on Rabies Control in Asia, Djakarta, Indonesia, 27-30 April 1993. Lyon: Foundation Marcel Mérieux, 213-221.

Spink W (1978). Infectious diseases. Minneapolis: University of Minnesota Press, 417.

Steele JH, Fernandez PJ (1991). History of rabies and global aspects. In: Baer G, ed. The natural history of rabies. Ed 2. Boston: CRC Press, 1-24.

Teulieres L (1993). New trends in rabies vaccines. In: Proceedings of the Symposium on Rabies Control in Asia, Jakarta, Indonesia, 27-30 April 1993. Lyon: Foundation Marcel Merieux, 203-211.

Valery-Radot, René. The life of Pasteur. (Devonshire RL, transl.) Garden City, NY: Garden City Publishing Co., Inc. (N.B.—Valery-Radot wrote this book in 1901, but the English edition used was published in 1971).

Vodopija I, Clark HF (1991). Human vaccination against rabies. In: Baer G, ed. The natural history of rabies. Ed 2. Boston: CRC Press, 571-595.

4

Recent Advances in the Study of Rabies

Zhen Fang Fu and Bernhard Dietzschold

Despite significant progress in the study of rabies since the time of Pasteur, particularly in the pre- and post-exposure treatment for rabies, the disease remains an important health problem today. It is estimated that 50,000 to 70,000 people worldwide die from rabies and four million people require post-exposure prophylaxis each year (Anonymous 1993). In addition, there is accumulating epidemiological evidence that new forms of human rabies are emerging in North America, which may represent a potentially serious health problem. Unlike classical rabies, which is inflicted by the bite of an infected animal, recent rabies cases in the United States have no known history of exposure. Most of these cases were caused by a variant associated with silver-haired bats (CDC 1994, 1995). In this chapter, we review the most recent advances in rabies research, including progress in the molecular pathogenesis of rabies, the mechanisms of clearance of rabies virus from the central nervous system (CNS), and the emerging rabies virus infections.

Molecular Pathogenesis

During the past several years, significant progress has been made toward a better understanding of rabies virus invasion into and spread within the CNS and the interactions between rabies virus and host cells. The role of inflammatory responses in the pathogenesis of rabies has also been studied.

Peripheral Entry of Rabies Virus to the Nervous System

One of the questions concerning the early events of rabies virus infection is whether virus enters peripheral nerves directly or after local replication in non-neuronal tissues (e.g., muscle cells). Studies carried out so far have yielded controversial results due to differences in virus strains and animal species used, as well as differences in the methods of virus detection. Murphy et al (1973) and Charlton and Casey (1979, 1981) reported the detection of rabies virus antigens by immunofluorescence in myocytes at the inoculation site of hamsters and skunks before they were detected in the spinal cord and dorsal root ganglia. Using the reverse transcriptase-polymerase chain reaction (RT-PCR) technique, Shankar et al (1991) detected rabies virus RNA in the trigeminal ganglia as early as 18 hours and in the brain stem 24 hours after inoculation of rabies virus CVS strain into the masseter muscle of mice. Analysis of masseter muscle tissue demonstrated viral RNA at one hour post-infection, but not thereafter, indicating that the virus had invaded the nerve ending directly, without prior replication in the muscle. The results from this study are consistent with observations of Coulon et al (1989) who demonstrated rabies virus antigen by immunocytochemistry in ventral motor neurons and primary sensory neurons at 18 hours after injection of mice with the CVS strain in the forelimb muscle. Early studies with neurectomy or leg amputation have also shown that rabies virus enters nerve endings within a few hours of inoculation (Baer et al 1965).

SPREAD OF RABIES VIRUS WITHIN THE CNS

After entering peripheral nerves, rabies virions are transported to the CNS via axons. Neurons are the primary targets for rabies virus infection (Charlton 1994). The spread of rabies virus within the CNS is mainly effected by intra-axonal transport and direct transfer of the virus between neurons, as demonstrated in rabies-infected mice (Iwasaki and Clark 1975) and skunks (Charlton and Casey 1979), respectively. Viral budding was observed, usually with simultaneous endocytosis by the neighboring axon terminal (Charlton and Casey 1979). In a study of rabies virus spread in Lewis rats infected intranasally with the CVS strain of rabies virus (Gosztonyi et al 1993), there was evidence that the viral nucleocapsid, rather than the complete virus, may be transported within the CNS. In the early stages of rabies virus infection, the viral nucleocapsid is synthesized in the infected neurons and appears in axons in association with tubular filaments. Staining with anti-nucleocapsid antibody confirmed the presence of this structure in the absence of complete viral particles. Apparently, the nucleocapsid is transported through synapses where it accumulates in the postsynaptic compartment. Only in the late stage of infection are the bullet-shaped, enveloped viral particles assembled on the membrane of the endoplasmic reticulum and on the plasma membrane of neurons (Gosztonyi et al 1993).

EFFECT OF RABIES VIRUS INFECTION ON GENE EXPRESSION IN NEURONAL CELLS

Neuropharmacological studies have revealed functional modifications of neurons during rabies virus infection, including decreased affinity of the opiate receptors for agonists (Munzel and Koschel 1981), a reduction in scopolamine binding to muscarine acetylcholine receptors (Tsiang 1982), and a decrease in γ-amino-n-butyric acid uptake (Ladogana et al 1994). Analyses of the electrophysiological and sleep alterations in mice experimentally infected with rabies virus indicated the disappearance of rapid eye movement, the persistence of the hippocampal rhythmic slow activity late in infection, the occurrence of myoclonus, and brain death (Gourmelon et al 1986, 1991). Brain electrical activity disappeared completely about 30 minutes before cardiac arrest, which was usually believed to be the cause of death. These observations indicate that fatal rabies results from functional alteration rather than structural damage of neurons. Recent studies have shown that rabies virus infection affects the expression of immediate-early response genes and late-response genes in neuronal cells of the host (Fu et al 1993). As shown in Figure 1, the appearance of rabies virus mRNA was accompanied by the increased mRNA expression of host immediate-early response gene egr-1 mRNA, and rabies virus mRNA was co-distributed with egr-1 mRNA in the hippocampus and the cortex. The correlation between activated mRNA expression and the strong increase in viral RNA raised the possibility that the egr-1 product induces phenotypic changes in neurons that render them more susceptible to viral replication.

By contrast, the expression of mRNAs transcribed from late-response genes, such as preproenkephalin (pENK, Figure 1) and neuronal constitutive nitric oxide synthase (Akaike et al 1995), was significantly decreased during rabies virus infection. Ceccaldi et al (1993) also observed a down-regulation of 5-hydroxytryptamine receptor in rat brain infected with rabies virus. The expression of a housekeeping gene glyceraldehyde-3-phosphate dehydrogenase (G3PDH) mRNA was also dramatically decreased in rabies

FIGURE 1. EFFECTS OF RABIES VIRUS INFECTION on expression of selected neuronal genes. Rats were infected with rabies virus CVS-11. Adjacent coronal sections through caudate putamen (x) and hippocampus (xx) of normal rats (a) and rabies virus-infected rats at 2 (b), 3 (c), 4 (d), 5 (e), and 6 (f) days post-infection were hybridized with rabies virus nucleocapsid (R-N), egr-1, pENK (ENK), and G3PDH cRNA probes as described (Fu et al 1993).

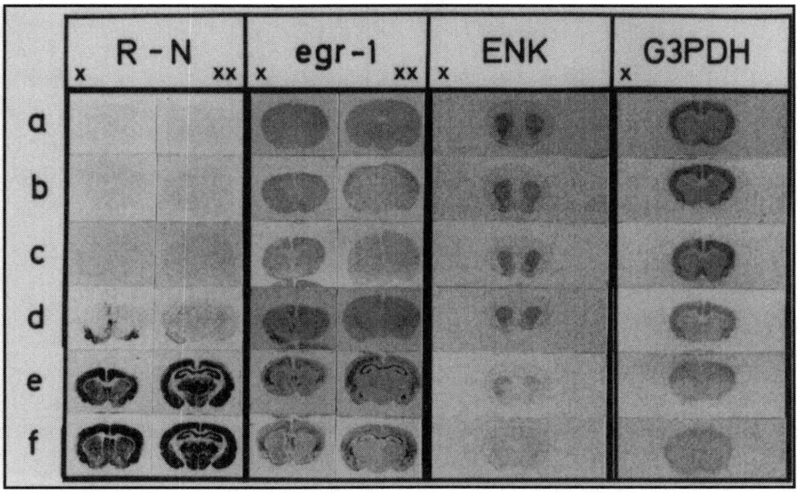

virus-infected rat brain (Figure 1). Together, the data suggest that rabies virus replication in neurons leads to a general inhibition of host gene expression, perhaps due to the recruitment of the host transcriptional machinery by the virus and the preferential production of rabies virus mRNA. The decreased expression of important neurotransmitters and neuropeptides such as pENK are likely to underlie the impairment of neuronal functions, as observed by Gourmelon et al (1986, 1991), and the death of infected animals.

ROLE OF INFLAMMATION AND NITRIC OXIDE IN THE PATHOGENESIS OF RABIES

Although inflammatory reactions are often mild in rabies virus-infected brains, with relatively little cell destruction (Miyamoto and Matsumoto 1967), studies in animal models demonstrate that immune mechanisms are involved in the neuropathogenesis of rabies. Both CD4 and CD8 T-cell subsets have been implicated in both rabies virus neuritic paralysis and fatal encephalo-pathogenic rabies (Sugamata et al 1992; Weiland et al 1992). However, it is not known whether immune-mediated neuronal injury in rabies is due to lytic activity of T cells or is mediated by factors (cytokines) released from activated T cells. Accumulation of inflammatory cells in the CNS, particularly cells of the monocyte lineage, is often seen in rabies virus-infected brains. Following rabies virus infection, the expression of proinflammatory cytokines such as interleukin (IL)-6, tumor necrosis factor (TNF)-α and interferon (IFN)-γ is markedly increased in rat brain and correlated with the development of disease symptoms (Figure 2). Recently, nitric oxide (NO) produced by inducible nitric oxide synthase (iNOS) in activated monocytes has been implicated in the pathogenesis of a variety of diseases of both infectious and autoimmune origin (Koprowski et al 1993). The expression of mRNA for iNOS as well as iNOS enzyme activity was demonstrated in brains of rats infected with rabies virus (Akaike et al 1995; Koprowski et al 1993). Since the nature of the possible pathogenic effects of NO is clearly dependent upon the levels generated locally, spin-trapping and electron paramagnetic resonance spectrometry was used to directly measure NO in the brains of animals infected with rabies virus CVS strain (Hooper et al 1995). NO, which can be detected in normal brain at concentrations less than 1 μM, increased approximately 30-fold in brains of rats with clinical rabies. The presence of such high levels of NO raises the possibility of a causal link between NO production and neuronal dysfunction in rabies.

CLEARANCE OF RABIES VIRUS FROM THE CNS

ANTIBODY-MEDIATED CLEARANCE OF RABIES VIRUS

Anti-rabies virus immunoglobulin has long been used in combination with rabies virus vaccines for treatment immediately after exposure, and it was believed that the antibody neutralizes virus before it enters the nervous system. However, anti-rabies virus antibodies have also been shown to protect animals from rabies when the virus has already invaded the CNS

(Dietzschold et al 1992). Eighty percent of hamsters were protected from rabies when a cocktail of mouse monoclonal antibodies (MAb) was administered 36 hours after infection (Schumacher et al 1989), and 60% of hamsters were protected when a human MAb was given as late as three days after infection (Dietzschold 1994). Eighty percent of rats were also protected when a mouse MAb was given one day after infection, whereas all nontreated rats succumbed to rabies (Dietzschold et al 1992). RT-PCR analysis revealed that by 12 hours, rabies virus had already invaded the CNS of all infected rats and confirmed the elimination of rabies virus RNA in 80% of the animals at 24 hours after treatment with MAb (Dietzschold 1993).

Clearance of rabies virus from the CNS has also been reported in adult mice infected with mutant virus variants. All of these mutant variants, which have a mutation at amino acid position 333 of the rabies glycoprotein (from arginine to isoleucine or glutamine) are apathogenic for adult mice but are still pathogenic for newborn mice (Dietzschold et al 1983; Seif et al 1985). Apathogenic variants invade the CNS at the same rate and with a similar topographical distribution in the CNS as the pathogenic viruses (Dietzschold et al 1985; Jackson 1991; Lafay et al 1991). The only differences between the

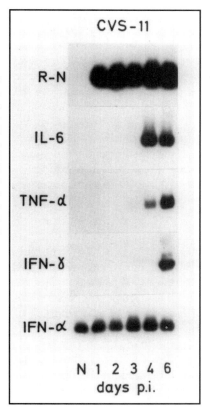

FIGURE 2. EFFECTS OF RABIES VIRUS INFECTION on the expression of proinflammatory cytokines. Rats were infected with rabies virus CVS-11. One day before and various days after infection, rats were sacrificed and perfused with phosphate-buffered saline. Brains were removed and RNA was isolated from the cerebrum. RT-PCR was performed on these samples using primers specific for rabies virus nucleocapsid (R-N) and for the cytokines, IL-6, TNF-α, IFN-γ and IFN-α. Amplified cDNA was resolved in agarose gels and analyzed by Southern blotting as described (Shankar et al 1991; 1992).

61

pathogenic and apathogenic viruses are that the pathogenic viruses infect many more neurons than do the apathogenic variants (Dietzschold et al 1985; Jackson 1991) and that infection of apathogenic variants is restricted to certain categories of neurons such as mitral cells (Lafay et al 1991). Apathogenic variants invade and replicate in the CNS, but are completely cleared from the CNS by 17 days after infection. Animals infected with apathogenic virus (F3) (Dietzschold et al 1983) produce virus-neutralizing antibodies at a faster rate than those infected with pathogenic virus (CVS strain), suggesting that such antibodies might play a role in the clearance of rabies virus from the CNS.

MECHANISMS OF RABIES VIRUS CLEARANCE FROM THE CNS

Antibody-mediated neutralization of virus is the result of several different mechanisms, including inhibition of virus attachment, release of bound virus, and inhibition of the fusion step that leads to virus uncoating (Dietzschold et al 1987). However, the protective activity *in vivo* of a particular antibody does not correlate with its virus-neutralizing activity *in vitro* (Schumacher et al 1989; Dietzschold et al 1992), indicating that antibody-mediated clearance of rabies virus is facilitated by a mechanism that is distinct from previously described antiviral effects mediated by antibody, such as neutralization of extracellular virus, complement-mediated lysis, or antibody-dependent cellular cytotoxicity. The *in vivo* protective activity of a particular MAb correlates with its ability to prevent virus spread from cell to cell and with the capacity to inhibit transcription of viral RNA *in vitro* (Dietzschold et al 1992). Possibly, MAbs exert their protective activity after being internalized in infected neurons. Another unanswered question is how MAbs administered intramuscularly can pass the blood-brain barrier and exert their antiviral function in the CNS. It has been suggested that rabies virus infection may stimulate neurons to produce vasoactive substances (e.g., substance P), which can mediate vasodilation and plasma extravasation (Dietzschold 1993). Thus, activation of neuronal function in early stages of virus infection may be an essential factor in the antibody-mediated virus clearance from the CNS.

Rabies as an Emerging Infection

EPIZOOTICS OF RACCOON RABIES IN THE EASTERN UNITED STATES

Raccoon rabies was first observed in Florida in 1947 (Scatterday et al 1960); by 1977, raccoon rabies had extended into the neighboring states of Georgia, Alabama, and South Carolina (CDC 1978). In 1977, a second focus of raccoon rabies was reported in West Virginia and, in 1978, in Virginia. These few index cases have led to the second major epizootic of raccoon rabies in the mid-Atlantic states, and by 1994, raccoon rabies was reported along the entire eastern seaboard with the exception of Maine. More than 20,000 cases of raccoon rabies have been recorded since 1977, about 6,000 cases in

1993 alone (Krebs et al 1994; Rupprecht and Smith 1994). The second epizootic of raccoon rabies may have been due to the translocation of rabid raccoons from Florida to West Virginia (Rupprecht and Smith 1994). Nucleotide sequence analysis showed that rabies viruses isolated from these two epizootics share more than 99% homology but show significant differences from isolates found elsewhere or in other species in the United States (Smith et al 1984; Rupprecht and Smith 1994). These raccoon epizootics present a major health threat to humans, although to date, no human fatalities have been documented. No doubt, improved human post-exposure treatment with human rabies immunoglobulin and human diploid cell vaccine (Wiktor et al 1964) has played a major role in preventing human fatalities associated with raccoon rabies. Indeed, the number of human post-exposure treatments has escalated in proportion to the increased number of rabid raccoons. For example, the number of human post-exposure treatments given in New York State rose from 84 in 1989 to more than 1000 by 1992 (Rupprecht and Smith 1994). The cost for biologicals alone exceeded one million dollars in 1992.

A recombinant vaccinia virus expressing rabies virus glycoprotein has been developed (Kieny et al 1984) that is effective against rabies virus infection under laboratory conditions (Wiktor et al 1984; Blancou et al 1986; Rupprecht et al 1986). Field trials with this recombinant vaccine in foxes have resulted in large-scale elimination of rabies in vaccinated areas in Europe (Brochier et al 1991). Field trails with the same vaccine in raccoons in the United States have also demonstrated its safety and efficacy (Rupprecht et al 1992). With the licensure of this recombinant vaccine now in the United States, the epizootics of raccoon rabies should be brought under control in the near future.

EMERGING RABIES VIRUS INFECTIONS IN HUMANS IN THE UNITED STATES

Recent trends in human rabies indicate that a new form of rabies is emerging in the United States. In the 14 years prior to 1994, only nine indigenous cases of rabies in humans in the United States were reported, five of which were caused by a variant of street rabies virus associated with the silver-haired bat *(Lasionycteris noctivagans)* (Childs et al 1994). In 1994 alone, five more indigenous cases of human rabies were described and the silver-haired bat variant was identified in brain tissue from four of these (CDC 1994, 1995). None of these recent cases of human rabies were identified with any history of exposure to conventional risks such as animal bites, scratches, licks, or aerosol in bat-infested caves. Furthermore, rabies was not diagnosed antemortem in most of these victims (CDC 1994, 1995), suggesting that considerable numbers of cases may escape scrutiny. The association of most human rabies cases in the United States with a rabies virus variant associated with silver-haired bats raises the possibility that the silver-haired bat variant has unique biological properties such as increased neuro-invasiveness. Alternatively, this variant could be transmitted by an unidentified vector(s). With the development of reverse genetics to clone infectious rabies virus (Schnell et al 1994), it will be possible to characterize biologically the rabies virus variant associated with silver-haired bats.

Acknowledgments

This work was supported by Public Health Service Grants AI-33029 and AI-09706. The authors express their gratitude to Dr. H. Koprowski for encouragement and support.

References

Akaike T, Weihe E, Schäfer MKH, Fu ZF, Vogel WH, Schmidt HHHW, Koprowski H, Dietzschold B (1995). Effect of neurotropic virus infection on neuronal and inducible nitric oxide synthase activity in rat brain. J Neuro Virol 1: 118-125.

Anonymous (1993). World Survey of Rabies 27, Veterinary Public Health Unit, WHO, Geneva.

Baer GM, Shanthaveerppa TR, Bourne GH (1965). The pathogenesis of street rabies virus in rats. Bull WHO 33: 783-794.

Blancou J, Kieny MP, Lathe R, Lecocq JP, Pastoret PP, Soulebot JP, Desmettre P (1986). Oral vaccination of the fox against rabies using a live recombinant vaccinia virus. Nature 322: 373-375.

Brochier B, Kieny MP, Costy F, Coppens P, Bauduin B, Lecocq JP, Languet B, Chappuis G, Desmettre P, Afiademanyo K, Libois R, Pastoret PP (1991). Large-scale eradication of rabies using recombinant vaccinia-rabies vaccine. Nature 354: 520-523.

CDC (1978). Rabies surveillance annual summary 1977. Atlanta: US Department of Health, Education and Welfare.

CDC (1994). Human rabies—California, 1994. MMWR 43: 455-457.

CDC (1995). Human rabies—West Virginia, 1994. MMWR 44: 86-93.

Ceccaldi PE, Fillion MP, Ermine A, Tsiang H, Fillion G (1993). Rabies virus selectively alters 5-HT$_1$, receptor subtypes in rat brain. Eur J Pharmacol 245: 129-138.

Charlton KM (1994). The pathogenesis of rabies and other lyssaviral infections: Recent studies. Curr Top Microbiol Immunol 187: 95-119.

Charlton KM, Casey GA (1979). Experimental rabies in skunks: Immunofluorescent, light and electron microscopic studies. Lab Invest 41: 36-44.

Charlton KM, Casey GA (1981). Experimental rabies in skunks: persistence of virus in denervated muscle at the inoculation site. Can J Comp Med 45: 357-362.

Childs JE, Trimarchi CV, Krebs JW (1994). The epidemiology of bat rabies in New York State, 1988-92. Epidemiol Infect 113: 501-511.

Coulon P, Derbin C, Kucera P, Lafay F, Prehaud C, Flamand A (1989). Invasion of the peripheral nervous system of adult mice by the CVS strain of rabies virus and its avirulent derivative Av01. J Virol 63: 3550-3554.

Dietzschold B (1993). Antibody-mediated clearance of viruses from the mammalian central nervous system. Trends Microbiol 1: 63-66.

Dietzschold B (1994). Monoclonal antibodies in rabies therapy. Clin Immunother 1: 245-249.

Dietzschold B, Wunner WH, Wiktor TJ, Lopes AD, Lafon M, Smith CL, Koprowski H (1983). Characterization of an antigenic determinant of the glycoprotein that correlates with pathogenicity of rabies virus. Proc Natl Acad Sci USA 80: 70-74.

Dietzschold B, Wiktor TJ, Trojanowski JQ, MacFarlan RI, Wunner WH, Torres-Anjel MJ, Koprowski H (1985). Differences in cell-to-cell spread of pathogenic and apathogenic rabies virus in vivo and in vitro. J Virol 56: 12-18.

Dietzschold B, Tollis M, Lafon M, Wunner WH, Koprowski H (1987). Mechanisms of rabies virus neutralization by glycoprotein-specific monoclonal antibodies. Virology 161: 29-36.

Dietzschold B, Kao M, Zheng YM, Chen ZY, Maul G, Fu ZF, Rupprecht CE, Koprowski H (1992). Delineation of putative mechanisms involved in the antibody- mediated clearance of rabies virus from the central nervous system. Proc Natl Acad Sci USA 89: 7252-7256.

Fu ZF, Weihe E, Zheng YM, Schäfer MKH, Sheng H, Corisdeo S, Rauscher FJ, Koprowski H, Dietzschold B (1993). Differential effects of rabies and Borna disease viruses on immediate-early and late-response gene expression in brain tissue. J Virol 67: 6674-6681.

Gosztonyi G, Dietzschold B, Kao M, Rupprecht CE, Ludwig H, Koprowski H (1993). Rabies and borna disease. A comparative pathogenetic study of two neurovirulent agents. Lab Invest 68: 285-295.

Gourmelon P, Briet D, Court L, Tsiang M (1986). Electrophysiological and sleep alterations in experimental mouse rabies. Brain Res 398: 128-140.

Gourmelon P, Briet D, Clarencon D, Court L, Tsiang H (1991). Sleep alterations in experimental street rabies virus infection occur in the absence of major EEG abnormalities. Brain Res 554: 159-165.

Hooper DC, Ohnishi ST, Kean R, Numagami Y, Dietzschold B, Koprowski H (1995). Local nitric oxide production in viral and autoimmune diseases of the central nervous system. Proc Natl Acad Sci USA 92: 5312-5316.

Iwasaki Y, Clark HF (1975). Cell to cell transmission of virus in the central nervous system. II. Experimental rabies in mouse. Lab Invest 28: 142-148.

Jackson AC (1991). Biological basis of rabies virus neurovirulence in mice: comparative pathogenesis study using the immunoperoxidase technique. J Virol 65: 537-540.

Kieny MP, Lathe R, Drillen R, Spehner D, Skory S, Schmitt D, Wiktor TJ, Koprowski H, Lecocq JP (1984). Expression of rabies virus glycoprotein from a recombinant vaccinia virus. Nature 312: 163-166.

Koprowski H, Zheng YM, Heber-Katz E, Rorke L, Fu ZF, Hanlon CA, Dietzschold B (1993). In vivo expression of inducible nitric oxide synthase in experimentally-induced neurological diseases. Proc Natl Acad Sci USA 90: 3024-3027.

Krebs JW, Strine TW, Smith JS, Rupprecht CE, Childs JE (1994). Rabies surveillance in the United States during 1993. JAVMA 205: 1695-1709.

Ladogana A, Bouzamondo E, Pocchiari M, Tsiang H (1994). Modification of tritiated gamma-amino-n-butyric acid transport in rabies virus-infected primary cortical cultures. J Gen Virol 75: 623-627.

Lafay F, Coulon P, Astic L, Saucier D, Riche D, Holley A, Flamand A (1991). Spread of the CVS strain of rabies virus and of the avirulent mutant Av01 along the olfactory pathways of the mouse after intranasal inoculation. Virology 183: 320-330.

Miyamoto K, Matsumoto S (1967). Comparative studies between pathogenesis of street and fixed rabies infection. J Exp Med 125: 447-456.

Muenzel P, Koschel K (1981). Rabies virus decreases agonist binding to opiate receptors of mouse neuroblastoma-rat glioma hybrid cells for 108 CC15. Biochem Biophys Res Commun 101: 1241-1250.

Murphy FA, Bauer SP, Harrison AK, Winn WC (1973). Comparative pathogenesis of rabies and rabies-like viruses. Viral infection and transit from inocultion site to the central nervous system. Lab Invest 28: 361-376.

Rupprecht CE, Smith JS (1994). Raccoon rabies: the re-emergence of an epizootic in a densely populated area. Semin Virol 5: 155-164.

Rupprecht CE, Hanlon CA, Koprowski H, Hamir AN (1992). Oral wildlife rabies vaccination: development of a recombinant virus vaccine. Trans. 57th NA Wildl & Nat Res Conf, 439-452.

Rupprecht CE, Wiktor TJ, Johnston DH, Hamir AN, Dietzschold B, Wunner WH, Glickman LT, Koprowski H (1986). Oral immunization and protection of raccoons (Procyon lotor) with a vaccinia-rabies glycoprotein recombinant virus vaccine. Proc Natl Acad Sci USA 83: 7949-7952.

Scatterday JE, Schneider NJ, Jennings WL, Lewis AL (1960). Sporadic animal rabies in Florida. Publ Hlth Rep 75: 945-953.

Schnell MJ, Mebatsion T, Conzelmann KK (1994). Infectious rabies viruses from cloned cDNA. EMBO J 13: 4195-4203.

Schumacher CL, Dietzschold B, Ertl HCJ, Niu HS, Rupprecht CE, Koprowski H (1989). Use of mouse anti-rabies monoclonal antibodies in postexposure treatment of rabies. J Clin Invest 84: 971-975.

Seif I, Coulon P, Rollin PE, Flamand A (1985). Rabies virulence: effect of pathogenicity and sequence characterization of rabies virus mutations affecting antigenic site III of the glycoprotein. J Virol 53: 926-935.

Shankar V, Dietzschold B, Koprowski H (1991). Direct entry of rabies virus into the central nervous system without prior local replication. J Virol 65: 2736-2738.

Shankar V, Koa M, Harmir AN, Sheng H, Koprowski H, Dietzschold B (1992). Kinetics of virus spread and changes in levels of several cytokine mRNAs in the brain after intranasal infection of rats with Borna disease virus. J Virol 66: 992-998.

Smith JS, Sumner JW, Roumillat LF, Baer GM, Winkler WG (1984). Antigenic characteristics of isolates associated with a new epizootic of raccoon rabies in the United States. J Infect Dis 149: 769-774.

Sugamata M, Miyazawa M, Mori S, Spangrude GJ, Ewalt LC, Lodmell DL (1992). Paralysis of street rabies virus-infected mice is dependent on T lymphocytes. J Virol 66: 1252-1260.

Tsiang H (1982). Neuronal function impairment in rabies infected rat brain. J Gen Virol 61: 277-281.

Weiland F, Cox JM, Meyer S, Dahme E, Reddehase MJ (1992). Rabies virus neuritic paralysis: Immunopathogenesis of nonfatal paralytic rabies. J Virol 66: 5096-5099.

Wiktor TJ, Fernandes MV, Koprowski H (1964). Cultivation of rabies virus in human diploid strain WI-38. J Immunol 93: 353-366.

Wiktor TJ, MacFarlan RI, Reagan KJ, Dietzschold B, Curtis PJ, Wunner WH, Kieny MP, Lathe R, Lecocq JP, Mackett M, Moss B, Koprowski H (1984). Protection from rabies by a vaccinia virus recombinant containing the rabies virus glycoprotein gene. Proc Natl Acad Sci USA 81: 7194-7198.

Measles Virus

Measles virus is among the most contagious of viruses and the disease it produces has been attributed in large part to successful European conquests in the New World. The development of a highly successful vaccine and its initial usage is chronicled by Samuel Katz, who participated in its development. Although the use of the vaccine has led to a dramatic control of disease, much remains to be done, especially in Third World countries where measles is believed to infect over 40 million persons and kills nearly one million persons a year.

Measles was the first virus described to cause immunosuppression, but how this occurs is still not clear. Major efforts are now being made to establish the genetics of measles and to evaluate virus-monocyte/lymphocyte interactions. The mechanism(s) that allows the virus to commonly cause an acute infection but rarely a subacute post-infectious encephalitis or a chronic viral infection of the brain (subacute sclerosing encephalitis) has also not been resolved. Recently, the receptor for measles virus has been identified and is being used to establish transgenic mice. Since humans are the natural host for measles virus and only subhuman primates have been used as models to study viral disease, the availability of transgenic small animal (mouse) models should advance knowledge in this field. This recent development in receptor chemistry and the derivation of transgenic mice is described by Manchester, Scheiflinger, and Oldstone.

5

The History of Measles Vaccine and Attempts to Control Measles

Samuel L. Katz

Although Rhazes, a Persian physician, is credited with the first written description of measles (Abu Becr 1748) and perhaps with distinguishing between it and smallpox, earlier Hebrew physicians (such as Al Yehudi) had recognized the illness but without a distinction between it and other rash disorders. As urbanization occurred in subsequent centuries, the proximity of larger populations permitted epidemics with continued circulation of virus in the cities. By the early seventeenth century, measles was more clearly recognized as a distinct entity as described by the London physician, Thomas Sydenham (1922). Nearly 40 years before Jenner's description of smallpox vaccine, a Scottish physician, Francis Home, appreciating the gravity of the disease, attempted to produce mild measles by mimicking the variation process used to mitigate smallpox (Enders 1964). Although, in contrast to smallpox, there were no vesicles or pustules, he chose to inoculate blood from an infected patient and succeeded in passing infection with rash to 10 of his 12 childhood subjects. He thereby demonstrated the presence of viremia more than a century before the concept of viruses had even been set forth.

A young Danish medical graduate, Peter Panum, observed and accurately described an outbreak of measles on the Faroe Islands in 1846 and was able to define the incubation period, the infectiousness of the illness, and the lifelong duration of immunity among individuals who had contracted measles more than 50 years previously (Panum 1939). The infectivity for susceptible monkeys was demonstrated in 1911 by Goldberger and Anderson who transmitted the infection to monkeys using blood and pharyngeal washings from measles patients. One unsustained series of experiments by Rake, Shaffer, and Stokes in the 1940s suggests that those investigators were able to cultivate measles virus, at least transiently, in chick embryos.

ATTENUATED VIRUS VACCINE

It remained for John Enders (1954) and his colleagues to propagate the virus successfully in cell cultures of human and simian renal tissue in the mid-1950s. The cytopathic effects observed were those of cell fusion with large syncytia containing multiple nuclei which revealed eosinophilic nuclear and cytoplasmic inclusions when fixed and stained. These were identical to those observed in lungs, gastrointestinal tract, and other organs of patients who had died of measles. This isolate, named for the youngster from whom it had been cultured, David Edmonston, was subsequently passed in human kidney, human amnion, fertilized hen's eggs (Milovanovic 1957), and eventually in chick embryo cell culture (Katz 1958). This became the progenitor for measles vaccines used subsequently and today throughout the world. Its attenuation was first demonstrated in susceptible monkeys who developed antibodies after inoculation with the chick cell virus but no detectable viremia or illness, in contrast to other monkeys inoculated with the early passaged material. After intracerebral inoculation of susceptible monkeys revealed no histopathology, the attenuated virus was administered to immune adults (Enders 1960).

EARLY VACCINE STUDIES

With the demonstrated innocuity of the attenuated virus, studies were then undertaken in a small group of institutionalized youngsters where measles occurred annually with high morbidity and mortality. In parallel with the earlier cell culture and animal studies, serologic assays had been developed to test complement-fixing, virus-neutralizing, and hemagglutination-inhibiting antibodies to measles virus. Susceptible children were chosen on the basis of absent antibodies and, after parental permission, were the first children to receive the attenuated virus subcutaneously (Katz 1960a). The success of these studies led to further trials among home-dwelling children in five United States cities (Katz 1960b). Although a significant number of the recipients developed fever, and some a mild rash, this occurred with apparent well-being and without discomfort. Nevertheless, the vaccine was later given along with a small dose of immunoglobulin to further attenuate its clinical reactivity. Subsequent laboratory passages at lower temperatures produced a "further attenuated" level of virus which was administered successfully without globulin. This vaccine or similar material has now been used throughout the world for more than 30 years in attempts to control measles (Markowitz 1994).

HOST FACTORS IN MEASLES AND ITS PATHOGENESIS

The clinical manifestations of measles virus infection are greatly influenced by host factors. Healthy, well-nourished children usually have a self-limited illness which may be accompanied by complications (otitis media, pneumonia, gastroenteritis) but carries a low mortality rate (1 per 500). In contrast, malnourished children, particularly those with protein deficiencies, are subject to severe illness with mortality rates that may reach 20% or higher (Morley 1974). In contrast to the self-limited disease of healthy children, they may have prolonged gastroenteritis, desquamation, negative nitrogen balance for weeks or months thereafter, cutaneous bacterial abscesses, cancrum oris, and other debilitating complications. In general, patients at either extreme of age are apt to be more ill, infants in the first 18 months of life and adults. Recent studies have demonstrated the efficacy of oral vitamin A administration in preventing the severe keratoconjunctivitis as well as lower respiratory tract infections accompanying measles (Hussey 1990). Rare central nervous system (CNS) complications have produced some of the most dread aspects of measles (Johnson 1984). A post-infectious encephalitis occurs in roughly 1 per 1000 cases and a delayed subacute sclerosing panencephalitis (SSPE) in 1 per 100,000. The latter is of particular interest because it may not become apparent until 7 to 10 years after the acute measles (Modlin 1977) and reveals defective measles virions within the neuronal and glial cells. An intermediate form has been observed in children with severe immune compromise. Likewise, a progressive giant cell pneumonia has occurred in such immune-deficient patients (Mitus 1959).

Early observations of the loss of delayed cutaneous hypersensitivity to tuberculoproteins among infected patients who developed measles correlated with an exacerbation of the underlying mycobacterial disease (von Pirquet 1908). This loss of cell-mediated immunity continues to be one of the provocative aspects of measles pathogenesis (McChesney 1989; Griffin 1989).

INACTIVATED VIRUS VACCINE

Simultaneously with the licensure of live attenuated virus vaccines in the early 1960s, several pharmaceutical firms marketed a "killed" measles vaccine which was prepared by formalin inactivation. Two or three injections of this vaccine produced a detectable antibody response, but when such patients subsequently were exposed to wild-type measles virus, they developed a severe atypical illness with CNS obtundation, marked pneumonia, and a centrifugal rash that was quite unlike that of natural measles (Fulginiti 1967). In addition to the acute nodular pneumonia with effusion that many suffered, they were later found to have persistent pulmonary nodules that remained for years after this syndrome (Annunziato 1982). After approximately four years in the United States (1963–1967), the inactivated vaccine was removed from the market. Studies suggest that formalin inactivation denatured the fusion protein of the virion, and patients with atypical measles lacked antibodies to this protein.

MEASLES CONTROL IN THE VACCINE ERA

Various countries have approached measles control in differing fashions, but the most common approach has been to administer vaccine shortly after the first birthday, by which time most maternal transplacental measles antibody has been catabolized so that the attenuated virus replicates successfully. With its high immunogenicity and prophylactic efficacy, a single exposure to measles vaccine produces serum antibody in 95 to 97% of susceptible individuals and field efficacy of at least 90%. Nonetheless, with birth cohorts of millions of children in various countries, even if all received measles vaccine, there would still be a significant annual increase in the number of children who failed to respond to that initial dose. As a result, many countries have now instituted two-dose schedules with a second dose given at varying times or ages after the initial inoculation. With such a strategy, it has been possible to eliminate measles from some countries (Finland, Sweden) and to reduce the annual number of reported cases in the United States from more than two million to less than 400 in 1995. Because the virus is so highly communicable, importations from countries where measles vaccination is not widely practiced continue to provide the introduction of virus among those few who remain susceptible.

GENETIC STABILITY OF MEASLES VIRUS

One reassuring aspect of measles virus has been its relative genetic stability. Viruses recovered from patients throughout the world and at various times from the 1950s to the present are all neutralized by antibodies induced by the vaccine. Genomic analysis has revealed changes in a few nucleotides of the hemagglutinin (HA) and nucleoprotein (N) genes, but this minor genetic drift has not altered overall antigenicity. The entire genome of the Edmonston strain has been sequenced and demonstrated to contain six genes that encode the six major structural proteins of the virus. Using genomic analyses, it is possible to identify strains from various parts of the world (Rota 1992) by such molecular epidemiology. Bellini (1994) and others have described in detail the replicative biology and biochemistry of the virus. They have demonstrated that the sequences among vaccine strains used throughout the world differed by no more than 0.5 to 0.6% at the nucleotide level.

GLOBAL ASPECTS OF MEASLES PREVENTION

The World Health Organization (WHO) estimates that in the early 1980s, as many as 2.5 million children died annually from measles (Assaad 1983). By 1994, with 78% global coverage by measles vaccine before age one year, reported cases had dropped significantly and deaths were judged to be less than 750,000. At least three different vaccination protocols have been utilized. Most commonly, an initial age of administration, varying from 9 months (the Expanded Program on Immunization [EPI] of WHO) to 15 months (the United States), has been added to the schedule for children. A number of nations, in attempts to eliminate completely the indigenous circulation of measles virus, have added a second dose of vaccine (Bottiger 1987; Peltola 1994) for those children who were primary vaccine failures or the unusual secondary failures. A third, completely different approach has been that of mass immunization during several days annually in which all children in varying age cohorts (usually up to age 15 years) have been immunized nationwide. This last approach has been successful in Cuba and Brazil (Anonymous 1992).

An unsolved problem remains the infection of infants with wild measles before nine months of age in nations where the virus continues to circulate widely, particularly some of the sub-Saharan African nations. The "window of susceptibility" between the loss of protection by maternal transplacental antibody and the age at which attenuated vaccine will successfully replicate leaves a hiatus in which wild measles continues to occur. In attempts to overcome the resistance to earlier immunization, various measles vaccines were administered at unusually high titers to children at four to six months of age. The successful seroconversion and protection by this approach led WHO in 1989 to recommend administration of vaccine at high titer to children as young as six months of age. Although it was technologically difficult to produce large amounts of such high-titered vaccine, these programs were initiated. However, in 1990–1991 it became apparent that there was an increased mortality rate later in infancy among recipients of these high-titered materials, not from measles itself but from other infections (respiratory, gastrointestinal, malarial). Moreover, this increased mortality occurred most often in girls. The boys in contrast remained well, protected by the vaccine without enhanced morbidity or mortality (Gellin 1994). The biologic basis of these epidemiologic observations remains incompletely explained, but the immune suppression that transiently follows natural measles and measles virus vaccines has been implicated. As a result, the early use of high-titered vaccines has been abandoned. In many countries, it appears that vaccination at age 9 to 12 months with 85 to 90% coverage has been successful in markedly reducing measles cases and subsequent spread of the virus within communities (Cutts 1994). The presence of successfully immunized older infants and children reduces the likelihood of younger infants' exposure.

One feature of current live attenuated measles virus vaccines that adds to its difficulties in field use has been its heat sensitivity. Because the vaccine is both light- and heat-sensitive, it loses infectivity rapidly when reconstituted in tropical lands. Attempts continue to provide a more stable product.

Current Diagnostic Dilemmas

As measles has become less frequent in many areas, physicians and other health care workers have lost familiarity with the illness so that diagnosis of rash disease is less reliable. In attempts to assist with field diagnosis, Bellini and others have devised rapid measles-specific IgM antibody tests that can be done on heel stick blood as a slide agglutination test (personal communication 1995). Further studies explore the possibility of similar tests on salivary and/or urine samples (Gresser 1960). Field availability of such rapid, sensitive, and specific tests will greatly facilitate the epidemiological surveillance of measles.

Conclusions

It remains to be determined whether the circulation of measles viruses can be totally controlled by the currently available vaccines or whether newer products will be required that permit immunization of young infants under the cover of maternal antibody. A number of laboratories have returned to the original monkey models to investigate the possibility of such approaches using experimental vaccines prepared in recombinant or other more current preparations. The EPI of WHO has announced a global target of reduction of measles incidence by 90% and measles mortality by 95% from pre-EPI (1976) levels by the end of this year. Any approaches to these goals will require the establishment of a sustained effort as new susceptible cohorts are born annually. Until the virus has been totally eliminated, its high degree of communicability will threaten any programs that are not sustained.

A number of unanswered questions remain as challenges to current and future investigators. These include the development of successful strategies to overcome transplacental immunity, a better understanding of measles-induced immunosuppression (Griffin 1994), the identification of the molecular basis for virus attenuation, and the explanation of gender differences in the pathogenesis of high-titered vaccine administration in the youngest infants.

References

Abu Becr M (1748). A discourse on the smallpox and measles (Mead R, trans.). London: J Brindley.

Annunziato D, Kaplan MH, Hull WW, Ichinose H, Lin JH, Balsam D, Paladino VS (1982). Atypical measles syndrome: Pathologic and serologic findings. Pediatrics 70: 203-209.

Anonymous (1992). Measles elimination in the Americas. Bull Pan Am Health Organ 26: 271-275.

Assaad F (1983). Measles: Summary of worldwide impact. Rev Infect Dis 5:452-459.

Bellini WJ, Rota JS, Rota P (1994). Virology of measles virus. J Infect Dis 170:(suppl 1), S15-S23.

Bottiger M, Christenson B, Romanus V, Taranger J, Strandell A (1987). Swedish experience of two dose vaccination programme aiming at eliminating measles, mumps and rubella. Brit Med J 295: 1264-1267.

Cutts FT, Markowitz LE (1994). Successes and failures in measles control. J Infect Dis 170: (suppl 1),S32-S41.

Enders JF, Peebles TC (1954). Propagation in tissue cultures of cytopathogenic agents from patients with measles. Proc Soc Exp Biol Med 86: 277-286.

Enders JF, Katz SL, Milovanovic MV, Holloway A (1960). Studies on an attenuated measles-virus vaccine. I. Development and preparation of the vaccine: Technics for assay of effects of vaccination. N Engl J Med 263: 153-159.

Enders JF (1964). Francis Home and his experimental approach to medicine. Bull Hist Med 38: 101-112.

Fulginiti VA, Eller JJ, Downie AW, Kempe CH (1967). Altered reactivity to measles virus. Atypical measles in children previously immunized with inactivated measles virus vaccines. JAMA 202: 1075-1080.

Gellin BG, Katz SL (1994). Measles: State of the art and future directions. J Infect Dis 170: (suppl 1), S3-14.

Goldberger J, Anderson JF (1911). An experimental demonstration of the presence of the virus of measles in the mixed buccal and nasal secretions. JAMA 57: 476-478.

Gresser I, Katz SL (1960). Isolation of measles virus from urine. N Engl J Med 263: 452-454.

Griffin DE, Ward BJ (1989). Differential CD4 T cell activation in measles. J Infect Dis 168: 275-281.

Griffin DE, Ward BJ, Esolen LM (1994). Pathogenesis of measles virus infection: An hypothesis for altered immune responses. J Infect Dis 170: (suppl 1), S24-S3 1.

Hussey GD, Klein M (1990). A randomized, controlled trial of Vitamin A in children with severe measles. N Engl J Med 323: 160-164.

Ikic DM (1983). Edmonston-Zagreb strain of measles vaccine: Epidemiologic evaluation in Yugoslavia. Rev Infect Dis 5: 558-563.

Johnson RT, Griffin DE, Hirsch RL, Wolinsky JS, Roedenbeck S, Lindo de Soriano I, Vaisberg A (1984). Measles encephalomyelitis—Clinical and immunologic studies. N Engl J Med 310: 137-141.

Katz SL, Milovanovic MV, Enders JF (1958). Propagation of measles virus in cultures of chick embryo cells. Proc Soc Exp Biol Med 97: 23-29.

Katz SL, Enders JF, Holloway A (1960a). Studies on an attenuated measles-virus vaccine. II. Clinical, virologic and immunologic effects of vaccine in institutionalized children. N Engl J Med 263: 159-161.

Katz SL, Kempe HC, Black FL, Lepow UL, Krugman S, Haggerty RJ, Enders JF (1960b). Studies on an attenuated measles-virus vaccine. VIII. General summary and evaluation of the results of vaccination. N Engl J Med 263: 180-184.

Markowitz LE, Katz SL (1994). Measles vaccine. In: Plotkin SA, Mortimer EA Jr, eds. Vaccines. Ed. 2. Philadelphia: WB Saunders Co, 229-276.

McChesney MB, Oldstone BA (1989). Virus-induced immunosuppression: Infections with measles virus and human immunodeficiency virus. Adv Immunol 45: 335-380.

Milovanovic MV, Enders JF, Mitus A (1957). Cultivation of measles virus in human amnion cells and developing chick embryo. Proc Soc Exp Biol Med 95: 120-127.

Mitus A, Enders JF, Craig JM, Holloway A (1959). Persistence of measles virus and depression of antibody formation in patients with giant cell pneumonia after measles. N Engl J Med 261: 882-889.

Modlin JF, Jabbour FT, Witte JJ, Halsey NA (1977). Epidemiologic studies of measles, measles vaccine, and sub-acute sclerosing panencephalitis. Pediatrics 59: 505-512.

Morley DC (1974). Measles in the developing world. Proc R Soc Med 67: 1112-1115.

Panum PL (1939). Observation made during the epidemic of measles on the Faroe Islands in the year 1846. Med Classics 3: 839-886.

Peltola H, Heinonen OP, Valle M et al (1994). The elimination of indigenous measles, mumps, and rubella from Finland by a 12 year, two-dose vaccination program. N Engl J Med 331: 1397-1402.

Rota JS, Hummel KB, Rota PA, Bellini WJ (1992). Genetic variability of the glyco-protein genes of current wildtype measles isolates. Virology 188: 135-142.

Sydenham T (1922). The works of Thomas Sydenham. Vol 2. London: Sydenham Society, 250-251.

von Pirquet CE (1908). Das Verhalten der kutanen Tuberkulin-reaktion wahrend der Masern. Dtsch Med Wochenschr 34: 1297-1300.

6

Recent Progress in Measles Research: Pathogenesis and the Virus Receptor CD46

MARIANNE MANCHESTER, FRIEDRICH SCHEIFLINGER,
AND MICHAEL B.A. OLDSTONE

Measles, one of the most contagious infections of humans, has a complex pathogenesis. In addition to causing an acute infection, measles virus (MV) is associated with a number of additional syndromes that cause morbidity and mortality, including immunosuppression, secondary infections, encephalitis, and neurological complications. While these syndromes have been closely studied, a clear understanding of their pathogenesis remains elusive. The first association of a viral infection with immunosuppression was described in 1908 by von Pirquet, who noted loss of the skin tuberculin reaction as well as remission of kidney degeneration (now known to be of an autoimmune pathogenesis) during acute MV infection (von Pirquet 1908). Suppression of the immune system by MV can lead to complications from secondary infections such as pneumonia, bacterial and fungal infections, and diarrhea. In addition, 0.1% of measles patients suffer from an encephalitis that appears a few weeks post-infection (Anonymous 1989a, 1989b). The most insidious of the measles-associated syndromes, subacute sclerosing panencephalitis (SSPE), is poorly understood. SSPE occurs in about 1 in 100,000 cases of measles and is characterized by a chronic neurodegeneration that is inevitably fatal (Bloom 1989; Clements and Cutts 1995).

Most of our knowledge about the pathogenesis of measles comes from observations of humans infected with the virus. The evidence that immunity to measles is primarily cell-mediated is derived from studies of genetically immune-deficient children. Those who lack an antibody response (agammaglobulinemic) can be vaccinated for measles or are able to recover from natural measles infection with lifelong immunity. By contrast, children who are deficient in cell-mediated responses display high morbidity and mortality when infected with measles, even when they receive anti-MV antibody therapy (Burnet 1968). Nevertheless, anti-measles antibody levels play an important role in reducing viral load and thus also have an important, although not essential, role in controlling MV. Hence any successful measles vaccine should induce both a strong cell-mediated and neutralizing antibody immune response.

Our incomplete understanding of the precise contributions of the immune system and the role of the various viral gene products in pathogenesis accounts for many of the difficulties with the current MV vaccine. The attenuated vaccine is not effective in the presence of maternal antibody and thus is given after the first birthday. However, very young children (usually age 3 to 12 months) protected by maternal neutralizing antibody are increasingly becoming infected with MV; as a result, it is estimated that more than one million children die of measles each year (Bloom 1989; Clements and Cutts 1995; Clemens et al 1988). In addition, while immunity following natural measles infection is lifelong, evidence is mounting that immunity derived from vaccination can wane in adulthood (Mulholland 1995; McLean 1995). More must be learned about passive and waning immunity to measles before new or current vaccines can be developed and used to their full potential.

Like many negative-strand RNA viruses, MV has been difficult to study in the absence of a system to introduce mutations into the MV genome and study their effects on replication and pathogenesis. Recent findings suggest that this difficulty may soon be addressed. Martin Billeter and colleagues have developed a system using an infectious MV cDNA clone to introduce mutations into complete infectious MV genomes (Radecke et al 1995). The recent discovery of the MV receptor also opens new avenues in the understanding of MV pathogenesis and in the development of new models for studying MV. Our group and others have characterized the MV receptor, CD46, on human cells (Naniche et al 1993a; Dorig et al 1993; Manchester et al 1994). We have begun to use those data to generate mice expressing human CD46 to provide the small animal model needed for careful dissection of the host immune response to MV.

Membrane Cofactor Protein (CD46): The MV Receptor

Transmitted by aerosol, MV enters the body through the respiratory epithelium and is subsequently disseminated through the lymphoid system. The MV envelope contains two viral glycoproteins, the hemagglutinin (HA) and fusion (F) proteins (Norrby and Oxman 1990). Infection of susceptible cells is a two-step process, with the HA protein mediating initial receptor binding and the F protein directing fusion of virus and host cell membranes (Nussbaum et al 1995; Choppin and Scheid 1980; Gerlier et al 1994b; Richardson et al 1980). By using monoclonal antibodies to block virus infection, Naniche et al (1993a, 1993b) showed that a surface glycoprotein, CD46, was the receptor for MV. Dorig et al (1993) and our group (Manchester et al 1994) confirmed these results. The MV receptor, membrane cofactor protein (MCP or CD46), is a member of the regulators of complement activation (RCA) superfamily of proteins (Liszewski et al 1991), which regulate complement deposition and activation on host tissue and protect host cells from complement-mediated damage (reviewed in Liszewski et al 1991). CD46 binds the complement components C3b and C4b and acts as cofactor for their cleavage by the serine protease factor 1, thus preventing deposition of these proteins on CD46-expressing tissues (Liszewski et al 1991).

CHARACTERISTICS OF CD46

CD46 is expressed on most human cells and tissues except erythrocytes (Liszewski et al 1991). CD46 is a type I membrane glycoprotein of isoforms ranging from 57,000 to 67,000 daltons in mass (Liszewski et al 1991) (Figure 1). The extracellular domain largely consists of four short consensus repeat (SCR) modules, which are the hallmark motif in the RCA superfamily; proteins in the family have between 4 and 56 SCRs (Table 1). An increasing number of proteins in the RCA family have been identified as virus receptors (Table 1)

FIGURE 1. PRIMARY ISOFORMS OF CD46. 67kD (BC1 and BC2) and 57kD (C1 and C2) isoforms are depicted. SCR, short consensus repeat (1-4); STP, serine-threonine-proline-rich domain. Cytoplasmic tails are designated 1 (cyt-1) or 2 (cyt-2); U, region of unknown function.

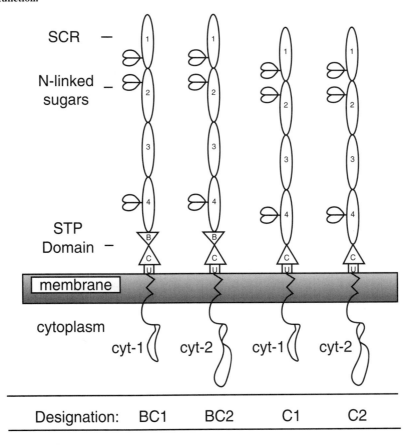

(Nemerow et al 1985; Ward et al 1994; Bergelson et al 1994, 1995). The structure of the SCR is highly conserved and contains four invariant cysteines that form two disulfide bonds (C1 pairs with C3; C2 with C4). SCR domains have approximately 30% homology between members of the RCA family. In CD46, SCRs 1, 2, and 4 contain sites for N-linked glycosylation; these sugars may be important for interaction of CD46 with MV (Maisner et al 1994). In addition to the SCRs, the CD46 extracellular domain contains a region that is rich in serine, threonine, and proline residues (designated STP) and is heavily O-glycosylated. The STP region is encoded by three separate exons, designated A, B, and C (Figures 1 and 2). Following a small region of unknown function, the transmembrane region links the extracellular domain to a cytoplasmic tail (Post et al 1991).

The isoforms of CD46 arise by alternative splicing of the STP domains near the membrane and splicing of two different cytoplasmic tails (cyt-1 and cyt-2) (Post et al 1991). Four of those isoforms are commonly found in human cells and tissues; they consist of either the B plus C or only the C STP regions (the 67-kilodalton and 57-kilodalton forms, respectively) and cyt-1 or cyt-2 (Figure 1) (Post et al 1991). The tails contain putative sites for phosphorylation by protein kinase C (cyt-1) and casein kinase 2 (cyt 1 and 2),

TABLE 1. REGULATORS OF COMPLEMENT ACTIVATION (RCA) superfamily

RCA family proteins	No. of SCRs	STP	Membrane anchor	Receptor for:
Membrane bound:				
MCP (CD46)	4	+	cyt	MV streptococcus
DAF (CD55)	4	+	gpi	echovirus Coxsackievirus B3
CR1 (CD35)	30	-	cyt	-
CR2 (CD21)	16	-	cyt	EBV
Fluid phase:				
C4bpα	56	-	-	-
C4bpβ	3	-	-	-
Factor H	20	-	-	-

MCP, membrane cofactor protein; DAF, decay accelerating factor; CR1, complement receptor 1; CR2, complement receptor 2; cyt, cytoplasmic tail; gpi, glycosyl-phosphatidylinositol membrane anchor; MV, measles virus; EBV, Epstein-Barr virus. Proteins in the family have been identified as receptors for measles (Naniche et al 1993a; Dorig et al 1993; Manchester et al 1994), echoviruses (Ward et al 1994; Bergelson et al 1994), coxsackievirus B3 (Bergelson et al 1995), Epstein-Barr virus (Nemerow et al 1985) and streptococcus (Okada et al 1995).

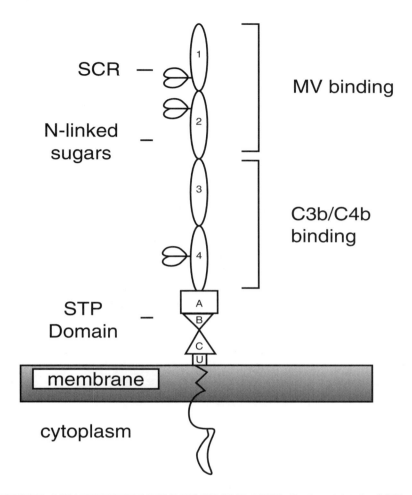

FIGURE 2. LIGAND BINDING DOMAINS ON CD46. MV binding is restricted to SCR1 and SCR2 (Manchester et al 1995), while SCRs 3 and 4 bind complement. Some cofactor activity is also provided by SCR2 (Adams et al 1991).

phosphorylation by src kinase (cyt-2) and nuclear localization (Liszewski et al 1994). The different tails affect the rate of isoform transport to the cell surface in the exocytic pathway (Liszewski et al 1994). Infection of cells expressing any of the four isoforms leads to productive infection, expression of viral proteins, and syncytium formation (Manchester et al 1994; Gerlier et al 1994a). In addition, the cytoplasmic tail of CD46 is not required for MV receptor function, since a lipid-anchored form of CD46 functions equally as wild-type (Manchester et al 1995; Varior-Krishnan et al 1995).

MV BINDING TO CD46

A hallmark of RCA proteins is the presence of multiple SCRs, with clusters of SCRs involved in ligand binding. For example, the natural ligands for CD46, i.e., C3b and C4b, interact with SCR3 and SCR4 of CD46 (Adams et al 1991) (Figure 2). This phenomenon has been extended to virus binding; for example, the Epstein-Barr virus (EBV) binding region has been mapped to SCR1 and SCR2 of CR2 (Moore et al 1991). For MV, inhibition of infection by monoclonal antibodies directed against the various SCRs of CD46 indicated that the virus binding site is located in a portion of the molecule distal from the membrane (Manchester et al 1995). Chimeric molecules between CD46 and CD55 (decay accelerating factor [DAF], a closely related member of the RCA superfamily) indicated that both SCR1 and SCR2 of CD46 were required for MV receptor function (Manchester et al 1995). Deletions of SCR1 or SCR2 of CD46 blocked receptor function, while deletions of SCR3 did not (Manchester et al 1995) (M. Manchester, K. Liszewski, J.P. Atkinson, and M.B.A. Oldstone, unpublished observations). Taken together, these results indicated that the MV binding site is located within SCR1 and SCR2 of CD46 (Figure 2). These results have been confirmed by Iwata et al (1995). Current studies with small CD46-DAF chimeras and CD46 peptides are underway to determine whether the virus interacts with distinct portions of these modules or with a region that spans the junction between the modules. Preliminary studies indicate that there are two distinct regions, one each in SCR1 and 2 and each smaller than 20 amino acids, that interact with MV (M. Manchester, J. Gairin, and M.B.A. Oldstone, unpublished observations).

MV HA: Viral Component of the Virus-Cell Interaction

MV HA is a type II transmembrane glycoprotein of 617 amino acids with an apparent molecular mass of 78,000 to 80,000 daltons (Alkhatib and Briedis 1986). HA is synthesized in the endoplasmic reticulum and processed to a highly glycosylated disulfide-linked homodimer as it moves along the exocytic pathway. The intracellular transport rate is relatively low as compared to other viral membrane proteins (Hu et al 1994). Cross-linking studies suggest that HA is expressed on the cell surface in an oligomeric, probably tetrameric form (Ogura et al 1991; Malvoisin and Wild 1993), and is closely associated with the MV F protein (Wild et al 1991). Both proteins together form spike-like projections from the viral envelope. Close proximity of HA and F is a requirement for the fusion of viral envelope and cell membrane (Wild et al 1991). Because fusion only occurs in the presence of the receptor (Nussbaum et al 1995), the ternary complex between HA, F, and CD46 is also probably necessary, indicating a highly organized interaction. However, based on the use of separate HA or F genes, it is clear that CD46 binding function can be assigned to the MV HA protein.

Although MV exists as a single immunotype, sequence data accumulated over the past few years suggest the existence of at least six different lineage groups containing a variety of strains. Most of the variations in the amino acid sequence have been identified in the region near the potential glycosylation sites, indicating a selective pressure (Rima et al 1995).

There is now ample evidence that single amino acid changes might influence the binding properties of viral attachment proteins (Sheehan and Iorio 1992; Shioda et al 1988). Changes in the primary protein sequence, leading to changes in the folding pattern, might have a profound impact on the biological function by influencing the kinetics, avidity, and rate of receptor binding. In terms of tertiary structure, the carbohydrate moiety and the position of cysteine residues are thought to have a particularly important role. Molecular analysis of HA revealed that four out of five potential N-linked glycosylation sites are actually used (Cattaneo and Rose 1993). Being concentrated over a short range of 70 amino acids in the extracellular domain, the glycosylation sites are in a typical arrangement for morbillivirus HA proteins. In addition, comparison of the predicted amino acid sequences of different morbillivirus HA proteins also reveals the conservation of at least 12 cysteine and approximately half of the glycine and proline residues. This indicates similar tertiary structure and biological function (Hu and Norrby 1994).

In a more detailed analysis using a large panel of N-glycosylation mutants, Hu et al (1994) evaluated the relative importance of individual N-linked glycosylation sites. There is indication that two of the four sites have a large influence on the antigenicity of HA and that alternative glycosylation at site 4 (position 215, Edmonston strain) is responsible for the appearance of two populations of HA glycoproteins (Hu et al 1994; Ogura et al 1991). Also, at least two carbohydrate chains are necessary for transport, folding, and oligomerization of HA. RNA sequencing of HA transcripts derived from a recent measles outbreak in the United States identified a new potential glycosylation site at position 416 (Rota et al 1992). Whether the site is actually N-glycosylated and whether glycosylation of this site correlates with low hemagglutination activity remains to be established (Sato et al 1995; Sakata et al 1993).

Recently, the role of individual cysteine residues in the HA ectodomain was investigated (Hu and Norrby 1994). Cysteine residues at position 287, 300, 381, 394, 494, 579, and 583 were found to be necessary for oligomerization, processing, cell surface expression, and antigenicity of HA protein. In that study, no particular role could be assigned to cysteine 139, and deletion of cysteine 154 yielded a dimerized molecule, although with an aberrant electrophoretic mobility. In contrast, results by Sato et al (1995) suggest that cysteines 139 and 154 participate in intermolecular disulfide binding.

HA protein activity is measured by two biological assays related to morbillivirus infections: hemagglutination and syncytia formation (cell-cell fusion). Whereas cell fusion requires F and HA protein interaction (Wild et al 1991; Nussbaum et al 1995; Stern et al 1995), hemagglutination of primate red blood cells *in vitro* is thought to be mediated by HA alone. Hemagglutination likely occurs by binding of HA to the primate homologue of CD46 found on red blood cells from certain primate species (Liszewski et al 1991).

Neutralizing antibodies are mainly, although not exclusively, directed against conformational epitopes on HA. Giraudon and Wild (1985) established the biological importance of hemagglutination-inhibiting antibodies, since only these antibodies have been able to protect MV-infected mice from MV-induced encephalitis. The HA gene sequences of several monoclonal antibody escape mutant viruses have been determined and the corresponding epitopes identified (Hu et al 1993). Similarly, swap mutants between monoclonal antibody-reactive and -nonreactive HA molecules allowed the identification of epitopes involved in different HA activities, i.e., fusion helper function or receptor binding (Hummel and Bellini 1995).

CD46 IN ANTIVIRAL STRATEGIES TO BLOCK MV BINDING AND ENTRY

Since it is possible to functionally separate the complement-binding and MV-binding regions of CD46, it might be possible to design soluble mediators that mimic the MV-binding regions of CD46 without interfering with complement binding. A soluble form of CD46 (sCD46) that contains only the extracellular region is capable of inhibiting MV infection and cell-cell fusion in a dose-dependent manner (M. Manchester, J. Atkinson, and M.B.A. Oldstone, unpublished) (Seya et al 1995). sCD46 also can bind C3b and C4b, so that anti-MV therapy using whole soluble CD46 might have immunosuppressive effects. Deletion of SCR1 from CD46 does not affect C3/C4 binding or cofactor activity *in vitro* (Adams et al 1991). Thus, it may be possible to design small molecules that block the MV-CD46 interaction without compromising the complement-CD46 interaction. Preliminary studies have shown that soluble SCR1 alone can inhibit MV infection in tissue culture (M. Manchester, L. Daniels, and M.B.A. Oldstone, unpublished observations). These results suggest that while both SCR1 and 2 of CD46 are necessary to bind HA and result in infection, a single SCR can also interact with MV HA.

MV PATHOGENESIS: CD46 DOWN-MODULATION

Down-modulation of virus receptors has been seen in a number of viral infections (Moscona and Peluso 1992; Hoxie et al 1986; Crise et al 1990; Garcia and Miller 1991; Sleckman et al 1992). Naniche et al (1993c) first observed that down-modulation of MV receptor from the cell surface occurred following infection of cells in culture by the Halle strain of MV. Down-modulation occurred approximately 12 hours post-infection and could be attributed to intracellular expression of the MV HA protein. Since the role of surface CD46 is to protect against complement deposition, down-modulation of surface CD46 might lead to increased complement deposition. Indeed, 13 years before the discovery of the MV receptor, Sissons et al (1980) showed that C3b builds up specifically on MV-infected cells. Recently Schneider-Schaulies et al (1995) showed that modulation of CD46 on infected cells leads to increased sensitivity to complement deposition and lysis. The modulation apparently occurs primarily with vaccine strains and not with wild-type lymphotropic isolates (Schneider-Schaulies et al 1995).

The mechanism and clinical significance of CD46 down-modulation is not clear. Since down-modulation is associated with intracellular expression of HA, the phenomenon may result from complexing and retention of CD46 in the secretory pathway by HA. While CD46 is an important contributor to complement regulation on both epithelial cells and lymphoid cells, both DAF (CD55) and CD59, proteins with similar and arguably redundant functions, are also expressed on these tissues (Liszewski et al 1991). Nevertheless, infection with vaccine strains might lead to down-modulation of CD46 on infected cells and killing by complement, thus curtailing the infection in respiratory epithelium by the vaccine strain and providing a basis for attenuation (Schneider-Schaulies et al 1995). Alternatively, the enhanced deposition by complement on infected cells could lead to the increased lysis and thus dissemination of cell-associated MV. Since several pathogens utilize the complement regulatory proteins as receptors (Table 1) (Nemerow et al 1985; Ward et al 1994; Bergelson et al 1994, 1995; Okada et al 1995), it is possible that pathogens gain a selective advantage in the host by disturbing or modulating complement regulation.

ANIMAL MODELS OF CD46 RECEPTOR EXPRESSION

Studies of MV pathogenesis have been hampered by the lack of a small animal model for MV. Humans are the only natural host for MV, but certain primates can be infected experimentally. In the cynomolgus monkey, MV infection leads to a rash, fever, and lymphopenia. Other primates such as chimpanzees, marmosets, gibbons, and rhesus macaques can be infected and show varied levels of illness (reviewed in Van Binnendijk et al 1995). Both the cost and difficulty of studying the primate models place severe limitations on their use for studying MV pathogenesis.

Murine models for disease are attractive because of the availability of genetically inbred strains and of the powerful tools that allow dissection of the immune system in the mouse. In a mouse model for natural measles infection, important issues such as immunosuppression, virus tropism, and pathogenesis might be resolved. In addition, such a model would be useful for testing new types of measles vaccines or anti-MV therapies. Mice can mount both cytotoxic T-cell and antibody responses to wild-type MV, indicating that infection of mice does occur. However, no pathology or illness is seen, so that infection presumably occurs only at a very low level. Among rodent models for MV infection, the most successful were generated by successive passage of MV strains in neonatal rat brains (Liebert and Finke 1995). These "rodent-adapted" MV strains cause severe disease in mice of the H-2^k and H-2^b histocompatibility haplotypes (Liebert and terMeulen 1987; Niewiesk et al 1993). The disease is characterized by a marked encephalitis and neurodegeneration. However, rodent-adapted MV strains do not cause any of the other facets of MV-associated pathology normally seen in humans or primates and hence the usefulness of this mouse model is limited.

Mouse cell lines that are normally nonpermissive for MV infection can be rendered permissive by expression of human CD46 (Naniche et al 1993a; Manchester et al 1994). These results have been demonstrated in murine cells of either fibroblast or lymphoid origin and indicate that the main cellular contributor to MV tropism is the presence of the CD46 receptor. Nevertheless, replication of MV in mouse cells is less efficient in murine cells, showing a 10- to 100-fold decrease in virus titers compared to human- or primate-derived cell lines. Thus host factors other than CD46 must play some role in determining species specificity of MV replication (Naniche et al 1993a; Manchester et al 1994). Mice do not express a genetic homologue of human CD46 and instead have a complement regulatory protein called Crry/p65 that performs functions of both the DAF/CD55 and CD46 proteins found in humans (Kim et al 1995). Crry/p65 does contain SCRs but does not interact with human complement components or, presumably, with MV (Kim et al 1995).

An auxiliary receptor for MV, moesin, has been identified that has low but apparent specific affinity for MV (Maisner et al 1994; Dunster et al 1994). Moesin is a widely expressed protein involved in cell membrane organization and structure at cell-cell junctions, cleavage furrows, and microvilli (Sato et al 1992). Moesin is highly conserved and the mouse and human homologues are 98% identical (Sato et al 1992). New evidence suggests that low-level MV infection of mouse cells is mediated by binding to mouse moesin and not to Crry/p65 (Dunster et al 1995). Thus expression of the human CD46 receptor in transgenic mice, combined with the natural expression of murine moesin, would be expected to enhance mouse susceptibility to MV.

MOUSE MODELS FOR MEASLES AND SSPE

Mice have been generated that express CD46 (M. Manchester, L. Daniels, and M.B.A. Oldstone, unpublished observation) using a β-actin promoter in order to mimic the broad expression pattern of wild-type human CD46 (Liszewski et al 1991). Expression of the CD46 protein can be detected in peripheral blood mononuclear cells isolated from transgenic mice; these cells are infectable with MV *in vitro*. In preliminary studies, MV replication has been observed in lymphoid tissues of transgenic mice *in vivo* (M. Manchester and M.B.A. Oldstone, unpublished data).

SSPE is characterized by a persistent MV infection primarily of neurons and also glia of the central nervous system (CNS). SSPE patients show high levels of MV expression, and much of the virus isolated from brain tissue is defective, containing many mutations throughout the virus genome (reviewed in Rima et al 1995; Cattaneo and Billeter 1992). In addition, SSPE patients have high levels of anti-MV antibodies and evidence of progressive CNS disease. To study MV-neuron interactions *in vivo*, we have generated transgenic mice that express CD46 under the control of the neuron-specific enolase promoter (NSE) (G. Rall and M.B.A. Oldstone, unpublished observations), a promoter that restricts expression to CNS neurons in the brains of transgenic mice (Sakimura et al 1987; Forss-Petter et al 1987; Rall et al 1995). In NSE-CD46 mice, the receptor can be detected in neurons isolated from the hippocampus of transgenic mice (G. Rall and M.B.A. Oldstone, unpublished data). When these mice are inoculated intracerebrally with Edmonston MV, viral replication as evidenced by the presence of viral antigen is found specifically in neurons (M. Manchester, G. Rall, L. Daniels, and M.B.A. Oldstone, unpublished). In contrast, MV antigens and MV replication are not found in similarly inoculated nontransgenic control mice. These results provide the first indication that expression of CD46 on murine tissue may indeed enhance the ability of MV to replicate in the mouse.

CONCLUSIONS

While much has been learned about the pathogenesis of measles, many questions remain unanswered. The discovery of the measles receptor, CD46, has opened new avenues for study of the pathogenesis and tropism of measles. Characterization of the MV-CD46 interaction has led to a better understanding of the virus-host relationship and has provided an opportunity to develop a small animal model for MV. Further study of the interaction between MV and CD46 should provide crucial insights into MV pathogenesis and immunity that might lead to the development of new and specific anti-MV therapies.

ACKNOWLEDGMENTS

Our work described herein was supported by the National Institutes of Health (grants AI09484 and AI36222 to M.B.A. Oldstone and by postdoctoral training grant MH19185 to M. Manchester). F. Scheiflinger was supported by a fellowship from ImmunoAG, Vienna, Austria. This is neuropharmacology department publication 9569-NP from The Scripps Research Institute.

References

Adams EM, Brown MC, Nunge M, Krych M, Atkinson JP (1991). Contribution of the repeating domains of membrane cofactor protein (MCP;CD46) of the complement system to ligand binding and cofactor activity. J Immunol 147: 3005-3011.

Alkhatib G, Briedis DJ (1986). The predicted primary structure of the measles virus hemagglutinin. Virology 150: 479-490.

Anonymous (1989a). Measles, United States, 1st 26 weeks 1989. MMWR 38: 863-871.

Anonymous (1989b). Measles–United States. MMWR 38: 601-605.

Bergelson JM, Chan M, Solomon KR, St. John NF, Lin H, Finberg RW (1994). Decay-accelerating factor (CD55), a glycosylphosphatidylinositol-anchored complement regulatory protein, is a receptor for several echoviruses. Proc Natl Acad Sci USA 91: 6245-6248.

Bergelson JM, Mohanty JG, Crowell RL, St. John NF, Lublin DM, Finberg RW (1995). Coxsackievirus B3 adapted to growth in RD cells binds to decay-accelerating factor (CD55). J Virol 69: 1903–1906.

Bloom BR (1989). Vaccines for the third world. Nature 342: 115-120.

Burnet FM (1968). Measles as an index of immunological function. Lancet 2: 610-613.

Cattaneo R, Billeter MA (1992). Mutations and A/I hypermutations in measles virus persistent infections. Curr Top Microbiol Immunol 176: 63-74.

Cattaneo R, Rose JK (1993). Cell fusion by the envelope glycoprotein of persistent measles virus which caused lethal human brain disease. J Virol 67: 1493-1502.

Choppin P, Scheid A (1980). The role of viral glycoproteins in absorption, penetration, and pathogenicity of viruses. Rev Infect Dis 2: 40-61.

Clemens JD, Stanton BF, Chakraborty J, Chowdhury S, Rao MR, Ali M, Zimicki S, Wojtyniak B (1988). Measles vaccination and childhood mortality in rural Bangladesh. Am J Epidemiol 128: 1330-1339.

Clements CJ, Cutts FT (1995). The epidemiology of measles: thirty years of vaccination. Curr Top Microbiol Immunol 191: 13-33.

Crise B, Buonocore L, Rose JK (1990). CD4 is retained in the endoplasmic reticulum by the human immunodeficiency virus type I glycoprotein precursor. J Virol 64: 5585-5593.

Dorig R, Marcel A, Chopra A, Richardson CD (1993). The human CD46 molecule is a receptor for measles virus (Edmonston strain). Cell 75: 295-305.

Dunster LM, Schneider-Schaulies J, Dehoff MH, Holers VM, Schwartz-Albiez R, terMeulen V (1995). Moesin, and not the functional homologue Crry/p65) of membrane cofactor protein (CD46), is involved in the entry of measles virus (strain Edmonston) into susceptible murine cell lines. J Gen Virol 76: 2085-2089.

Dunster LM, Schneider-Schaulies J, Loffler S, Lankes W, Schwartz-Albiez R, Lottspeich F, terMeulen V (1994). Moesin: a cell membrane protein linked with susceptibility of measles virus infection. Virology 198: 265-274.

Forss-Petter S, Danielson PE, Catsicas S, Battenberg E, Price J, Nerenberg M, Sutcliffe JG (1987). Transgenic mice expressing beta-galactosidase in mature neurons under neuron-specific enolase control. Neuron 5: 187-197.

Garcia JV, Miller D (1991). Serine phosphorylation-independent downregulation of cell-surface CD4 by nef. Nature 350: 508-511.

Gerlier D, Loveland B, Varior-Krishnan G, Thorley B, McKenzie IF, Rabourdin-Combe C (1994a). Measles virus receptor properties are shared by several CD46 isoforms differing in extracellular regions and cytoplasmic tails. J Gen Virol 75: 2163-2171.

Gerlier D, Trescol-Biemont MC, Varior-Krishnan G, Naniche D, Fugier-Vivier I, Rabourdin-Combe C (1994b). Efficient major histocompatibility complex class II-restricted presentation of measles virus relies on hemagglutinin-mediated targeting to its cellular receptor human CD46 expressed by murine B cells. J Exp Med 179: 353-358.

Giraudon P, Wild F (1985). Correlation between epitopes on hemagglutinin of measles virus and biological activities: passive protection by monoclonal antibodies is related to their hemagglutination activity. Virology 144: 45-58.

Hoxie JA, Alpers JD, Rakowski JL, Huebner K, Haggarty BS, Cedarbaum AJ, Reed JC (1986). Alterations in T4 (CD4) protein and mRNA synthesis in cells infected with HIV. Science 234: 1123-1127.

Hu A, Cattaneo R, Schwartz S, Norrby E (1994). Role of N-linked oligosaccharide chains in the processing and antigenicity of measles virus haemagglutinin protein. J Gen Virol 75: 2173-2181.

Hu A, Norrby E (1994). Role of individual cysteine residues in the processing and antigenicity of the measles virus hemagglutinin protein. J Gen Virol 75: 2173-2181.

Hu A, Sheshberadaran H, Norrby E, Kovamees J (1993). Molecular characterization of epitopes on the measles virus hemagglutinin protein. Virology 192: 351-354.

Hummel BK, Bellini WJ (1995). Localization of antibody epitopes and functional domains in the hemagglutinin protein of measles virus. J Virol 69: 1913–1916.

Iwata K, Seya T, Yanagi Y, Pesando J, Johnson P, Okabe M, Ueda S, Ariga H, Nagasawa S (1995). Diversity of sites for measles virus binding and for inactivation of complement C3b and C4b on membrane cofactor protein CD46. J Biol Chem 270: 15148-15152.

Kim YU, Kinoshita T, Molina H, Hourcade T, Seya T, Wagner LM, Holers VM (1995). Mouse complement regulatory protein Crry/p65 uses the specific mechanisms of both human decay-accelerating factor and membrane cofactor protein. J Exp Med 181: 151-159.

Liebert UG, Finke D (1995). Measles virus infections in rodents. Curr Top Microbiol Immunol 191: 149-166.

Liebert UG, terMeulen V (1987). Virological aspects of measles virus induced encephalitis in Lewis and BN rats. J Gen Virol 68:1715-1722.

Liszewski MK, Post TW, Atkinson JP (1991). Membrane cofactor protein (MCP or CD46): newest member of the regulators of complement activation gene cluster. Ann Rev Immunol 9: 431-455.

Liszewski MK, Tedja I, Atkinson JP (1994). Membrane cofactor protein (CD46) of complement: Processing differences related to alternatively spliced cytoplasmic domains. J Biol Chem 269: 10776-10779.

Maisner A, Schneider-Schaulies J, Liszewski MK, Atkinson JP, Herrler G (1994). Binding of measles virus to membrane cofactor protein (CD46): importance of disulfide bonds and N-glycans for the receptor function. J Virol 68: 6299-6304.

Malvoisin E, Wild TF (1993). Measles virus glycoproteins: studies on the structure and interaction of the hemagglutinin and fusion proteins. J Gen Virol 74: 2365-2372.

Manchester M, Liszewski MK, Atkinson JP, Oldstone MBA (1994). Multiple isoforms of CD46 (membrane cofactor protein) serve as receptors for measles virus. Proc Natl Acad Sci USA 91: 2161-2165.

Manchester M, Valsamakis A, Kaufman R, Liszewski MK, Atkinson JP, Lublin DM, Oldstone MBA (1995). Measles virus and C3 binding sites are distinct on membrane cofactor protein (MCP; CD46). Proc Natl Acad Sci USA 92: 2303-2307.

McLean A (1995). After the honeymoon in measles control. Lancet 345: 272.

Moore MD, Cannon MJ, Sewall A, Finlayson M, Okimoto M, Nemerow GR (1991). Inhibition of Epstein-Barr virus infection in vitro and in vivo by soluble CR2 (CD21) containing two short consensus repeats. J Virol 65: 3559-3565.

Moscona A, Peluso RW (1992). Fusion properties of cells infected with human parainfluenza virus type 3: receptor requirements for viral spread and virus-mediated membrane fusion. J Virol 66: 6280-6287.

Mulholland K (1995). Measles and pertussis in developing countries with good vaccine coverage. Lancet 345:305-307.

Naniche D, Varior-Krishnan G, Cervino F, Wild TF, Rossi B, Rabourdin-Combe C, Gerlier D (1993a). Human membrane cofactor protein (CD46) acts as a cellular receptor for measles virus. J Virol 67: 6025-6032.

Naniche D, Wild TF, Rabourdin-Combe C, Gerlier D (1993b). A monoclonal antibody recognized a human cell surface glycoprotein involved in measles virus binding. J Gen Virol 73: 2617-2624.

Naniche D, Wild TF, Rabourdin-Combe C, Gerlier D (1993c). Measles virus haemagglutinin induces down-regulation of gp57/67, a molecule involved in virus binding. J Gen Virol 74: 1073-1079.

Nemerow GR, Wolfert R, McNaughton ME, Cooper NR (1985). Identification and characterization of the Epstein-Barr virus receptor on human B lymphocytes and its relationship to the C3d complement receptor (CR2). J Virol 55: 347-351.

Niewiesk S, Brinckmann U, Bankamp B, Sirak S, Liebert UG, terMeulen V (1993). Susceptibility of measles virus-induced encephalitis in mice correlated with impaired antigen presentation to cytotoxic T lymphocytes. J Virol 67: 75-81.

Norrby E, Oxman M (1990). Measles virus. In: Fields BN, Knipe DM, eds. Virology. New York: Raven Press, 1013-1045.

Nussbaum O, Broder CC, Moss B, Bar-Lev Stern L, Rozenblatt S, Berger EA (1995). Functional and structural interaction between measles virus hemagglutinin and CD46. J Virol 69: 3341-3349.

Ogura H, Sato H, Kamiya S, Nakamura S (1991). Glycosylation of measles virus hemagglutinin protein in infected cells. J Gen Virol 72: 2679-2684.

Okada N, Liszewski MK, Atkinson JP, Caparon M (1995). Membrane cofactor protein (CD46) is a keratinocyte receptor for the M protein of the group A streptococcus. Proc Natl Acad Sci USA 92: 2489-2493.

Post TW, Liszewski MK, Adams EM, Tedja I, Miller EA, Atkinson JP (1991). Membrane cofactor protein of the complement system: alternative splicing of serine/threonine/proline-rich exons and cytoplasmic tails produces multiple isoforms which correlate with protein phenotype. J Exp Med 174: 93-102.

Radecke F, Spielhofer P, Schneider H, Kaelin K, Huber M, Dotsch C, Christiansen G, Billeter MA (1995). Rescue of measles viruses from cloned DNA. EMBO J 14: 5773-5784.

Rall G, Mucke L, Oldstone MBA (1995). Consequences of cytotoxic T-lymphocyte interaction with major histocompatibility complex type I expressing neurons in vivo. J Exp Med 182: 1201-1212.

Richardson C, Scheid A, Choppin PW (1980). Specific inhibition of paramyxovirus and myxovirus replication by oligopeptides with amino acid sequences similar to those at the N-termini of the R1 or HA2 viral polypeptides. Virology 105: 205-222.

Rima BK, Earle JAP, Baczko K, Rota P, Bellini WJ (1995). Measles virus strain variations. Curr Top Microbiol Immunol 191: 65-83.

Rota JS, Hummel KB, Rota PA, Bellini WJ (1992). Genetic variability of the glycoprotein genes of current wild-type measles isolates. Virology 188: 135-142.

Sakata H, Kobune F, Sato TA, Tanabayashi K, Yamada A, Sugiura A (1993). Variation in field isolates of measles virus during an 8-year period in Japan. Microbiol Immunol 37: 233-237.

Sakimura K, Kushiya E, Takahashi Y, Suzuki Y (1987). The structure and expression of the neuron specific enolase gene. Gene 60: 103-113.

Sato N, Funayama N, Nagafuchi A, Yonemura S, Tsukita S, Tsukita S (1992). A gene family consisting of ezrin, radixin and moesin. J Cell Sci 103: 131-143.

Sato TA, Enami M, Kohama T (1995). Isolation of measles virus hemagglutinin protein in a soluble form by protease digestion. J Virol 69: 513-516.

Schneider-Schaulies J, Schnorr J-J, Brinckmann U, Dunster LM, Baczko K, Schneider-Schaulies S, terMeulen V (1995). Receptor usage and differential downregulation of CD46 by measles virus wild type and vaccine strains. Proc Natl Acad Sci USA 92: 3943-3947.

Seya T, Kurita M, Tara T, Iwata K, Semba T, Hatanaka M, Matsumoto M, Yanagi Y, Ueda S, Nagasawa S (1995). Blocking measles virus infection with a recombinant soluble form of, or monoclonal antibodies against, membrane cofactor protein of complement (CD46). Immunology 84: 619-625.

Sheehan JP, Iorio RM (1992). A single amino acid substitution in the hemagglutinin-neuraminidase of Newcastle disease virus results in a protein deficient in both functions. Virology 189: 778-781.

Shioda T, Wakao S, Suzu S, Shibuta H (1988). Differences in bovine parainfluenza 3 virus variants studied by sequencing of the genes of envelope proteins. Virology 162: 388-396.

Sissons JGP, Oldstone MBA, Schreiber RD (1980). Antibody-independent activation of the alternative complement pathway by measles virus-infected cells. Proc Natl Acad Sci USA 77: 559-562.

Sleckman BP, Shin J, Igras VE, Collins TL, Strominger JL, Burakoff SJ (1992). Disruption of the CD4-p56lck complex is required for rapid internalization of CD4. Proc Natl Acad Sci USA 89: 7566-7570.

Stern LB, Greenberg M, Gershoni JM, Rozenblatt S (1995). The hemagglutinin envelope protein of canine distemper virus (CDV) confers cell tropism as illustrated by CDV and measles virus complementation analysis. J Virol 69: 1661-1668.

Van Binnendijk RS, van der Heijden RWJ, Osterhaus ADME (1995). Monkeys in measles research. Curr Top Microbiol Immunol 191: 135-148.

Varior-Krishnan G, Trescol-Biemont MC, Naniche D, Rabourdin-Combe C, Gerlier D (1995). Glycosyl-phosphatidylinositol-anchored and transmembrane forms of CD46 display similar measles virus receptor properties: virus binding, fusion, and replication; down-regulation by hemagglutinin; and virus uptake and endocytosis for antigen presentation by major histocompatibility complex class II molecules. J Gen Virol 68: 7891-7899.

von Pirquet CE (1908). Das Verhalten der kutanen Tuberkulin-reaktion wahrend der Masern. Dtsch Med Wochenschr 34: 1297-1300.

Ward T, Pipkin PA, Clarkson NA, Stone DM, Minor PD, Almond JW (1994). Decay-accelerating factor CD55 is identified as the receptor for echovirus 7 using CELICS, a rapid immune-focal cloning method. EMBO J 13: 5070-5074.

Wild TF, Malvoisin E, Buckland R (1991). Measles virus: both the hemagglutinin and fusion glycoproteins are required for fusion. J Gen Virol 72: 439-442.

Vignette

Yellow Fever

Quarantined patients of Greenwich Hospital in England who wore a jacket with yellow patches to forewarn other patients and outsiders about contagion were nicknamed "Yellow Jacks." A yellow-colored flag flying over quarantined areas was referred to as "Yellow Jack." But the "yellow jack" everyone feared was yellow fever. First described in the seventeenth century, this dreaded plague of mankind did not spare cities, countries, or continents. Perhaps only epidemics of bubonic plague claimed more victims than yellow jack. Scourge of man for his sins? But how delivered? It was the genius of Finlay to describe the mosquito as a death messenger carrying the disease agent from one human to another. The fight against *Aedes aegypti* as the carrier of yellow fever was initiated in cities where yellow jack was raging. The great Brazilian epidemiologist Osvaldo Cruz led the campaign for extermination of mosquitoes in the cities and succeeded in the control of yellow fever in urban areas of Brazil. But the persistence of yellow fever among rural populations in Brazil led to the discovery of "jungle" yellow fever, where man was an accidental host of a viral infection acquired from jungle monkeys which remain the source of infection transmitted through mosquito species other than *Aedes aegypti*. Laboratory infection through direct contact with the virus by personnel working with yellow fever at the Rockefeller Foundation Laboratories were not uncommon. This situation changed drastically after the discovery by Theiler and Lloyd of live attenuated virus as a vaccine source. Although the principle of using laboratory-modified infectious agents as live vaccine was established by Pasteur in 1880, it was Theiler who first attenuated yellow fever sufficiently to produce a variant called 17D virus which became the basis for prophylactic vaccination against this disease throughout the world. His vaccine is still used successfully today. The Nobel Prize awarded to Theiler in 1951 honored a man who opened new vistas in the protection of man against infection.

93

7

Milestones in the Conquest of Yellow Fever

Thomas P. Monath

In 1995, world attention was riveted by the events in Kikwit, Zaire, where an epidemic of viral hemorrhagic fever claimed 100 to 200 lives. The cause of this outbreak, Ebola virus, an enigmatic "emerging" virus disease with a very high case-fatality rate, conjured up fears of spreading contagion. Little is known about the natural history of the virus, the origin of periodic outbreaks, the mode of transmission, or the means to prevent future epidemics, and no treatment is available to stop the inexorable course of the disease. In many respects, the Kikwit episode is a paradigm of the fear, ignorance, and helplessness surrounding yellow fever for nearly 250 years since it was first described and a poignant reminder that this original viral hemorrhagic fever, fatal in half of the cases with full-blown infection, still occurs as an epidemic disease with an impact on humanity and a risk of global contamination far greater than Ebola virus disease. The difference, of course, is that we now have a rather complete understanding of the natural history and mode of transmission of yellow fever, epidemic control measures based on this knowledge, and a highly effective vaccine to prevent the disease. These accomplishments, which occurred mainly during the first half of this century, were the results of efforts by a relatively short list of dedicated and visionary men. The personal courage of the early yellow fever workers is noteworthy, since no other virus has taken so high a toll in fatal laboratory infections.

The earliest probable account of yellow fever was during an epidemic in the Yucatan Peninsula in 1648. In Africa, the disease was first recognized during an epidemic in 1778. In the eighteenth and nineteenth centuries, yellow fever was a major threat to human health in the Americas and Africa, and it invaded Europe on numerous occasions. Yellow fever is characterized by severe liver necrosis, renal failure, and gastrointestinal hemorrhage. The impact of the disease has been described many times, and a litany of epidemics and enumeration of deaths will not be repeated here. It is important, however, to understand that societies affected by epidemic yellow fever faced such a threat to life that normal commerce and human activities virtually ceased. In the Philadelphia epidemic of 1793, for example, 10% of the population perished, and in the New Orleans epidemics of 1853 and 1878, 28% and 7% of the city's inhabitants died of yellow fever, respectively. Under such conditions, there was enormous pressure on the medical profession and on governments for information and intervention.

Controversy centered on the origin and mode of spread of the disease, and theories abounded. The origin of yellow fever, like that of malaria, was attributed to miasmas arising from swamps or detritus unloaded by sailing vessels, and early debates raged over the question of communicability of the disease and the local origin of epidemics *vs* introduction of an exotic agent. Efforts to control the disease centered on improving general hygiene and sanitation, and on the principle of quarantine espoused by state and federal health authorities. By the 1870s, most physicians agreed that the disease was not directly transmitted by person-to-person contact or by food or drinking water, and by 1880, there was general agreement that a "germ" was responsible for the disease. Attempts to demonstrate the germ of yellow fever in blood or black vomit had, however, universally failed. The lack of contagion and inability to demonstrate the agent of yellow fever in the patient led many to conclude that yellow fever had a primary environmental source, while others proposed a human origin, with subsequent transformation of the infection in the environment before it could be transmitted to others. Since epidemics often followed introduction of yellow fever cases or ships carrying such cases, fomites—infected inanimate materials—were suspected sources of infection. The "germ" apparently multiplied under unidentified conditions and in local sites external to the human body, especially in unsanitary areas with decomposing organic matter. Climatic, meteorological conditions and other factors were widely thought to affect such replication. Once introduced, yellow fever appeared to spread by inhalation of infected air. After the discovery of anthrax and tubercle bacilli in the 1870s, the principles of microbiology were applied to the study of yellow fever. This work led to many blind alleys between 1897 and the early 1920s, particularly regarding the etiologic role of *Leptospira* in yellow fever.

CARLOS J. FINLAY AND THE MOSQUITO THEORY OF YELLOW FEVER TRANSMISSION

In the midst of the confusion surrounding the origin and transmission of yellow fever during the later nineteenth century, a singular hypothesis was put forth by the perspicacious Cuban physician, Carlos Juan Finlay (Figure 1). Born in Camaguey, Cuba, in 1833, Finlay was the son of Scottish and French parents and went to college in France. He entered Thomas Jefferson Medical College in Philadelphia in 1853, a year in which yellow fever raged in the city. In 1855, he graduated, and two years later he began practicing medicine in Havana. Finlay was inclined towards research and published his first paper

FIGURE 1. CARLOS J. FINLAY, a Cuban physician who first developed the theory of yellow fever transmission by mosquitoes.

in 1864. His publications range widely across the fields of medicine, including surgery, ophthalmology, and infectious diseases, as well as natural and physical sciences. These latter works, which encompassed studies on climate, gravity, and chemistry, indicate the breadth of Finlay's interests in fields allied loosely to medicine and foretold his inclination to consider novel approaches and undertake entomological studies in his work on yellow fever. For example, Finlay published a report on the possible relationship between atmospheric alkalinity and yellow fever activity. In 1879 and 1880, Finlay's interest in yellow fever was greatly stimulated during a visit by an American Commission sent to Cuba, where yellow fever was endemo-epidemic, to investigate the disease following the devastating outbreak in the Mississippi basin in 1877–1878.

In 1880, Finlay formulated the theory that mosquitoes played a role in yellow fever transmission, and in August of 1881, he read a paper at the Royal Academy of Havana titled "The mosquito hypothetically considered as the agent of transmission of yellow fever." In this landmark work, Finlay concluded that, because yellow fever affected vascular endothelium clinically and histologically, a bloodsucking insect may be the likely intermediary host responsible for transmission. Three conditions were defined as necessary for transmission:

1. The existence of a yellow fever patient into whose capillaries the mosquito is able to drive its sting and to impregnate it with the virulent particles, at an appropriate stage of the disease.

2. That the life cycle of the mosquito be spared after its bite upon the patient until it has a chance of biting the person in whom the disease is to be reproduced.

3. The coincidence that some of the persons whom the same mosquito happens to bite thereafter shall be susceptible of contracting the disease.

In drawing these conclusions, Finlay provided a plausible solution to the cabal of yellow fever transmission that explained a role for infected (but not directly contagious) human beings in the introduction of the disease and for a "transformation" in the local environment into a transmissible infection. How did Finlay arrive at his theory? There is dispute among historians as to the originality of Finlay's hypothesis, based on the possibility that he had read the reports (published in *The Lancet* in 1878) of Manson's work on the mosquito transmission of filiariasis (Delaporte 1991). Finlay's 1881 hypothesis makes no mention of Manson. Whatever the truth in this matter, one must still conclude that Finlay had a prepared mind and the vision to link scientific information into a hypothesis of immense importance to the scientific community. In 1884, Finlay presented a paper in which he recommended prevention of yellow fever based on preventing mosquito bites and on destroying mosquitoes. He also suggested that once infected, mosquitoes remain so for life.

Finlay suspected the *Culex cubensis* mosquito as the vector of yellow fever. From his description, it is clear that this is the species now known as *Aedes aegypti*. In an attempt to prove his theory, in 1880, Finlay conducted a series of experimental infections in which wild-caught adult mosquitoes were allowed to feed on yellow fever patients and refeed on "healthy, non-immunes." In the initial studies, five transmission attempts were made, and in four cases, the recipient contracted a febrile illness within 5 to 15 days of the bite. In one of these cases, the clinical illness was characterized as mild yellow fever with jaundice and albuminuria, whereas the other cases fit the description of abortive yellow fever. Doubt surrounds these studies, since none of the cases developed classical (severe) disease and the interval between mosquito infection and transfer to the recipients was two to six days, an extrinsic incubation period presumably too short for the mosquitos to have developed a disseminated infection. Moreover, although none of Finlay's 15 non-immune subjects became ill at the time, one could not exclude the possibility that the recipients had been exposed to natural infection from another source. It is also conceivable that Finlay's wild-caught mosquitoes had been infected before the experiments. In his defense, it is worth noting that while Finlay's cases were not pathognomonic of yellow fever, they were certainly consistent with the diagnosis, and the extrinsic incubation period of arboviruses may in fact be quite variable. For example, Hardy et al (1978) showed that some mosquitoes in a population demonstrate a "leaky" midgut, in which the virus in ingested blood may bypass the requirement for replication in the midgut epithelium prior to dissemination of the virus to salivary gland, resulting in a greatly shortened extrinsic incubation period. Under such conditions, transmission is possible in as few as four days after ingestion of a blood meal containing virus.

Finlay also sought approaches for the prevention of yellow fever. Many non-immune immigrants had entered Cuba, and morbidity was high. In studies conducted between 1882 and 1886, Finlay concluded that individuals exposed to yellow fever through bites of mosquitoes that had recently fed on yellow fever patients became immune to regular attacks of the disease (Finlay 1881). This work, as well as the Reed experiments, led to an unfortunate series of studies performed in 1901 by Juan Guiteras (see below).

Finlay's hypothesis on mosquito transmission was not given much weight by his contemporaries. However, in the final analysis he was right, and he has been widely acclaimed by the Spanish-speaking world as deserving primary credit for this discovery. Finlay was a strong local influence in Cuba, and his views also influenced George Sternberg and members of the Reed Commission. As described below, Reed and his colleagues proved the mosquito theory by direct experimentation about 20 years after Finlay proposed it. Finlay's hypothesis paved the way for mosquito control and the eventual conquest of yellow fever as a major threat to civilization.

THE REED COMMISSION AND THE PROOF OF MOSQUITO TRANSMISSION

The contribution of Major Walter Reed (Figure 2) and his colleagues Jesse Lazear, James Carroll, and Aristedes Agramonte to yellow fever research has been reviewed many times, and only highlights are provided here. Reed,

FIGURE 2. MAJOR WALTER REED, leader of the United States Army Yellow Fever Commission to Cuba in 1900. Reed and his colleagues performed human studies demonstrating mosquito transmission, the absence of transmission by fomites, and the presence of a virus in human blood.

born in 1851 in Virginia, was at age 17 the youngest graduate of the University of Virginia Medical School. Reed was commissioned in 1875 as Assistant Surgeon in the United States Army. After various field assignments, he made a successful application to enter the field of scientific research, and in 1890 was transferred to Baltimore, where, as Attending Surgeon and Examiner of Recruits, he began to work part-time in the pathology and bacteriology laboratories at Johns Hopkins. In 1893, he assumed the post of curator of the Army Medical Museum and Professor of Bacteriology at the Army Medical School. Over the next six years, he was engaged in a number of projects, the most important of which was a series of investigations on typhoid fever in military camps during the Spanish American War.

In 1897, at the urging of George Sternberg, then Surgeon General of the Army Medical Corps, Reed and James Carroll began investigations to confirm or refute the Italian microbiologist Sanarelli's recent report of *Bacillus icteroides (Leptospira)* as the cause of yellow fever. Reed and Carroll failed to confirm Sanarelli's work in the laboratory. Neither a Commission sent to Cuba to study the problem nor Dr. Agramonte, who was sent subsequently, was able to provide a final answer to the *Leptospira* question. In 1900, Sternberg appointed Reed to lead a new Commission to Havana to study yellow fever. In part, this decision was stimulated by the morbidity due to yellow fever among American soldiers who were occupying Cuba in 1899. Reed had several objectives, including further study of the *Leptospira* etiology, on the possibility of mosquito transmission, and on blood as a source of yellow fever infection.

The mosquito theory proposed by Finlay in 1881 had been given new credibility by the work of Major Ronald Ross (British Army) on malaria transmission by mosquitoes in India. Those studies, conducted in 1898, showed clearly that an infectious agent could have two independent life cycles in the human host and in mosquito vectors. In the same year, Henry Rose Carter, a United States Public Health Service physician, conducted a careful epidemiological study of yellow fever in two Mississippi villages, and concluded that an "extrinsic incubation period" of approximately two weeks was required after the introduction of a yellow fever case before new cases could occur (Carter 1900). Carter did not identify the nature of this extrinsic part of the transmission pathway, but his observation was consistent with a role for mosquitoes in the external "transformation" of the disease. This fact was appreciated by others soon after. The work of Ross and Carter provided a clear direction for studies on the role of mosquitoes in yellow fever transmission.

The Reed Commission, with the exception of Dr. Lazear (Figure 3), was not at first heavily invested in studies on mosquito transmission. However, shortly after Reed and his colleagues arrived in Havana in June, 1990, they met a British delegation from Liverpool, constituted by Herbert Durham and Walter Myers. The latter scientists had learned about extrinsic incubation from Carter, had discussed mosquito transmission with Finlay, and were well

FIGURE 3. DR. JESSE LAZEAR, hero and martyr of the Reed Commission. Lazear was instrumental in designing and conducting the first trials showing that yellow fever was transmissible by mosquito bite and that the extrinsic incubation period was 12 days. Lazear was the only person experimentally infected who succumbed to yellow fever in Reed's studies.

versed in the discoveries of Manson and Ross on mosquito transmission of filariasis and malaria, respectively (Delaporte 1991). Their visit may have created a scientific rivalry that propelled Reed toward the study of mosquito transmission. Indeed, in August 1900, Reed and his party met with Carlos Finlay and received some of Finlay's *Ae. aegypti* mosquitoes.

Lazear took up the work on mosquito transmission in earnest. Reed was called back to Washington and was not personally present during Lazear's experiments. Mosquitoes reared from eggs were allowed to feed on yellow fever patients, eliminating the possibility of a previous contamination from a source of infection other than the yellow fever donor. At first, Finlay's approach was taken, with refeeding upon recipients immediately after mosquitos had fed on a yellow fever case. The results were negative in nine cases. James Carroll then received a bite from a mosquito that Lazear had fed 12 days earlier on a patient in the second day of illness, and within four days, he developed yellow fever. The observation was repeated in the case of volunteer Pvt. W.H. Dean. It appeared that mosquitoes fed early in infection and allowed to become infected themselves over a period of 12 days could efficiently transmit the disease. While these studies were in progress, Lazear developed severe yellow fever and died within a week. There appears to be little question that Lazear had allowed one of his experimentally infected mosquitoes to bite him, in an attempt to extend his observations (Truby 1943; Franklin and Sutherland 1984). Fearing that this act might be declared a suicide and Lazear's insurance withheld from his family, Reed and co-workers reported that Lazear's infection was inflicted by a wild mosquito. In any case,

Lazear deserves the major credit for the experimental work proving the mosquito theory, which was initially published as a preliminary report in late 1900 (Reed et al 1900).

The Reed Commission went on to perform further studies on the transmission of yellow fever at a strict quarantine facility, dubbed in memoriam, Camp Lazear. The purpose of those studies was to extend the observations of Lazear and to conclusively rule out sources of infection other than the experimental mosquito. Other studies addressed the possible role of fomites in transmission and the presence of the yellow fever agent in the blood. In all, 16 volunteers held in quarantine were subjected to the bites of mosquitoes that had previously fed on yellow fever cases, and 14 became ill and recovered. The extrinsic incubation period was shown to be 10 to 12 days, and infected mosquitoes were shown to survive for at least 71 days, providing an explanation for the observation that the "contagion of yellow fever may cling to a building, although it has been vacated for two or more months" (United States Senate Document 1911). Conclusive evidence was obtained against fomite transmission, as none of the volunteers who were exposed to bedding and clothing soiled with fresh black vomit and excreta of yellow fever patients fell ill. Finally, 8 of 10 volunteers injected with minute amounts of blood taken from yellow fever cases during the first two days of illness became infected, proving that blood was the source of infection for mosquitoes. Blood passed through a bacteria-proof filter was infectious, providing the first evidence that yellow fever was caused by a virus (although they did not rigorously exclude—by performing a serial transmission experiment—the possibility of a filterable bacterial toxin).

These detailed studies provided the basis for a strategy of prevention and control of yellow fever that was later implemented in Cuba, the Panamanian isthmus, and port cities in South America by Maj. William C. Gorgas (then Chief Sanitary Officer of the city of Havana) and by Dr. Oswaldo Cruz in Brazil. Finlay's hypothesis, the Lazear experiment, and Reed's meticulous studies had created the framework for a strategic attack on one of the most feared plagues of humankind. The days of ignorance and controversy about yellow fever transmission were over, and millions would be saved by cessation of the useless practices of disinfection based on the fomite and contagion theories.

A Sidelight of Mosquito Research:
The Misfortunes of Juan Guiteras

In 1901, Juan Guiteras, Professor of Pathology and Tropical Medicine, University of Havana and Director of the Inoculation Station, Las Animas Hospital, Havana, together with other Cuban physicians and Maj. William C. Gorgas, conducted a series of studies in recently arrived immigrants who were exposed to the bites of mosquitoes that had fed more than 12 days previously on acute yellow fever cases. Their objective was not to reproduce the Reed Commission's work, but to "propagate the disease in a controllable form" as a means of producing immunity against naturally acquired yellow fever. In this respect, Guiteras and Gorgas were clearly following the lead of Finlay, who in 1886 claimed that volunteers bitten by infected mosquitoes had acquired immunity to subsequent attacks (Finlay 1881). Gorgas was apparently also motivated by doubt about the practical reality of mosquito control.

Nineteen volunteers were exposed to the bites of infected mosquitoes on 42 occasions. Guiteras was able to elicit clinical yellow fever in eight volunteers, three of whom died. His results stood in stark contrast to the total experience of the Reed Commission, in which none of 16 persons[1] who developed the disease died (p=.03, Fisher's exact test, two-tailed). Among the dead was Clara Louise Maass, a United States Army nurse from New Jersey who, throughout her career, exhibited great personal courage and dedication and was the only woman volunteer for this or Reed's study (Tigertt and Tigertt 1983).

The Guiteras experiments put a halt to the mosquito immunization approach. One intriguing possibility to explain why his results were different from those of the Reed Commission is that Guiteras inadvertently used a highly virulent strain of yellow fever virus. For the 42 experimental infection attempts, mosquitoes were fed on 11 different yellow fever cases (Guiteras 1901). Only eight clinical infections occurred, and seven were the result of infection with the same (Alvarez) virus strain. Of nine individuals infected with Alverez virus, seven fell ill. It is now well known that wild-type yellow fever strains vary in virulence, gene sequence, and epitopes recognized by monoclonal antibodies. The fact that only one in two persons sustaining primary infection develops symptoms of yellow fever (Monath 1989) and that 20 to 50% of such cases have a fatal outcome may be explained both by variation in virulence of the infecting strain and in host susceptibility. The outcome of infection in the studies by Reed and Guiteras fit the current concepts of virus and host factors. Guiteras was apparently unlucky to have used a virus strain (Alvarez) that produced an infection:case ratio of only 1.2:1 and a case-fatality ratio of 43%.

FRED L. SOPER: DISCOVERY OF JUNGLE YELLOW FEVER
AND THE IMPERATIVE OF *AE. AEGYPTI* ERADICATION

The Rockefeller Foundation's involvement in yellow fever research and control had been initiated as early as 1915, when the Foundation's International Health Commission resolved "to give aid in the eradication of yellow fever in those areas where the disease is endemic and where conditions would seem to invite cooperation for its control." The ultimate result of these beginnings was a global effort on yellow fever research and disease control (Strode

---

[1] Includes Carroll, Dean, and the 14 volunteers intentionally infected in Camp Lazear, but not Dr. Lazear himself. Lazear suffered a fatal infection, but its source is definitely known.

1951). In 1927, as part of this effort, Drs. A.H. Mahaffy and J.H. Bauer, members of the Rockefeller's West African Yellow Fever Commission succeeded in isolating the yellow fever virus from blood of a mild yellow fever case (named Asibi). Efforts at the Rockefeller Yellow Fever Laboratory by M. Theiler, W. Sawyer and W. Lloyd led to the development of a convenient neutralization test in mice, and in 1931, immunity surveys were initiated in Africa and South America. The immunity survey in South America began to reveal an astonishing fact: yellow fever activity was considerably more widespread than expected and was prevalent over large areas of the Amazon region where yellow fever cases were not reported. Indeed, the prevalence of immunity in rural and jungle regions was considerably higher than in the cities where outbreaks of the disease occurred. Yellow fever surveillance, based on histopathological examination of post-mortem liver samples (so-called viscerotomy), had been initiated by national yellow fever services and the Rockefeller Foundation in 1930 and enhanced the discovery of yellow fever cases in remote areas.

In 1932, Dr. Fred L. Soper (Figure 4) and his colleagues Penna, Cardoso, Serafim, Frobisher and Pinheiro, investigated an outbreak of yellow fever in a rural area of Brazil—Valle do Chanaan, Espiritu Santo—where *Ae. aegypti* was absent (Soper et al 1933). The human cases were widely scattered, and there was a preponderance of cases and immune individuals among adult males. The full significance of this outbreak was not immediately appreciated, and Soper concluded that "This epidemic...was apparently self-limited and it is possible that there are no rural regions in America which are truly endemic for yellow fever."

During 1932 and 1933, further outbreaks of yellow fever in remote rural areas were investigated, including those at Caparrapi in the Magdalena Valley of Colombia; at Lauro Sodre along the Amazon River near the borders of Brazil and Peru; and in San Ramon, Nullo de Chaves, Bolivia. These outbreaks in the absence of the classical vector *Ae. aegypti,* as well as the immunity surveys showing virus activity in the Amazon, led Soper to conclude that yellow fever had a distinct transmission cycle in rural areas and affected a different population (adult males engaged in forest clearing), which he termed "jungle yellow fever." During 1933 and 1934, further reports from multiple locations supported this conclusion, and in 1935, Soper and colleagues conducted extensive epidemiological studies in Minas Geraes, Brazil, that convinced him of the importance of the jungle cycle.

In a paper read in Portuguese before the Brazilian National Academy of Medicine in 1935 (Soper 1936), Soper considered historical evidence for jungle yellow fever. A report from Bolivia as early as 1907 was suggestive of jungle yellow fever. In that year, Franco and colleagues described an outbreak in Muzo, Colombia, and concluded that the disease was contracted in the forest. In 1930, Adolfo Lutz described outbreaks in remote rural areas of Sao Paulo, Brazil, affecting forest workers in the absence of *Ae. aegypti.* Unlike Carlos

FIGURE 4. DR. FRED L. SOPER (ca. 1960). Director of the Rockefeller Foundation Laboratory in Brazil and later of the Pan American Sanitary Bureau. Soper elucidated the jungle cycle of yellow fever in South America and became a champion of disease vector and eradication, including *Aedes aegypti.*

Finlay, who may have concealed the precedents of his personal hypothesis, Soper enquired into the precedents of his theory. His great contribution was to link key observations on the epidemiology of yellow fever and to form a hypothesis that was to revolutionize thinking about the ecology of yellow fever and the strategies for its control.

In 1935, the elements involved in the jungle transmission cycle were not clear to Soper. Forest-dwelling mosquitoes were responsible for transmission, but their identity was uncertain. Laboratory studies as early as 1929 had shown that mosquitoes other than *Ae. aegypti* could transmit yellow fever virus, but the identity of the jungle vector (*Haemagogus* sp.) was confirmed only in 1938, when Shannon, Whitman, and Franco isolated the virus from wild-caught mosquitoes and successfully demonstrated transmission to susceptible hosts in the laboratory. Monkeys were suspect hosts in the jungle cycle, based on their susceptibility in the laboratory and immunity tests performed on wild monkeys, but it was not clear at the time whether monkeys were only a temporary element in the cycle or whether humans could sustain jungle transmission (Soper 1935).

Despite the lack of clarity of yellow fever ecology in 1935, Soper drew several insightful conclusions from the information available at the time. He identified the jungle cycle as the natural form of yellow fever endemnicity, whereas "...urban *aegypti*-transmitted yellow fever is a highly exotic form maintained with more difficulty than the jungle type," and concluded that the jungle cycle was a source of introduction to *Ae. aegypti*-infected areas. On this basis, he concluded that "…the idea of eliminating yellow fever…has no chance of immediate fulfillment," while emphasizing the necessity for eradicating the urban vector and for developing vaccines for prevention of jungle yellow fever (Soper 1935). Soper appreciated the great importance of further scientific discovery for yellow fever ecology and foresaw the possibility that there may be animal reservoirs and alternative vectors of yellow fever.

Others would contribute to endeavors in this area during the late 1930s and 1940s, revealing many of the secrets of the natural transmission of yellow fever in South America and Africa. Soper was a politically astute, tireless and ambitious worker with a great vision: disease eradication. He concentrated his personal efforts on the total elimination of *Ae. aegypti* in Brazil as a means of preventing urban epidemics and international spread of yellow fever through port cities (Soper 1955). He promulgated eradication on a hemisphere-wide basis. His vision of the feasibility of vector eradication for disease control was applied to other problems, and he successfully led a campaign against *Anopheles gambiae* in Brazil and against typhus in Italy and north Africa during World War II. His philosophy of eradication ultimately extended to non-vector-borne diseases, and he played an important role in smallpox eradication and in various immunization programs. In 1950 Soper became Director, Pan American Sanitary Bureau (the predecessor of the Pan American Health Organization). Under his leadership, *Ae. aegypti* eradication continued as a priority and by 1972 the mosquito had been eliminated from 19 countries, representing 73% of the landmass originally inhabited by the vector.

In the 1980s, the recognition of dengue hemorrhagic fever in the Americas renewed an interest in Soper's approach to intensified and directed efforts to control *Ae. aegypti*. Once again with leadership from the Rockefeller Foundation (in the person of Dr. Scott Halstead), Soper's principles were followed in selected countries in an effort to reinvigorate sanitation and mosquito control, this time at a grassroots level rather than by use of a vertically organized program.

MAX THEILER AND HUGH H. SMITH, DISCOVERERS OF YELLOW FEVER 17D VACCINE

Max Theiler was the principal architect of yellow fever 17D, perhaps the world's safest and most effective vaccine. Theiler (Figure 5) was born in 1899 in South Africa to parents of Swiss descent. His father, a veterinarian, worked on another arbovirus disease (African horse sickness) and gained acclaim for his work on tick-borne protozoa. Max studied medicine in Cape Town and then at St. Thomas's in London. He subsequently took the six-month course at the London School of Tropical Medicine & Hygiene in 1923, where he was recruited to work in Andrew Watson Sellards' laboratory in the Department of Tropical Medicine at Harvard. Yellow fever virus had been isolated in 1927 in Accra and Dakar, and Sellards was collaborating with the

FIGURE 5. MAX THEILER, American physician awarded the Nobel Prize in 1951 for the development of yellow fever vaccine

French investigators in West Africa. In 1928, Theiler and Sellards proved that the yellow fever virus strains in Africa and South America were immunologically identical, a finding of obvious importance to vaccine development and seroepidemiology.

In 1930, Theiler made a major breakthrough that would accelerate yellow fever research and would provide the basis for work on vaccines: he reported that the French yellow fever virus caused encephalitis in weaned mice (Theiler 1930). He was able to pass the virus serially in mice and to reproduce hepatic disease in monkeys inoculated with infected mouse brain. Theiler's work came to the attention of Wilbur Sawyer, Director of the Rockefeller Foundation Yellow Fever Laboratory in New York, who encouraged him to develop a protection test for immunity based on the mouse model. In 1931, Theiler described the first yellow fever neutralization test (Theiler, undated), which was immediately put into practice in large-scale population surveys in Africa and South America that uncovered the unexpected rural distribution of immunity.

Theiler noted that serial passage of the French virus in mouse brain led to an increased neurovirulence for mice but a loss of hepatotropism and virulence for rhesus monkeys. In collaboration with Sellards, Laigret and his colleagues in Dakar began to develop the French virus as a human vaccine and in 1932 reported the first human immunizations. However, Theiler and other American and British researchers believed that the French virus had safety problems as a vaccine and pursued a different path.

The relationship between Theiler and Sellards had always been tenuous, and in 1930, Theiler accepted a position in the Rockefeller's International Health Division Laboratory in New York. He began to work with the Asibi strain of yellow fever virus that had been isolated in 1927 by Mahaffy and Bauer in West Africa. Working first with Dr. Wray Lloyd and then with Dr. Hugh H. Smith, Theiler infected cultures of minced mouse embryo and, after 18 passages, began adaptations in minced chick embryo cultures. The neurovirulence for mice was monitored during the serial passages. To Hugh Smith (Figure 6) goes the credit of having first noticed that a decrease in neurovirulence occurred at the 176th passage level of a passage lineage designated 17D. Subsequent studies confirmed that this strain had also lost its ability to induce clinical infection in monkeys; no virus could be demonstrated in the liver or brain of these animals. However, monkeys inoculated by the subcutaneous route were protected against challenge with virulent yellow fever virus. The 17D strain was also markedly less neurotropic for mice and monkeys than wild-type virus, and intracranially inoculated monkeys survived challenge. The virus thus had the characteristics of a human vaccine.

Theiler conducted a series of studies, showing that the alteration in virulence of 17D occurred between the 89th and 114th subcultures. Theiler and Smith published their results in 1937 (Theiler and Smith 1937a). They also were the first volunteers to take the candidate vaccine. Both were immune, having sustained laboratory infections with yellow fever virus—Theiler by accidental infection while at Harvard, and Smith by intentional injection of immune goat serum and partially attenuated virus as a means of immunization. After the 17D vaccine proved safe in the two immune investigators, their colleagues at the Rockefeller Laboratory, Drs. Andrew Warren and Hugo Muench, volunteered and were successfully immunized without any attendant side effects. Theiler and Smith reported the first human immunizations with 17D virus in 1937 (Theiler and Smith 1937b).

That same year, Smith traveled to Brazil in order to extend the data on human immunization. Smith, born in 1902 in Pickens County, South Carolina, was educated at Davidson College, Johns Hopkins School of Medicine, and, as a resident physician, the University of Rochester (Smith 1978). Toward the end of his residency, he obtained an interview with Rockefeller Foundation staff and was nearly selected to join the first Rockefeller Foundation Yellow Fever Commission to go to West Africa (1918). However, his first overseas assignment with the Foundation was in Jamaica, where he undertook studies on tuberculosis vaccines. In 1935, Smith was picked by Wilbur Sawyer—who had a clear ability for identifying men of talent and who had himself made a major contribution to the epidemiology of yellow fever—to join the research staff in New York.

Arriving in Brazil in 1937, Smith found himself in the heyday of research on the ecology and control of yellow fever. Fred Soper was Director of the Rockefeller staff. Soper appreciated the important role that immunization

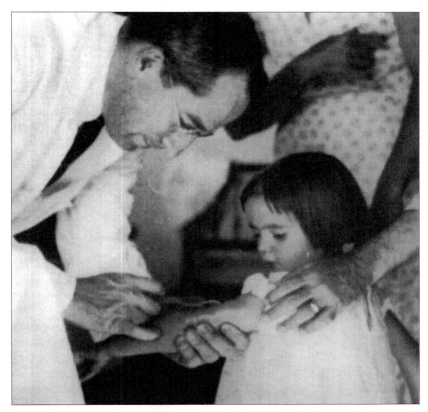

FIGURE 6. DR. HUGH H. SMITH (ca. 1937). Dr. Smith made key contributions to the discovery of 17D vaccine and conducted the first field trials of the vaccine in Brazil.

could play in the prevention of the jungle disease, which was not amenable to mosquito control. Loring Whitman, Raymond Shannon, and Nelson Davis were conducting entomological studies that would elucidate the vectors of jungle yellow fever. A new laboratory for yellow fever research was under construction at the Oswaldo Cruz Institute in Rio de Janeiro. Smith linked up with a talented Brazilian virologist, Dr. Henrique A. Penna. Within the first six months of Smith's arrival, he and Penna had selected non-immune individuals for vaccine trials based on Theiler's neutralization test, had inoculated over 100 volunteers, and had established 17D vaccine production at the laboratory in Rio. They took 17D vaccine to a carefully selected field site in Minas Geraes, where the local population, mainly peasants working on coffee plantations, had a healthy respect for yellow fever as a disease problem. A system of cold storage and transport had to be set up to preserve the vaccine during the field studies. In the first month of these trials, approximately 1,000 persons were vaccinated, without untoward effects. Smith and Penna defined the incidence of mild reactions to the vaccine and demonstrated that 95% of those vaccinated under field conditions developed neutralizing antibodies by the 14th day after inoculation. The stage was set for widescale use of the vaccine, and, under the leadership of Fred Soper, vaccine production and immunization campaigns were stepped up in 1938. By the time Smith departed Brazil to take up the study of yellow fever in Colombia in 1938, over 500,000 persons had received 17D vaccine, and the vaccine had been used for the first time in the control of a yellow fever epidemic in Sao Paulo and Minas Geraes states.

In 1951, Max Theiler was awarded the Nobel Prize in Medicine for the discovery of yellow fever 17D vaccine. Theiler's initial observation of viral attenuation during adaptation in mice, his derivation of 17D by *in vitro* culture, and his careful dissection of the biological activities of yellow fever virus during passage were the underlying achievements. Theiler's initial work at Harvard with Sellards had also identified the virus that would ultimately be used by French workers to make a mouse brain vaccine (the French neurotropic virus) widely used in francophone Africa. Theiler was personally embarrassed that Hugh Smith had not shared in the distinction accorded by the Nobel Committee and, when he spoke about his achievements after the award, was always quick to give credit to his colleague.

With Theiler's landmark achievement, so ends this brief resumé of a few selected yellow fever pioneers. Only the constraints of space excuse the omission of so many others who have forged the way to our present understanding of this interesting and terrible disease. Yellow fever has been a favorite subject of historians, to whom it is hoped the reader will turn for a more scholarly interpretation of events and personalities.

REFERENCES

Carter HR (1900). Note on interval between infecting and secondary cases of yellow fever from records at Orwood and Taylor, Mississippi, in 1898. New Orleans Med Surg J 52: 617-636.

Delaporte F (1991). The history of yellow fever. An essay on the birth of tropical medicine. Cambridge: MIT Press, 181 pp.

Finlay C (1891). Inoculations for yellow fever by means of contaminated mosquitoes. Am J Med Sci 102: 264-268.

Finlay CJ (1881). El mosquito hipoteticamente considerado como agente de la fiebre amarilla. Anal Real Acad Cienc Med Fis Natural 18: 147-169.

Franklin J, Sutherland J (1984). Guinea pig doctors. New York: Morrow.

Guiteras J (1901). Experimental yellow fever at the inoculation station of the Sanitary Department of Havana with a view to producing immunization. Am Med 2: 809-817.

Hardy JL, Apperson G, Asman SM et al (1978). Selection of a strain of Culex tarsalis highly resistant to infection following ingestion of western equine encephalomyelitis virus. Am J Trop Med Hyg 27: 313.

Monath TP (1989). Yellow fever. In: Monath TP, ed. The arboviruses: ecology and epidemiology, Vol V. Boca Raton, Fla.: CRC Press, 139-231.

Reed W, Carroll T, Agramonte A, Lazear JW (1900). The etiology of yellow fever—a preliminary note. Proc 28th Ann Meet Am Pub Health Assoc, Indianapolis, Oct. (Also published in United States Senate Document 1911, below).

Smith HH (1978). Life's a pleasant institution: The peregrinations of a Rockefeller doctor. Tucson, AZ, 254 pp.

Soper FL (1935). Recent extensions of knowledge of yellow fever. Q Bull League of Nations Health Organ, Geneva, 5: 19-68.

Soper FL (1936). Jungle yellow fever. A new epidemiological entity in South America. Rev Hyg Saude Pub 10: 1-39.

Soper FL (1955). The unfinished business with yellow fever. In: Yellow fever. A symposium in commemoration of Carlos Juan Finlay. Philadelphia: Jefferson Medical College, 79-88.

Soper FL, Penna H, Cardoso E, Serafim J, Frobisher M, Pinheiro J (1933). Yellow fever without *Aedes aegypti*. Study of a rural epidemic in the Valle do Chanaan, Espirito Santo, Brasil, 1932. Am J Hyg 18: 555-587.

Strode WK (ed) (1951). Yellow fever. New York: McGraw Hill, 710 pp.

Theiler M (1930). Susceptibility of white mice to the virus of yellow fever. Science 71: 367.

Theiler M (1931). Neutralization tests with immune yellow fever sera and strain of yellow fever virus adapted to mice. Ann Trop Med 25: 69.

Theiler M, Smith HH (1937a). Effect of prolonged cultivation in vitro upon pathogenicity of yellow fever virus. J Exp Med 65: 767-786.

Theiler M, Smith HH (1937b). Use of yellow fever virus modified by in vitro cultivation for human immunization. J Exp Med 65: 787-800.

Tigertt HB and Tigertt WD (1983). Clara Louise Maass: a nurse volunteer for yellow fever inoculations—1901. Milit Med 148: 252-253.

Truby AE (1943). Memoir of Walter Reed. The yellow fever episode. New York: Paul B Hoeber, 239 pp.

United States Senate Document (1911). Yellow fever. US 61st Congress, 3rd Session, Washington DC.

8

Current Advances
in Yellow Fever Research

James H. Strauss and Ellen G. Strauss

Yellow fever virus (YF) was the first virus shown to be transmitted to humans by the bite of an infected mosquito and one of the earliest viruses to be studied in the laboratory. However, molecular biological studies of YF and of its close relatives in the genus *Flavivirus* were slow in developing, due in part to the difficulties of working with these viruses in the laboratory. They grow to lower titer and have less pronounced effects on infected cells than do many other animal viruses. Moreover, many flaviviruses are serious human pathogens, which further complicates laboratory study. Within the last 10 years, however, knowledge of the molecular biology and molecular epidemiology of these viruses has expanded greatly and it seems an appropriate time to examine the rapid progress that has been made.

The modern era in the molecular biology of YF began with the publication of the complete sequence of the YF 17D vaccine strain 11 years ago (Rice et al 1985). This was quickly followed by the determination of complete sequences for a number of other flaviviruses (Castle and Wengler 1987; Cola et al 1988; Deubel et al 1988; Fu et al 1992; Hahn et al 1988; Irie et al 1989; Lee et al 1990; Mackow et al 1987; Mandl et al 1989, 1993; Osatomi and Sumiyoshi 1990; Sumiyoshi et al 1987), including that of the Asibi strain of YF (Hahn et al 1987c), the parent of the 17D strain, and of an insect virus related to the arthropod-borne flaviviruses (Cammisa-Parks et al 1992). Table 1 lists the major groups of flaviviruses for which complete nucleotide sequences are known.

The nucleotide sequences of flavivirus genomes and the deduced amino acid sequences of the encoded proteins make clear that all flaviviruses are closely related and descended from a common ancestor (Chambers et al 1990a; Lepiniec et al 1994; Lewis et al 1993; Marin et al 1995; Rice et al 1986). They share amino acid sequence identity throughout the genome, with the nonstructural proteins NS3 and NS5 being the most highly conserved. The relationships of the various flaviviruses to one another, which in the past were

TABLE 1. FLAVIVIRUSES for which the complete nucleotide sequence of the genome is known

Virus	Year of completion	Reference
Mosquito-borne		
Yellow fever		
17D vaccine	1985	Rice et al
Asibi	1987a	Hahn et al
Dengue		
Type 1	1992	Fu et al
Type 2		
Puerto Rico S1	1988	Hahn et al
Jamaica	1988	Deubel et al
New Guinea C	1989	Irie et al
Type 3	1990	Osatomi et al
Type 4	1987	Mackow et al
West Nile	1987	Castle et al
Japanese encephalitis	1987	Sumiyoshi et al
Kunjin	1988	Coia et al
Murray Valley encephalitis	1990	Lee et al
Cell-fusing agent (CFA)	1992	Cammisa-Parks et al
Tick-borne		
Tick-borne encephalitis	1989	Mandl et al
Langat	1992	Iacono-Connors et al
Powassan	1993	Mandl et al

Flaviviruses are grouped according to sequence relationships. The references shown either contain the entire sequence of a viral genome or are the last in a series of sequencing papers and report the completion of a viral genomic sequence.

deduced from antigenic cross-reactions among the viruses, are clearly shown in the evolutionary trees based upon sequence data (Figure 1). The mosquito-borne flaviviruses form a defined taxon that contains three distinct subgroups, the YF, dengue, and Japanese encephalitis subgroups. All mosquito-borne flaviviruses share a minimum of about 40% amino acid sequence identity in the E protein, the major surface antigen of the virus (Figure 1A). The tick-borne

Disc. cell hopson of tissue hop—

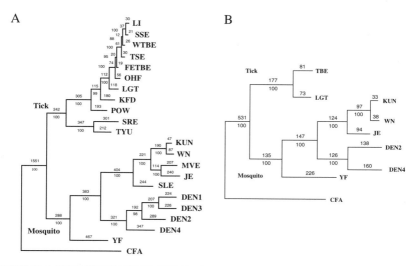

FIGURE 1. EVOLUTION OF FLAVIVIRUSES. Dendrograms illustrate the relationships of flaviviruses, based on the nucleotide sequences encoding the E protein (Panel A) or the NS5 protein (Panel B). Branch lengths have been drawn proportional to the number of parsimonious reconstructed changes (number above) along the branch. Bootstrap percent values are shown below the branches. From Marin et al (1995), with permission.

flaviviruses form a second defined taxon; the E proteins of these viruses share about 37% sequence identity with those of the mosquito-borne flaviviruses. A third taxon is represented by the cell-fusing agent (CFA), which is more distantly related to the mosquito-borne and tick-borne flaviviruses and may ultimately be classified as a distinct genus in the family Flaviviridae. The flavivirus replicase protein NS5 is more highly conserved than the envelope protein, but an examination of a phylogenetic tree based on NS5 sequences exhibits the same general groupings. Moreover, comparison of the trees in Figure 1A and 1B indicate that all flaviviruses have diverged from a single ancestral virus without recombination events.

Because of the close relationships of the various flaviviruses, the study of the molecular biology of virus replication is simplified in general by the direct applicability of findings with one virus to other viruses in the group. However, although the viruses are identical in genome organization and very similar in the details of their replication biology, they do differ in their interactions with whole organisms and cause very different diseases (Monath and Heinz 1996). The differences in disease syndromes are largely caused by the infection of a different spectrum of cells by the different viruses, but although we have learned much about flavivirus replication over the years, the mechanisms by which the viruses differ in their tissue tropisms are still not understood.

Organization of the Yellow Fever Genome

The YF genome is a single-stranded RNA molecule 10,862 nucleotides in length (Figure 2). It is capped at the 5' end but, unlike most animal RNA viruses, is not polyadenylated at the 3' end. Instead it has a conserved 3'-terminal secondary structure, reminiscent of the situation for a number of plant viruses (Hahn et al 1987b). There are relatively short 5' and 3' nontranslated regions with most of the genome present in one very large open reading frame (10,233 nucleotides) that is translated as a very large polyprotein processed by cellular and viral enzymes to produce at least 10 products. The genome is organized into two domains. The 5' region encodes the structural proteins required for construction of progeny virions, whereas the 3' region encodes proteins required for processing of the polyprotein and replication of the viral RNA.

The structural proteins include a nucleocapsid protein C and two envelope proteins M and E (Chambers et al 1990a). C is produced as a precursor, anC (anchored C), as is M (prM or precursor to M), although production of anC itself may not be the major pathway for the production of C, as described below. The nonstructural proteins include a glycoprotein NS1, four small hydrophobic peptides NS2A, 2B, 4A, 4B that are believed to be integral membrane proteins, and two large cytoplasmic proteins NS3 and NS5. The function of NS1, which has been reported to form homodimers (Winkler et al 1988), is obscure but it may be involved in virus assembly. More than one form of NS1 is produced (Chambers et al 1990b; Mason 1989; Mason et al 1987) and these multiple forms may have different functions in the virus life cycle.

NS3 is a multi-functional enzyme. The N-terminal 180 residues constitute a serine protease that cleaves the flavivirus polyprotein in as many as seven places (Rice and Strauss 1990). The central domain appears to be a helicase required to unwind RNA during replication. Although helicase activity has not been demonstrated *per se*, this region contains sequence motifs

FIGURE 2. SCHEMATIC ILLUSTRATION OF THE YELLOW FEVER VIRUS GENOME. The locations of the various encoded proteins are shown, drawn approximately to scale. The positions of several amino acid sequence motifs that have been used to establish the function of different domains in NS3 and NS5 are indicated. C, nucleocapsid protein; M and E, two envelope proteins; pr, precursor.

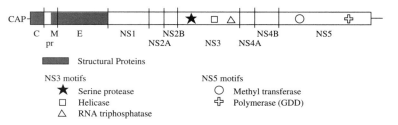

present in known cellular helicases (Gorbalenya et al 1989b; Laín et al 1989). Furthermore, NS3 from YF (Warrener et al 1993) and West Nile (Wengler and Wengler 1991) viruses have been shown to have ATPase activity that is stimulated by single-stranded RNA, which is a property of helicases (see, for example, Laín et al 1991). The CI protein of plum pox virus, which appears to be homologous to NS3, is capable of unwinding double-stranded RNA in a reaction that requires ATP or other nucleoside triphosphates (Laín et al 1990).

The C-terminal domain of NS3 has been shown to be an RNA triphosphatase that removes the γ phosphate from the 5' triphosphate on a single-stranded RNA molecule (Wengler and Wengler 1993). Removal of this phosphate is presumed to be the first step in capping the viral RNA.

NS5 has at least two functions. The N-terminal region contains motifs shared with methyltransferases (Koonin 1993), and this domain is believed to cap the viral RNA after dephosphorylation by NS3. The C-terminal domain of NS5 has motifs shared with RNA polymerases of a number of other RNA-containing viruses, and this domain is believed to be an RNA polymerase (Rice et al 1986).

NS2B is a component of the viral protease (Chambers et al 1991; Falgout et al 1991; Preugschat et al 1990), and NS2B and NS3 form a complex (Arias et al 1993; Chambers et al 1993; Jan et al 1995), referred to here as NS2B/NS3, that is the active form of the protease. NS4A and NS4B are probably components of the viral replicase complex, which is known to be membrane-associated. The function of NS2A is not known, although one function might be to present NS1 in an alternative conformation. Table 2 lists the different YF proteins and summarizes their functions where known.

PRODUCTION OF THE STRUCTURAL PROTEINS

The translation and processing of the structural proteins is complicated and requires both virus encoded and cellular enzymes. Figure 3 schematically diagrams this process. The capsid protein is N-terminal in the viral polyprotein, and following its C-terminus is an internal signal sequence that leads to insertion of the nascent polyprotein into the endoplasmic reticulum (ER) and the transport of the following protein, prM, into the lumen of the ER. Two cleavages are required to release the capsid protein from the precursor and to generate the N-terminus of prM; the first of these is accomplished by the viral NS2B/NS3 protease (Amberg et al 1994; Lobigs 1993; Yamshchikov and Compans 1994) and the second by signalase acting in the lumen of the ER. The order of these cleavages was first thought to be a signalase cleavage to separate a precursor of the capsid protein, anC, from prM, and anC has in fact been reported in infected cells, although not in large amounts. However, several observations now suggest that the cleavage to release the capsid protein is normally the first event and is followed closely by the signalase cleavage to produce

TABLE 2. FUNCTIONS of flavivirus proteins

Structural proteins		
anC (C)	121 (102) aa*	Nucleocapsid protein
prM (M)	164 (75) aa	Envelope protein
		prM forms heterodimer with E
		Cleaved during virion assembly
E	493 aa	Major envelope protein
		Hemagglutinin, Neutralizing epitope
		Receptor-binding domain
Nonstructural proteins		
NS1	352 aa	Nonstructural glycoprotein
		Possible role in virion assembly
NS2A	224 aa	Membrane-associated protein
		Possible role in anchoring NS1
NS2B	130 aa	Component of the nonstructural protease
		May affect prM-E cleavage by signalase
NS3	623 aa	Nonstructural serine protease
		Helicase
		RNA triphosphatase (required for cap-ping?)
NS4A	126 aa	Membrane-associated protein
		May be part of replicase complex
NS4B	250 (273) aa	Membrane-associated protein
		May be component of replicase
NS5	905 aa	Methyltransferase (capping reaction)
		RNA polymerase

* The sizes shown are for YF virus; aa = amino acids

prM. First, cleavage by signalase is delayed, unlike the removal of most signal sequences, and the polyprotein anC-prM can be isolated from infected cells early but not late after infection (Murray et al 1993). Second, efficient cleavage of anC-prM by signalase requires NS2B/NS3 (Amberg et al 1994; Lobigs 1993; Yamshchikov and Compans 1993). Two models have been presented to explain the requirement for NS2B/NS3. In one model, cleavage of C from the polyprotein occurs first and this results in a conformational change in an-prM so that signalase now cleaves prM from an. In a second model, cleavage of anC by NS2B/NS3 is not required *per se,* but the binding of NS2B/NS3, and more specifically NS2B, to anC causes the conformational change that allows signalase to cleave the anC-prM bond. In support of the second

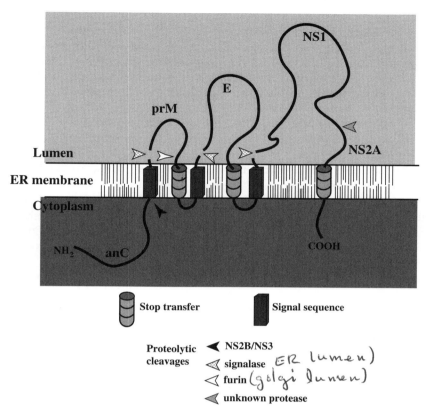

FIGURE 3. SCHEMATIC ILLUSTRATION OF PROCESSING OF FLAVIVIRUS STRUC-
TURAL PROTEINS. As described in the text, some of the cleavages are cotranslational,
whereas others are delayed. The protease that cleaves following NS1 has not been identified
but is believed to be a cellular enzyme that resides in the lumen of the endoplasmic reticulum or another intracellular vesicle.

model, Yamshchikov and Compans (1995) reported that mutations in the anC
cleavage site that block cleavage by the NS2B/NS3 protease do not prevent
cleavage of anC-prM by signalase. In either case, it appears likely that during
normal infection, the cleavage of C from the polyprotein anC-prM occurs first
and is followed very quickly by cleavage to release prM.

At the C-terminus of prM is a stretch of 37 amino acids that contains
only one charged residue, an arginine at residue 22; this arginine is conserved
in all flaviviruses. The N-terminal stretch of uncharged amino acids is thought
to act as a stop transfer signal that leads to the anchoring of prM in the ER,
with the arginine in the cytoplasm. The second uncharged stretch probably

acts as an internal signal that leads to the insertion of the following protein, E, into the lumen of the ER. Experiments in which one or the other of these uncharged domains was deleted and the production and orientation of the resulting prM and E in the ER was studied supported this interpretation (Markoff et al 1994). The situation is comparable to that found in alphaviruses in which such stop transfer and internal signal sequences have been thoroughly studied (reviewed in Strauss and Strauss 1994).

Separation of prM from E is once again believed to be effected by signalase, but in this case cleavage appears to be cotranslational. At the C-terminus of E, there is a hydrophobic domain very similar in organization to that found at the end of prM, with a single arginine residue at position 15 of an otherwise uncharged stretch of 39 amino acids. Once again it seems probable that the first uncharged stretch acts as a stop transfer signal to anchor E in the lipid bilayer, whereas the second uncharged stretch acts as an internal signal to transport the following protein, NS1, into the lumen of the ER. Consistent with this hypothesis, it has been shown that the second uncharged stretch can act as an N-terminal signal sequence for the insertion of NS1 (Fan and Mason 1990). Separation of E from NS1 is believed to be cotranslationally effected by signalase.

Following its synthesis and primary processing by signalase, prM is further processed at some stage preceding or during virus assembly. This cleavage of prM to M and a fragment (called "pr") released ultimately into the culture fluid (Murray et al 1993; Randolph et al 1990) is effected by a cellular protease, probably furin, present in the Golgi apparatus (reviewed in Strauss and Strauss 1994).

STRUCTURE OF YELLOW FEVER VIRUS

The YF virion is an enveloped particle with a diameter of about 50 nm as seen in the electron microscope (Murphy 1980). It is presumed to have icosahedral symmetry although conclusive evidence for this is lacking. The nucleocapsid, composed of the viral RNA and C, is about 30 nm in diameter. The viral envelope contains two integral membrane proteins, M and E. E has been found to be present in virions as a homodimer (Allison et al 1995). The structure of E has recently been solved to atomic resolution (Rey et al 1995) and, in contrast to the envelope proteins of other enveloped viruses, is proposed to lie flat along the surface of the virion so that its projection above the membrane is less than that seen for the spikes of other viruses. The dimensions of E are compatible with T=3 icosahedral symmetry (180 molecules on the surface) but not with T=4 (240 molecules) (Kuhn and Rossmann 1995).

Assembly of the capsid and of the virion are not well understood. In most flaviviruses, assembly occurs in association with intracellular membranes followed by subsequent release of the virus (Murphy 1980), although at least a few cases of flavivirus budding from plasma membranes have been described

(Ng et al 1994a). Intermediates in the assembly process are seldom observed (but see Ng et al 1994b), and it is assumed that assembly proceeds very rapidly once triggered by some process. As described above, cleavage to release C and the signalase cleavage to produce prM appear to be coordinated, and it is conceivable that these coordinated cleavages result in an assembly cascade.

prM and E form a heterodimer that is present in immature virions found inside the cell (Heinz et al 1994; Wengler and Wengler 1989), and cleavage of prM to M is believed to occur during transport and release of these immature virions. Because pr, the released portion of prM, is found only in the medium and not inside the cell, it has been postulated that cleavage of prM occurs late, just before or coincident with release of the virus (Murray et al 1993). Cleavage of prM can be blocked by treatment of the cells with lysosomotropic agents such as Tris or NH4Cl, and blockage of cleavage results in the release of prM-containing particles from the cell (Shapiro et al 1972, 1973; Randolph et al 1990; Guirakhoo et al 1991; Heinz et al 1994). These immature particles are noninfectious or have only low infectivity. Cleavage of prM results in the reorganization of the prM-E heterodimer to form E homodimers, and this is essential for infectivity (Allison et al 1995). The situation is analogous to that for the much better studied alphaviruses (Strauss and Strauss 1994), in which a PE2-E1 heterodimer forms which is converted into an E2-E1 heterodimer upon cleavage of PE2 (thought to be effected by the same enzyme that cleaves prM). prM-containing flaviviruses or PE2-containing alphaviruses form readily but possess low infectivity and are more stable to treatment with acidic pH. It is believed that heterodimers containing uncleaved glycoproteins are protected from disassembly during transport of the heterodimers or of particles containing the heterodimers through acidic compartments within the cell, and that cleavage is required to trigger disassembly of virus when it enters endosomes during infection. Infection by flaviviruses has been found to be sensitive to treatment of cells with agents that raise endosomal pH (Guirakhoo et al 1991; Heinz et al 1994; Randolph and Stollar 1990) and treatment of flaviviruses with acidic pH allows them to fuse with model membranes (Vorovitch et al 1991) or with cells (Guirakhoo et al 1991; Randolph and Stollar 1990), consistent with the hypothesis that when flaviviruses enter into endosomes during infection, the low pH triggers a rearrangement that leads to fusion of the viral membrane with the endosomal membrane. The pH required for fusion can be altered by mutation (Guirakhoo et al 1993). Subjecting the virions to the fusion pH leads to a rearrangement of the envelope proteins such that the E homodimers are converted to homotrimers (Allison et al 1995), and the reactivity of the particle with antibodies or proteases is changed (Guirakhoo et al 1989; Heinz et al 1994). Particles that contain prM are resistant to low pH-induced alterations and are unable to fuse cells upon low pH treatment, consistent with this model (Guirakhoo et al 1991, 1992; Heinz et al 1994; Allison et al 1995).

During flavivirus infection, heterogeneous particles are released that contain prM, M, and E or that might represent empty virus particles formed in the absence of capsids (Russell et al 1980). These particles are antigenic and since they can form when flavivirus structural genes are expressed by a vector such as vaccinia virus, it has been suggested that this approach might be useful for production of vaccines (Konishi et al 1991; Mason et al 1991; Yamshchikov and Compans 1993).

Yellow Fever NS1, NS2AB, NS3

NS1 follows E in the translated amino acid sequence. It is inserted into the lumen of the ER and glycosylated, and is found both cell-associated, in part on the cell surface, and extracellularly. At least two forms of NS1 have been described. For the most abundant form, cleavage from NS2A is effected in the lumen of the ER by an unknown protease (Figure 3). This cleavage is delayed, requires NS2A sequences (Falgout et al 1989) as well as the eight C-terminal amino acids of NS1 (Hori and Lai 1990), and results in an NS1 that lacks a membrane anchor. This NS1 is nonetheless membrane-associated, at least in part (Fan and Mason 1990; Mason 1989; Wengler et al 1990). A second form of NS1, called NS1' or NS1-2A, has been found in Japanese encephalitis virus-infected cells (Mason 1989; Mason et al 1987) and YF-infected cells (Chambers et al 1990b). NS1' migrates more slowly in gels than NS1 and is presumed to be larger. NS1' may possess a C-terminal anchor derived from NS2A, and might be produced by cleavage of a cryptic site in NS2A by the NS2B/NS3 protease (Figure 4); abolition of this cleavage site by site-specific mutagenesis is lethal for YF (Nestorowicz et al 1994). In this model, then, alternative cleavage sites are used to produce different forms of NS1.

NS2A and NS2B, which follow NS1, are very hydrophobic peptides that are membrane-associated. Their topology within membranes remains unclear, but one possible model for their orientation is shown in Figure 4. In this model, NS2B functions as a cofactor for the nonstructural protease, although only the central portion of NS2B, which is hydrophilic, is required for proteolytic activity and for complex formation with NS3. Thus the NS2B/NS3 protease is membrane-associated, and both NS2B and NS3 as well as the protease activity have indeed been reported to be membrane-associated in infected cells (Amberg et al 1994; Wengler et al 1990).

Cleavage between NS2A and NS2B, NS2B and NS3, and NS3 and NS4A are all effected by the NS2B/NS3 protease. The NS2B-NS3 cleavage occurs in *cis* and can readily be seen during cell-free translation of appropriate constructs. However, the NS3-NS4A cleavage is delayed and an NS3-NS4A polyprotein can be isolated from infected or transfected cells. From a study of the kinetics of cleavage, Lobigs (1992) suggested that this cleavage occurs primarily in *trans*. The kinetics of cleavage have also been found to be affected by the conformation of the polypeptide produced (Zhang and Padmanabhan 1993).

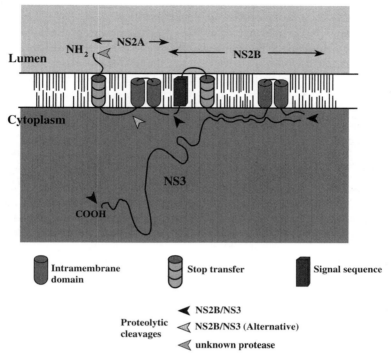

FIGURE 4. PROCESSING OF THE FLAVIVIRUS NS2A, NS2B, AND NS3 PROTEINS.
In this schematic illustration, an alternative site of cleavage within NS2A is shown that
might lead to an anchored form of NS1. NS2B and NS3 form a complex that involves the
central 40 amino acids of NS2B and that is required for expression of the proteolytic activity
in NS3; the interaction ties NS3 to the membrane. The orientation of NS2A and NS2B in
the membranes has not been shown directly, but the model illustrated is compatible with
the sequences of these two proteins and with all data derived to date.

THE NONSTRUCTURAL PROTEASE

NS3 was first postulated to be a serine protease on the basis of computer
folding studies performed by Bazan and Fletterick (1989) and by Gorbalenya
et al (1989a). This was confirmed by studies in a number of laboratories,
including our own, in which the N-terminal 180 residues of NS3 together with
NS2B were found to be an active protease that cleaved different sites in the
flavivirus polyprotein, and site-specific mutagenesis studies showed that the
serine, histidine, and aspartic acid proposed to form the catalytic triad of a
serine protease were indeed required for activity (Preugschat et al 1990; Chambers et al 1990; Falgout et al 1991; Wengler et al 1991). Since the protein
folding routines used by Bazan and Fletterick (1990) and by Gorbalenya et al

(1989a) are based in part upon the assumption that the protein folds in the same way as do serine proteases such as chymotrypsin, we have postulated that the flavivirus protein and the cellular proteins are evolutionarily related to one another, although crystal structures will be required to determine just how similar the flavivirus proteases are to cellular serine proteases. It seems probable that the viral protease was obtained from the host cell during the evolution of these viruses by recombination of a precursor flavivirus with a cellular mRNA encoding a serine protease.

Unlike many serine proteases, the flavivirus nonstructural protease requires a cofactor, NS2B, for cleavage to occur. NS2B and NS3 form a complex which can be readily visualized by their co-immunoprecipitation from infected cell lysates by antibodies to NS2B or NS3 (Arias et al 1993; Chambers et al 1993; Jan et al 1995) (Figure 5); this complex forms even when the proteins are expressed in *trans* from separate constructs. Deletion studies and mutagenesis studies have shown that only the central conserved domain of NS2B, consisting of 40 amino acids, is required for this cofactor activity (Chambers et al 1993; Falgout et al 1993). Furthermore, it has been found that certain heterologous combinations of NS2Bs and NS3s derived from different flaviviruses will form an active protease, whereas other combinations will not (Preugschat et al 1991; Jan et al 1995).

Some of the cleavages effected by the NS2B/NS3 protease occur in *cis* but many of them occur in *trans*. In general, such *trans* cleavages can be readily demonstrated in cells transfected with constructs expressing parts of the flavivirus polyprotein, but *trans* cleavage by the NS2B/NS3 protease has been much more difficult to demonstrate in cell-free translation systems (Amberg et al 1994). It has been suggested that the association of the protease with membranes through the hydrophobic regions of NS2B might accelerate such *trans* proteolytic processing events because diffusion is limited to two dimensions.

The consensus cleavage site recognized by the NS2B/NS3 protease is G/A-R-R $\downarrow$ S/G, where cleavage occurs at the position of the arrow. The double basic motif is not invariant in all cleavage sites of all flaviviruses, but cleavage always follows a basic residue. Site-specific mutagenesis has shown that, in some sites, alteration of either basic residue will kill the site, whereas other sites accept changes in the P2 position (Lin et al 1993b; Nestorowicz et al 1994; Chambers et al 1995). Similarly, cleavage at some sites is very sensitive to the identity of the P1' residue, whereas other sites will accept changes. Thus domains other than the P3-P2-P1-P1' sequence must also be important for cleavage site recognition. Studies with chimeric proteases have shown that proteases differ in their activity on the cleavage sites from different flaviviruses and that there is an interaction between the substrate binding domain identified by Bazan and Fletterick (1989, 1990) and the cleavage site (Preugschat et al 1991).

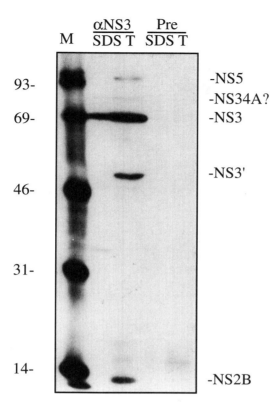

FIGURE 5. CO-IMMUNOPRECIPITATION OF NS3 AND NS2B. Mosquito cells were infected with dengue 2 virus and labeled with [³⁵S]methionine. Cell lysates were prepared under denaturing conditions (SDS) or under non-denaturing conditions, using Triton-X-100 (T), and were immunoprecipitated with anti-NS3 antiserum (α-NS3) or with pre-immune serum (Pre). Note that after SDS extraction, only NS3 is precipitated, whereas after Triton extraction, NS2B is also precipitated. There is also a fragment produced in the Triton extract labeled NS3' that is produced by cleavage of NS3 by the NS2B/NS3 protease (see text). From Arias et al (1993), with permission.

The NS2B/NS3 protease cleaves six sites in the polyprotein that are known to be important for virus replication. There is a seventh NS2B/NS3 cleavage site within the helicase domain of NS3 whose function is not known (Arias et al 1993). It may be irrelevant for the virus life cycle, representing an adventitious site that the protease recognizes, or it may represent an unknown control mechanism. The latter possibility is at least suggested by the fact that the extent of cleavage at this latter site is different in different cells.

YELLOW FEVER NS4A, NS4B, NS5

NS4A and NS4B follow NS3 in the polyprotein sequence. Both are very hydrophobic and associated with membranes. Although their topology in the membrane has not been directly determined, one possible model for their orientation is shown in Figure 6.

Cleavage between NS3 and NS4A is effected by the NS2B/NS3 protease. Cleavage between NS4A and NS4B is more complicated. The C-terminus of NS4A is generated by cleavage by the NS2B/NS3 protease; removal of NS4A liberates an N-terminal signal sequence that leads to the transfer of NS4B across the ER, where the signal is removed by signalase to generate the N-terminus of NS4B (Lin et al 1993a). This signal normally cannot function as an internal signal (but see Cahour et al 1992), but only as an N-terminal signal (Preugschat and Strauss 1991; Lin et al 1993a), and thus cleavage by the NS2B/NS3 protease to release NS4A is normally required before the

FIGURE 6. SCHEMATIC ILLUSTRATION OF THE SYNTHESIS, insertion, and processing of flavivirus NS4A, NS4B, and NS5 proteins. The orientation of NS4A and NS4B within the membrane has not been determined directly, but this model is consistent with the sequences of these peptides.

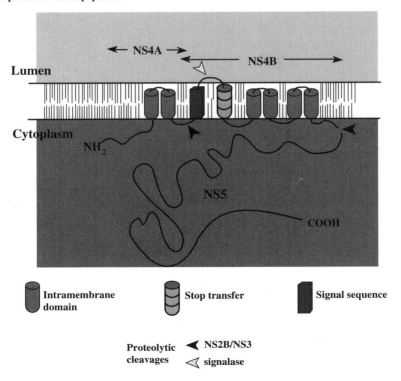

signalase cleavage to produce NS4B can occur. Cleavage between NS4B and NS5 is again effected by the viral NS2B/NS3 protease.

NS4A and NS4B are probably components of the viral replicase complex and may serve as a membrane-associated scaffold for the replicase. NS5, the RNA polymerase, is known to be membrane-associated in infected cells (Wengler et al 1990), and this membrane association could be effected by association with NS4A or NS4B.

VIRULENCE DETERMINANTS IN YELLOW FEVER

The 17D vaccine strain of YF was isolated more than 50 years ago by passage of YF in chicken cells, and it represents one of the most successful attenuated vaccines yet produced (the history of the selection of 17D YF is reviewed in Hahn et al 1987a). Complete sequencing of the parental Asibi strain of YF and of three different 17D strains has shown that there are 48 nucleotide differences between Asibi and the vaccine strains which are the same in all three 17D strains examined (shown in Table 3). These 48 substitutions result in 22 amino acid differences between Asibi and 17D (Hahn et al 1987a; Duarte dos Santos et al 1995).

Although it is widely assumed that attenuation results from multiple mutations in the 17D genome, certain changes are prime candidates for being responsible, at least in part, for attenuation. No alterations were found in the 5' nontranslated region, and the changes in the 3' nontranslated region do not appear to be important for attenuation. NS2 and NS4 are not particularly

TABLE 3. SEQUENCE DIFFERENCES between Asibi and the 17D vaccine strain

Region	Nucleotide differences	Amino acid differences
5' nontranslated	0	–
C	1	0
prM	1	1
E	11	8
NS1	4	2
NS2A	5	4
NS2B	3	1
NS3	5	1
NS4A	3	1
NS4B	2	2
NS5	9	2
3' nontranslated	4	–

The differences listed have been found in all of three different substrains of 17D sequenced.

conserved as to sequence and the changes in them do not alter the hydrophobicity of the proteins, suggesting that these substitutions may not be important for attenuation. Thus, the prime candidates for the substitutions that result in attenuation are the changes in E and in NS5. The changes in E are of particular interest, because 8 of 11 nucleotide substitutions result in a change in amino acid sequence, suggesting that many of these changes were positively selected during passage in chicken cells.

Figure 7 illustrates differences between Asibi and 17D in two domains of the E protein. In panel A, Ser-305 in Asibi became Phe-305 in 17D. A strain of YF isolated from a fatal case of vaccine-induced encephalitis has been sequenced through the E protein, and two changes between the parent 17D vaccine strain and the neurovirulent strain were noted (Jennings et al 1994). One of these, Lys-303 in 17D/Asibi to Gln in the neurovirulent variant, was hypothesized to be important for the virulence. Thus there is reason to postulate that changes at positions 303 and 305 change the tropism of the virus and that this domain in the YF E protein may be important for binding of virus to cells in order to initiate infection.

FIGURE 7. MUTATIONS IN YF and Murray Valley encephalitis (MVE) flavivirus that alter virulence. Two regions of the flavivirus E protein are shown. In A, a sequence following amino acid 300 has changed during selection of 17D and it is postulated that this change attenuates YF. A second change in this same region was found in the virus isolated from a case of fatal encephalitis induced by the vaccine; this second change appears to render the virus neurovirulent. In B, a sequence near residue 380 is illustrated for both YF and MVE. The attenuated 17D has acquired an RGD sequence, which may be important for attenuation. MVE has an RGD sequence in the same location; loss of the RGD attenuates the virus.

	Virus	Sequence with mutations	Phenotype
		300 303 305	
A.	Asibi YF	C T D K M S F V K	virulent
	17D YF	F	attenuated
	17D-(NV)	Q F	neurovirulent

	Virus	Sequence with mutations	Phenotype
		380	
B.	Asibi YF	I V G T G D S A L T Y Q W	virulent
	17D YF	R	attenuated
	MVE	V <u>V</u> <u>G</u> R <u>G</u> <u>D</u> K Q I N H H <u>W</u>	virulent
	MVE (Att.)	H	attenuated
		G	attenuated
		A	attenuated
		N	attenuated

A second difference between Asibi and 17D in the E protein, TGD in Asibi at amino acids 380-382 to RGD in 17D (Figure 7B), has received attention for three reasons. First, RGD is known to be a motif recognized by certain integrins, and the interaction between RGD-bearing ligands and integrins mediates cell adhesion (reviewed in D'Souza et al [1991]); this tripeptide motif has been shown to be an essential part of the cell attachment site on foot-and-mouth virus (Fox et al 1989) and on Coxsackie viruses (Roivainen et al 1991). Second, 17D was selected by passage in chicken cells, and it is of interest that the RGD sequence is present in all members of the Murray Valley encephalitis (MVE) group of flaviviruses, which are bird-associated and neurovirulent, suggesting that replication in birds selects for RGD. Third, when MVE was passaged in the SW13 human cell line, mutants were selected with alterations in the RGD sequence that were attenuated in mice (Lobigs et al 1990). The changes found were all at the D residue, which changed to H, G, A, or N (Figure 7B), indicating that changes in this domain may alter virulence.

To examine whether the RGD sequence might be important for virus entry, we tested the ability of RGD-containing peptides to interfere with the infection of cultured chicken cells by 17D YF. We found that such peptides had no detectable effect upon infection, suggesting that the selection of the RGD motif by passage on avian cells was not based upon this motif being part of a cell attachment site on the virus, but upon some other aspect of virus-chicken cell interaction, and suggesting that integrins are probably not used as receptors by 17D YF (F. Preugschat, unpublished).

cDNA CLONES OF FLAVIVIRUSES

Obtaining full-length cDNA clones of flaviviruses from which infectious virus can be resurrected has proven to be much more difficult than for most other RNA viruses. Portions of the genomes of almost all flaviviruses are toxic for bacteria for reasons that are not understood, and this results in unstable cDNA clones that undergo rearrangements and deletions to eliminate the toxic sequences. cDNA clones from which infectious RNA can be rescued now exist for at least four flaviviruses (Table 4). In two cases, a full-length clone has been assembled that is stable under the conditions used, but in two other cases, the genome has been assembled in two segments that can be ligated together *in vitro* to produce full-length cDNA.

Perhaps appropriately, the first "infectious" cDNA clone was obtained for 17D YF in 1989 (Rice et al 1989). This clone has been used to test the requirement for cleavage at many of the sites in the polyprotein for virus infectivity (Chambers et al 1995; Nestorowicz et al 1994). Using this clone, it will be possible to determine which of the differences between Asibi and 17D are responsible for attenuation, an approach that may make it possible to design an even better vaccine, since monkey neurovirulence tests have shown that the virus rescued from the cDNA

TABLE 4. INFECTIOUS CLONES of flaviviruses

Virus	Construct	Reference
Yellow fever	ligated	Rice et al 1989
Dengue type 4	full-length	Lai et al 1991
Japanese encephalitis	ligated	Sumiyoshi et al 1992
Kunjin	full-length	Khromykh and Westaway 1994

For YF and Japanese encephalitis virus, the genome is maintained in two pieces that can be ligated together *in vitro* to produce a full-length cDNA. For dengue 4 and Kunjin, a stable full-length cDNA clone has been obtained. The year of publication and the reference to the description of each clone is listed.

clone is attenuated (as is 17D) (Marchevsky et al 1995). Furthermore, because the cloned strain is attenuated, the cDNA clone represents a stable repository for the vaccine strain.

Equivalent clones now exist for dengue 4 (Lai et al 1991), Japanese encephalitis (Sumiyoshi et al 1992), and Kunjin (Khromykh and Westaway 1994) viruses. The dengue 4 clone has been used to construct a number of attenuated viruses as an approach to vaccine development. In one approach, deletions within both the 5' (Cahour et al 1995) and 3' (Lai et al 1992) nontranslated regions have been tested and many of these deletions were found to attenuate the virus. In a second approach, the dengue 4 clone was used to construct chimeric viruses between dengue 4 and other strains of dengue (Bray and Lai 1991) or between dengue 4 and tick-borne encephalitis (TBE) virus (Pletnev et al 1992). Viable chimeras were produced that were attenuated relative to the virulent parent; the dengue 4-TBE chimeras could be further attenuated by site-specific mutations (Pletnev et al 1993). In effect, this approach uses dengue 4 as a vector to express antigens of other flaviviruses. The results to date suggest that a good vaccine virus such as YF 17D or an attenuated dengue virus could be an effective vector for immunization against other flaviviruses.

PESTIVIRUSES AND HEPATITIS C VIRUS

Nucleic acid sequencing has revealed that there are similarities in genome organization between the flaviviruses, the pestiviruses, and hepatitis C virus (reviewed in Strauss et al 1991). These similarities include the presence of a nucleocapsid protein and structural glycoproteins at the N-terminus of the single long open reading frame, separated by a very hydrophobic, NS2A/NS2B-like domain from an NS3-like protein, which is in turn separated from an NS5-like polypeptide by a second very hydrophobic domain. The NS3-like domains of flaviviruses, pestiviruses, and hepatitis C virus all possess protease and helicase activities; flavivirus NS3, in addition, possesses a triphosphatase activity. In pestiviruses, the NS3 protein has been shown to possess ATPase activity characteristic of cellular helicases. The NS5 domain in the pestiviruses is longer than that of flaviviruses and probably encodes more functions in addition to those found in flavivirus NS5. Interestingly, the NS5 regions of pestiviruses and hepatitis C viruses, but not that of flaviviruses, are cleaved into two proteins, 5A and 5B. Nonetheless, the similarities in genome organization have led to the classification of pestiviruses and hepatitis C virus as genera in the family Flaviviridae.

CONCLUDING REMARKS

Much has been learned about the flaviviruses over the past 10 years in terms of their genome organization, the expression of their proteins, and the functions of many of these proteins. Yet gaps remain in the body of knowledge about flavivirus molecular biology, particularly about the interactions of these viruses with their hosts to cause disease. These viruses have proven difficult to work with over the years. The cleavage cascade to process the precursor polyprotein has proved to be much more complex than first anticipated and required major efforts to elucidate. This complex cleavage pathway is surely important for regulation of the virus life cycle in ways that are not yet completely understood. However, we now have in hand the tools and background information to make very rapid progress over the next 10 years.

ACKNOWLEDGMENTS

Our work with flaviviruses has been supported by Grant V22/181/62 from the World Health Organization and Grant AI 20612 from the National Institutes of Health.

REFERENCES

Allison SL, Schalich J, Stiasny K, Mandl CW, Kunz C, Heinz FX (1995). Oligomeric rearrangement of tick-borne encephalitis virus envelope proteins induced by an acidic pH. J Virol 69: 695-700.

Amberg SM, Nestorowicz A, McCourt DW, Rice CM (1994). NS2B-3 proteinase-mediated processing in the yellow fever virus structural region: in vitro and in vivo studies. J Virol 68: 3794-3802.

Arias CF, Preugschat F, Strauss JH (1993). Dengue 2 virus NS2B and NS3 form a stable complex that can cleave NS3 within the helicase domain. Virology 193: 888-899.

Bazan JF, Fletterick RJ (1989). Detection of a trypsin-like serine protease domain in flaviviruses and pestiviruses. Virology 171: 637-639.

Bazan JF, Fletterick RJ (1990). Structural and catalytic models of trypsin-like viral proteases. Sem Virol 1: 311-322.

Bray M, Lai C-J (1991). Construction of intertypic chimeric dengue viruses by substitution of structural protein genes. Proc Natl Acad Sci USA 88: 10342-10346.

Cahour A, Falgout B, Lai C-J (1992). Cleavage of the dengue virus polyprotein at the NS3/NS4A and NS4B/NS5 junctions is mediated by viral protease NS2B-NS3, whereas NS4A/NS4B may be processed by a cellular protease. J Virol 66: 1535-1542.

Cahour A, Pletnev A, Vazeille-Falcoz M, Rosen L, Lai C-J (1995). Growth-restricted dengue virus mutants containing deletions in the 5' noncoding region of the RNA genome. Virology 207: 68-76.

Cammisa-Parks H, Cisar LA, Kane A, Stollar V (1992). The complete nucleotide sequence of cell fusing agent (CFA): Homology between the nonstructural proteins encoded by CFA and the nonstructural proteins encoded by arthropod-borne flaviviruses. Virology 189: 511-524.

Castle E, Wengler G (1987). Nucleotide sequence of the 5'-terminal untranslated part of the genome of the flavivirus West Nile virus. Arch Virol 92: 309-313.

Chambers TJ, Grakoui A, Rice CM (1991). Processing of the yellow fever virus nonstructural polyprotein: a catalytically active NS3 proteinase domain and NS2B are required for cleavages at dibasic sites. J Virol 65: 6042-6050.

Chambers TJ, Hahn CS, Galler R, Rice CM (1990a). Flavivirus genome organization, expression, and replication. Ann Rev Microbiol 44: 649-699.

Chambers TJ, McCourt DW, Rice CM (1990b). Production of yellow fever virus proteins in infected cells: Identification of discrete polyprotein species and analysis of cleavage kinetics using region-specific antisera. Virology 177: 159-174.

Chambers TJ, Nestorowicz A, Amberg SM, Rice CM (1993). Mutagenesis of the yellow fever virus NS2B protein: effects on proteolytic processing, NS2B-NS3 complex formation, and viral replication. J Virol 67: 6797-6807.

Chambers TJ, Nestorowicz A, Rice CM (1995). Mutagenesis of the yellow fever virus NS2B/3 cleavage site: Determinants of cleavage site specificity and effects on polyprotein processing and viral replication. J Virol 69: 1600-1605.

Chambers TJ, Weir RC, Grakoui A, McCourt DW, Bazan JF, Fletterick RJ, Rice CM (1990). Evidence that the N-terminal domain of nonstructural protein NS3 from yellow fever virus is a serine protease responsible for site-specific cleavages in the viral polyprotein. Proc Natl Acad Sci USA 87: 8898-8902.

Coia G, Parker MD, Speight G, Byrne ME, Westaway EG (1988). Nucleotide and complete amino acid sequences of Kunjin virus: Definitive gene order and characteristics of the virus-specified proteins. J Gen Virol 69: 1-21.

Deubel V, Kinney RM, Trent DW (1988). Nucleotide sequence and deduced amino acid sequence of the nonstructural proteins of dengue type 2 virus, Jamaica genotype: comparative analysis of the full length genome. Virology 165: 234-244.

D'Souza SE, Ginsberg MH, Plow EF (1991). Arginyl-glycyl-aspartic acid (RGD): a cell adhesion motif. Trends Biochem Sci 16: 246-250.

Duarte dos Santos CN, Post PR, Carvalho R, Ferreira II, Rice CM, Galler R (1995). Complete nucleotide sequence of yellow fever virus vaccine strains 17DD and 17D-213. Virus Res 35: 35-41.

Falgout B, Chanock R, Lai C-J (1989). Proper processing of dengue virus nonstructural glycoprotein NS1 requires the N-terminal hydrophobic signal sequence and the downstream nonstructural protein NS2a. J Virol 63: 1852-1860.

Falgout B, Miller RH, Lai C-J (1993). Deletion analysis of dengue virus type 4 nonstructural protein NS2B: Identification of a domain required for NS2B-NS3 protease activity. J Virol 67: 2034-2042.

Falgout B, Pethel M, Zhang Y-M, Lai C-J (1991). Both nonstructural proteins NS2B and NS3 are required for the proteolytic processing of Dengue virus nonstructural proteins. J Virol 65: 2467-2475.

Fan W, Mason PW (1990). Membrane association and secretion of the Japanese encephalitis virus NS1 protein from cells expressing NS1 cDNA. Virology 177: 470-476.

Fox G, Parry NR, Barnett PV, McGinn B, Rowlands DJ, Brown F (1989). The cell attachment site on foot-and-mouth disease virus includes the amino acid sequence RGD (arginine-glycine-aspartic acid). J Gen Virol 70: 625-637.

Fu J, Tan B-H, Yap E-H, Chan Y-C, Tan YH. (1992). Full-length cDNA sequence of dengue type 1 virus (Singapore strain S275/90). Virology 188: 953-958.

Gorbalenya AE, Donchenko AP, Koonin EV, Blinov VM (1989a). N-Terminal domains of putative helicases of flavi- and pestiviruses may be serine proteases. Nucleic Acids Res 17: 3889-3897.

Gorbalenya AE, Koonin EV, Donchenko AP, Blinov VM (1989b). Two related superfamilies of putative helicases involved in replication, repair, and expression of DNA and RNA genomes. Nucleic Acids Res 17: 4713-4730.

Guirakhoo F, Bolin RA, Roehrig JT (1992). The Murray Valley encephalitis virus prM protein confers acid resistance to virus particles and alters the expression of epitopes within the R2 domain of E glycoprotein. Virology 191: 921-931.

Guirakhoo F, Heinz FX, Kunz C (1989). Epitope model of tick-borne encephalitis virus envelope glycoprotein E: Analysis of structural properties, role of carbohydrate side chain, and conformational changes occurring at acidic pH. Virology 69: 90-99.

Guirakhoo F, Heinz FX, Mandl CW, Holzmann H, Kunz C (1991). Fusion activity of flaviviruses: comparison of mature and immature (prM-containing) tick-borne encephalitis viruses. J Gen Virol 72: 1323-1329.

Guirakhoo F, Hunt AR, Lewis JG, Roehrig JT (1993). Selection and partial characterization of dengue 2 virus mutants that induce fusion at elevated pH. Virology 194: 219-223.

Hahn CS, Dalrymple JM, Strauss JH, Rice CM (1987a). Comparison of the virulent Asibi strain of yellow fever virus with the 17D vaccine strain derived from it. Proc Natl Acad Sci USA 84: 2019-2023.

Hahn CS, Hahn YS, Rice CM, Lee E, Dalgarno L, Strauss EG, Strauss JH (1987b). Conserved elements in the 3' untranslated region of flavivirus RNAs and potential cyclization sequences. J Mol Biol 198: 33-41.

Hahn CS, Rice CM, Strauss JH, Dalrymple JM (1987c). Comparison of the Asibi and 17D strains of yellow fever virus. In: Chanock R, Lerner R, Brown F, Ginsberg H, eds. Vaccines 87. Cold Spring Harbor, N.Y.: Cold Spring Harbor Laboratory Press, 316-321.

Hahn YS, Galler R, Hunkapiller T, Dalrymple J, Strauss JH, Strauss EG (1988). Nucleotide sequence of Dengue 2 RNA and comparison of the encoded proteins with those of other flaviviruses. Virology 162: 167-180.

Heinz FX, Stiasny K, Püschner-Auer G, Holzmann H, Allison SL, Mandl CW, Kunz C (1994). Structural changes and functional control of the tick-borne encephalitis virus glycoprotein E by the heterodimeric association with protein prM. Virology 198: 109-117.

Hori H, Lai C-J (1990). Cleavage of dengue virus NS1-NS2A requires an octapeptide sequence at the C terminus of NS1. J Virol 64: 4573-4577.

Iacono-Connors LC, Schmaljohn CS (1992). Cloning and sequence analysis of the genes encoding the nonstructural proteins of Langat virus and comparative analysis with other flaviviruses. Virology 188: 875-880.

Irie K, Mohan PM, Putnak R, Padmanabhan R (1989). Sequence analysis of cloned dengue virus type 2 genome (New Guinea C strain). Gene 75: 197-211.

Jan L-R, Yang C-S, Falgout B, Lai C-J (1995). Processing of Japanese encephalitis virus non-structural proteins: NS2B-NS3 complex and heterologous proteases. J Gen Virol 76: 573-580.

Jennings AD, Gibson CA, Miller BR, Mathews JH, Mitchell CJ, Roehrig JT, Wood DJ, Taffs F, Sil BY, Whitby SN, Whitby JE, Monath TP, Minor PD, Sanders PG, Barrett ADT (1994). Analysis of a yellow fever virus isolated from a fatal case of vaccine-associated human encephalitis. J Infect Dis 169: 512-518.

Khromykh AA, Westaway EG (1994). Completion of Kunjin virus RNA sequence and recovery of an infectious RNA transcribed from stably cloned full-length cDNA. J Virol 68: 4580-4588.

Konishi E, Pincus S, Fonesca BEL, Shope RE, Paoletti E, Mason PW (1991). Comparison of protective immunity elicited by recombinant vaccinia viruses that synthesize E or NS1 of Japanese encephalitis virus. Virology 185: 401-410.

Koonin EV (1993). Computer-assisted identification of a putative methyltransferase domain in NS5 protein of flaviviruses and $\lambda 2$ protein of reovirus. J Gen Virol 74: 733-740.

Kuhn RJ, Rossmann MG (1995). When it's better to lie low. Nature 375: 275-276.

Lai C-H, Zhao B, Hori H, Bray M (1991). Infectious RNA transcribed from stably cloned full-length cDNA of dengue type 4 virus. Proc Natl Acad Sci USA 88: 5139-5143.

Lai C-J, Men R, Pethel M, Bray M (1992). Infectious RNA transcribed from stably cloned dengue virus cDNA: Construction of growth-restricted dengue virus mutants. In: Brown F, Chanock RM, Ginsberg HS, Lerner RA, eds. Vaccines 92. Cold Spring Harbor, N.Y.: Cold Spring Harbor Laboratory Press, 265-270.

Laín S, Martín MT, Riechmann JL, García JA (1991). Novel catalytic activity associated with positive strand RNA virus infection: nucleic acid-stimulated ATPase activity of plum pox potyvirus helicaselike protein. J Virol 65: 1-6.

Laín S, Riechmann JL, García JA (1990). RNA helicase: a novel activity associated with a protein encoded by a positive strand RNA virus. Nucleic Acids Res 18: 7003-7006.

Laín S, Riechmann JL, Martín MT, García JA (1989). Homologous potyvirus and flavivirus proteins belonging to a superfamily of helicase-like proteins. Gene 82: 357-362.

Lee E, Fernon C, Weir RC, Rice CM, Dalgarno L (1990). Sequence of the 3' half of the Murray Valley encephalitis virus genome and mapping of the non-structural proteins NS1, NS3 and NS5. Virus Genes 4: 197-213.

Lepiniec L, Dalgarno L, Huong VTQ, Monath TP, Digoutte J-P, Deubel V (1994). Geographic distribution and evolution of yellow fever viruses based on direct sequencing of genomic cDNA fragments. J Gen Virol 75: 417-423.

Lewis JG, Chang G-J, Lanciotti RS, Kinney RM, Mayer LW, Trent DW (1993). Phylogenetic relationships of dengue-2 viruses. Virology 197: 216-224.

Lin C, Amberg SM, Chambers TJ, Rice CM (1993a). Cleavage at a novel site in the NS4A region by the yellow fever virus NS2B-3 proteinase is a prerequisite for processing at the downstream 4A/4B signalase site. J Virol 67: 2327-2335.

Lin C, Chambers TJ, Rice CM (1993b). Mutagenesis of conserved residues at the yellow fever virus 3/4A and 4B/5 dibasic cleavage sites: Effects on cleavage efficiency and polyprotein processing. Virology 192: 596-604.

Lobigs M (1992). Proteolytic processing of a Murray Valley encephalitis virus nonstructural polyprotein segment containing the viral proteinase: accumulation of a NS2-4A precursor which requires mature NS3 for efficient processing. J Gen Virol 73: 2305-2312.

Lobigs M (1993). Flavivirus premembrane protein cleavage and spike heterodimer secretion require the function of the viral proteinase NS3. Pro Natl Acad Sci USA 90: 6218-6222.

Lobigs M, Usha R, Nestorowicz A, Marshall ID, Weir RC, Dalgarno L (1990). Host cell selection of Murray Valley encephalitis virus variants altered at an RGD sequence in the envelope protein and in mouse virulence. Virology 176: 587-595.

Mackow E, Makino Y, Zhao B, Zhang Y-M, Markoff LI, Buckler-White A, Guiler M, Chanock RM, Lai C-J (1987). The nucleotide sequence of dengue type 4 virus: analysis of genes coding for nonstructural proteins. Virology 159: 217-228.

Mandl CW, Heinz FX, Stöckl E, Kunz C (1989). Genome sequence of tick-borne encephalitis virus (Western subtype) and comparative analysis of nonstructural proteins with other flaviviruses. Virology 173: 291-301.

Mandl CW, Holzmann H, Kunz C, Heinz FX (1993). Complete genomic sequence of Powassan virus: Evaluation of genetic elements in tick-borne versus mosquito-borne flaviviruses. Virology 194: 173-184.

Marchevsky RS, Mariano J, Ferreira VS, Almeida E, Cerqueira MJ, Carvalho R, Pissurno JW, Travassos da Rosa APA., Simoes MC, Santos CND, Ferreira II, Muylaert IR, Mann GF, Rice CM, Galler R. (1995). Phenotypic analysis of yellow fever virus derived from complementary DNA. Am J Trop Med Hyg 52: 75-80.

Marin MS, Zanotto PMdA, Gritsun TS, Gould EA (1995). Phylogeny of TYU, SRE, and CFA virus: different evolutionary rates in the genus *Flavivirus*. Virology 206: 1133-1139.

Markoff L, Chang A, Falgout B (1994). Processing of flavivirus structural glycoproteins: Stable membrane insertion of premembrane requires the envelope signal peptide. Virology 204: 526-540.

Mason PW (1989). Maturation of Japanese encephalitis virus glycoproteins produced by infected mammalian and mosquito cells. Virology 169: 354-364.

Mason PW, McAda PC, Dalrymple JM, Fournier MJ, Mason TL (1987). Expression of Japanese encephalitis virus antigens in *Escherichia coli*. Virology 158: 361-372.

Mason PW, Pincus S, Fournier MJ, Mason TL, Shope RL, Paoletti E (1991). Japanese encephalitis virus-vaccinia recombinants produce particulate forms of the structural membrane proteins and induce high levels of protection against JEV infection. Virology 180: 294-305.

Monath TP, Heinz FX (1996). Flaviviruses. In: Fields BN, Knipe DM, Howley PM, eds. Virology. Ed 3. New York: Raven Press, 961-1034.

Murphy FA (1980). Togavirus morphology and morphogenesis. In: Schlesinger RW, ed. The togaviruses: biology, structure, replication. New York: Academic Press, 241-316.

Murray JM, Aaskov JG, Wright PJ (1993). Processing of the dengue virus type 2 proteins prM and C-prM. J Gen Virol 74: 175-182.

Nestorowicz A, Chambers TJ, Rice CM (1994). Mutagenesis of the yellow fever virus NS2A/2B cleavage site: Effects on proteolytic processing, viral replication, and evidence for alternative processing of the NS2A protein. Virology 199: 114-123.

Ng ML, Howe J, Sreenivasan V, Mulders JJL (1994a). Flavivirus West Nile (Sarafend) egress at the plasma membrane. Arch Virol 137: 303-313.

Ng ML, Yeong FM, Tan SH (1994b). Cryosubstitution technique reveals new morphology of flavivirus-induced structures. J Virol Methods 49: 305-314.

Osatomi K, Sumiyoshi H (1990). Complete nucleotide sequence of Dengue type 3 virus genome RNA. Virology 176: 643-647.

Pletnev AG, Bray M, Huggins J, Lai C-J (1992). Construction and characterization of chimeric tick-borne encephalitis/dengue type 4 viruses. Proc Natl Acad Sci USA 89: 10532-10536.

Pletnev AG, Bray M, Lai C-J (1993). Chimeric tick-borne encephalitis and dengue type 4 viruses: Effects of mutations on neurovirulence in mice. J Virol 67: 4956-4063.

Preugschat F, Lenches EM, Strauss JH (1991). Flavivirus enzyme-substrate interactions studied with chimeric proteins: Identification of an intragenic locus important for substrate recognition. J Virol 65: 4749-4758.

Preugschat F, Strauss JH (1991). Processing of nonstructural proteins NS4A and NS4B of dengue 2 virus in vitro and in vivo. Virology 185: 689-697.

Preugschat F, Yao C-W, Strauss JH (1990). In vitro processing of dengue 2 nonstructural proteins NS2A, NS2B, and NS3. J Virol 64: 4364-4374.

Randolph VB, Stollar V (1990). Low pH-induced cell fusion in flavivirus- infected *Aedes albopictus* cell cultures. J Gen Virol 71: 1845-1850.

Randolph VB, Winkler G, Stollar V (1990). Acidotrophic amines inhibit proteolytic processing of flavivirus prM protein. Virology 174: 450-458.

Rey FA, Heinz FX, Mandl CW, Kunz C, Harrison SC (1995). The envelope glycoprotein from tick borne encephalitis virus at 2Å resolution. Nature 375: 291-298.

Rice CM, Lenches EM, Eddy SR, Shin SJ, Sheets RL, Strauss JH (1985). Nucleotide sequence of yellow fever virus: Implications for flavivirus gene expression and evolution. Science 229: 726-733.

Rice CM, Grakoui A, Galler R, Chambers TJ (1989). Transcription of infectious yellow fever RNA from full-length cDNA templates produced by in vitro ligation. The New Biologist 1: 285-296.

Rice CM, Strauss JH (1990). Production of flavivirus polypeptides by proteolytic processing. Sem Virol 1: 357-367.

Rice CM, Strauss EG, Strauss JH (1986). Structure of the flavivirus genome. In: Schlesinger S, Schlesinger MJ, eds. The Togaviridae and Flaviviridae. New York: Plenum Publishing Corp., 279-327.

Roivainen M, Hyypia T, Piirainen L, Kalkkinen N, Stanway G, Hovi T (1991). RGD-dependent entry of Coxsackievirus A9 into host cells and its bypass after cleavage of VP1 protein by intestinal proteases. J Virol 65: 4735-4740.

Russell PK, Brandt WE, Dalrymple JM (1980). Chemical and antigenic structure of flaviviruses. In: Schlesinger RW, ed. The Togaviruses. New York: Academic Press, 503-529.

Russell PK, Shapiro D, Brandt WE, (1972). Change involving a viral membrane glycoprotein during morphogenesis of group B arboviruses. Virology 50: 906-911.

Shapiro D, Kos KA, Russell PK (1973). Japanese encephalitis virus glycoproteins. Virology 56: 88-94.

Strauss JH, Preugschat F, Strauss EG (1991). Structure and function of the flavivirus and pestivirus genomes. In: Hollinger FB, Lemon SM, Margolis HS, ed. Viral hepatitis and liver disease. Baltimore: Williams and Wilkins, 333-344.

Strauss JH, Strauss EG (1994). The alphaviruses: Gene expression, replication, and evolution. Microbiol Rev 58: 491-562.

Sumiyoshi H, Hoke CH, Trent DW (1992). Infectious Japanese encephalitis virus RNA can be synthesized from in vitro-ligated cDNA templates. J Virol 66: 5425-5431.

Sumiyoshi H, Mori C, Fuake I, Morita K, Kuhara S, Kondou J, Kikuchi Y, Nagamatu H, Igarashi, A (1987). Complete nucleotide sequence of the Japanese encephalitis virus genome RNA. Virology 161: 497-5 10.

Vorovitch MF, Timofeev AV, Atanadze SN, Tugizov SM, Kushch AA, Elbert LB (1991). pH-dependent fusion of tick-borne encephalitis virus with artificial membranes. Arch Virol 118: 133-138.

Warrener P, Tamura JK, Collett MS (1993). RNA-stimulated NTPase activity associated with yellow fever virus NS3 protein expressed in bacteria. J Virol 67: 989-996.

Wengler G, Czaya G, Färber PM, Hegemann JH (1991). In vitro synthesis of West Nile virus proteins indicates that the amino-terminal segment of the NS3 protein contains the active centre of the protease which cleaves the viral polyprotein after multiple basic amino acids. J Gen Virol 72: 851-858.

Wengler G, Wengler G (1989). Cell-associated West Nile flavivirus is covered with E+Pre-M protein heterodimers which are destroyed and reorganized by proteolytic cleavage during virus release. J Virol 63: 2521-2526.

Wengler G, Wengler G (1991). The carboxy-terminal part of the NS3 protein of the West Nile flavivirus can be isolated as a soluble protein after proteolytic cleavage and represents an RNA-stimulated NTPase. Virology 184: 707-715.

Wengler, G., Wengler, G. (1993). The NS3 nonstructural protein of flaviviruses contains an RNA triphosphatase activity. Virology 197, 265-273.

Wengler G, Wengler G, Nowak T, Castle, E (1990). Description of a procedure which allows isolation of viral nonstructural proteins from BHK vertebrate cells infected with the West Nile flavivirus in a state which allows their direct chemical characterization. Virology 177: 795-801.

Winkler G, Randolf VB, Cleaves GR, Ryan TE, Stollar V (1988). Evidence that the mature form of flavivirus nonstructural protein NS1 is a dimer. Virology 162: 187-196.

Yamshchikov AF, Compans RW (1993). Regulation of late events in flavivirus protein processing and maturation. Virology 192: 38-51.

Yamshchikov AF, Compans RW (1994). Processing of the intracellular form of the West Nile virus capsid protein by the viral NS2B-NS3 protease: an in vitro study. J Virol 68: 5765-5771.

Yamshchikov AF, Compans RW (1995). Formation of the flavivirus envelope: Role of the viral NS2B-NS3 protease. J Virol 69: 1995-2003.

Zhang L, Padmanabhan R (1993). Role of protein conformation in the processing of dengue virus type 2 nonstructural polyprotein precursor. Gene 129: 197-205.

Vignette

Poliovirus

Poliomyelitis, once among the most dreaded diseases of children and adults, has been eliminated from North and South America through the development, use, and administration of vaccines. The challenge remaining is the elimination of poliomyelitis from the remainder of the earth, a goal set by the World Health Organization for the year 2000. The history of the poliomyelitis virus infection and the first successful vaccine trials in humans is described by H. Koprowski.

Over the last several years, knowledge of poliomyelitis virus replication, assembly, and three-dimensional crystallographic structure has been established and its receptor discovered. The limited tropism of the virus for its natural host, humans, and experimentally for monkeys, has made studies *in vivo* difficult and expensive. Further, the need to utilize both viral and host genetics in manipulating the immune response underlines the importance of a small animal model for an understanding of poliomyelitis replication *in vivo*. Some strains of poliomyelitis have been adapted to growth in mice, but this system has limitations. As an alternative, a transgenic mouse model has been established based on expression of the poliomyelitis virus receptor. The chapter by Akio Nomoto summarizes the recent information on polio and the use of the transgenic model.

9

Poliomyelitis:
Visit to Ancient History

HILARY KOPROWSKI

Comparison of the historical facts known about polio and rabies reveals a striking sparseness concerning polio. Both diseases have been recognized since ancient times. Rabies was probably known in India in the thirtieth century BC and was accurately described in the pre-Mosaic Neshuma Code in the twenty-third century BC (Koprowski 1995). A statue of a priest with a withered leg as evidence of polio infection on an Egyptian stele dates to the thirteenth to the fifteenth century BC (Paul 1971) (Table 1). But thereafter, it is rabies that remained of constant interest to naturalists, e.g., Celsus, to philosophers, e.g., Aristotle, to poets, e.g., Homer, and to savants, e.g., Fracastoro (Koprowski 1995). Apart from a dubious reference to a clubfoot in the writings of Hippocrates (Paul 1971), virtually no historical information about polio emerges until the description of paralytic poliomyelitis epidemics in the nineteenth century. What is the reason for this discrepancy between the records for these two diseases? Historians of polio claim that the paralytic form of the disease occurred only sporadically before the nineteenth century epidemic, so that it was difficult to arouse interest in distinguishing cases of paralytic polio from those of other origins. In other words, it was difficult to classify a disease that did not present a well-defined entity to the clinicians of those times.

Although rabies in humans rarely, if ever, assumed epidemic proportions, the nature of the disease may account for its interest to observers. Rabies can infect any warm-blooded animal, whereas polio has only one natural host—humans. There are terrifying descriptions of animals and humans suffering from rabies. Observations for centuries that humans become infected through the bite of a rabid animal have led to a universal dread of rabies which, in the course of centuries, stimulated interest in the disease. By contrast, polio produces a flaccid, certainly less dramatic paralysis and was transmitted among humans in a manner not known for centuries.

TABLE 1. SUPPOSITIONS AND FACTS in the history of poliomyelitis

Years	Events
1580–1350 BC	A suspected polio victim in an Egyptian stele.
After 460 BC	Hippocrates refers to clubfoot "Κυλλος," a malformation resulting from poliomyelitis.
138–201 AD	Galen refers to congenital and infancy-acquired clubfoot.
1789	Underwood: First description of paralytic poliomyelitis (Sir Walter Scott possible victim of polio).
1813	Monteggio: Paralysis and atrophy of lower extremities in infancy following a disease in infancy.
1840	Heine: Accurate systematic description of a paralytic polio pointing to the involvement of spinal cord.
1894	Caverly: Epidemic of polio in Rutland County, Vermont.
1905	Wiekman: Accurate description during an epidemic in Sweden of the contagious nature of the disease.

I personally have tried to make the case that polio caused the lameness of Lord Byron (Koprowski 1960b). Based on Hippocrates' cases of clubfoot, and Kyllos' description of "more than one variety of clubfoot," I postulated that Lord Byron may have been a victim of polio in infancy. However, according to one of Byron's biographers, Doris Langley Moore, one of Byron's boots that was located in the shop of a London saddlemaker was typically worn by people born with a clubfoot but without wastage of calf muscle tissue.

Underwood (1789), in his eighteenth century *Diseases of Children, Second Edition*, provided the first accurate clinical picture of paralytic polio but without reference to any epidemic of the disease. Underwood's clinical picture of polio was considerably expanded and refined by the Italian surgeon Monteggia (1813) in *Institutione Chirurgicale*, where he described the disease from its onset through the acute phase into chronic disability. Although it is still unclear who first noted that polio exists in epidemic form, it is possible that the British neurologist, Sir Charles Bell (1844), described one of the first polio epidemics on the island of St. Helena in his 1844 treatise on the nervous system. His account was preceded by Badham (1834–1835) in 1835 in England. Thereafter, the number of reports on polio epidemics increased rapidly, and the disease invoked as much fear as rabies for children at a vulnerable age. The question remains whether polio was really sporadic until the beginning of the nineteenth century or whether epidemics of the disease were simply missed or unrecorded.

During my stay in the Belgian Congo from 1956 to 1960, I was told that no cases of paralytic polio cases existed there, despite our observations of adults and adolescents with lameness typical of the withered legs of polio victims. Careful study of the early infancy of Congolese children revealed that polio epidemics did indeed occur in the Congo but as a true infantile paralysis. By the age of five years, almost every child in the Congo had antibodies that reacted with all three types of polio. Perhaps in the course of centuries such epidemics were overlooked because of high infant mortality in general and an acceptance of the loss of a young child as "God's will" rather than the result of a specific disease. However, this question is likely to remain unanswered.

Landsteiner and Popper (1909) opened a new era in polio by successfully transmitting the disease to monkeys through intraperitoneal infection with a suspension of spinal cords from two victims of paralytic poliomyelitis. Probably less well known is that shortly after C. Levaditti joined Landsteiner's efforts, it became possible to detect polio virus in many tissues other than the central nervous system on autopsy of paralytic polio victims.

Following the research breakthrough represented by the identification of an experimental animal system for polio virus, clinicians and scientists flocked to the polio field, and in the ensuing decades, many significant scientific publications resulted from these efforts. Another milestone in the history of polio in the twentieth century was provided by Burnet and MacNamara (1931) who demonstrated the existence of at least two serologically and immunologically different strains of polio. The National Foundation for Infantile Paralysis not only confirmed their results, but also clearly established the existence of three different types of polio (The Committee of Typing of the National Foundation for Infantile Paralysis 1951).

I began to recognize the early predictions of success and later fiascoes of developing a polio vaccine during my work at The Yellow Fever Research Laboratory in Brazil. Flexner in 1911 is credited with predicting that polio prevention was imminent. In the early 1930s, Brodie and Park (1935) in New York attempted to inactivate polio virus with formalin and boldly proceeded to vaccinate 3,000 children, several of whom developed paralytic polio. Another disaster occurred during an attempt by Kolmer (Leake 1935) to vaccinate children with a suspension of infected monkey spinal cord treated with sodium ricincolate; 12 children developed paralytic poliomyelitis and six of them died.

The excellent results of immunization with live attenuated yellow fever virus developed by Max Theiler of the Rockefeller Foundation were certainly impressive. After lengthy discussions with Theiler about the potential of this approach in immunization against polio, we concluded that live attenuated polio virus was the vaccine of choice. But how to proceed with the selection of attenuated virus particles in the limiting dilution assay for polio in the pre-tissue culture era? To obtain attenuated virus, we decided to use an unnatural host, the cotton rat, which was shown by Armstrong (1939) to be susceptible to polio infection, and mice, which support the propagation of Type II polio virus. Attenuation was determined based on the inability of the virus to induce signs of disease in intracerebrally injected rhesus monkeys. After many more tests and many more conferences with Theiler, we decided to produce a large pool of virus for further experimental use and, if the occasion arose, for human immunization. This pool was derived from cotton rat brain representing eight passages in mouse cord plus two in cotton rat brain—and it tasted like cod liver oil!

The occasion to use this viral pool in clinical trials arose when George Jervis, Director of Research at The Letchworth Village Institution for feeble-minded children, approached us with a request to immunize his wards since even a single case of polio in the Institution would spread the disease among the children because of intimate contact with feces. On February 27, 1950 (Koprowski et al 1952), we fed a formula containing the attenuated Type II virus to a child who had no antibodies to this type. The child showed no evidence of discomfort or central nervous system involvement. During 1950 and the beginning of 1951, attenuated Type II virus was fed to 19 more children, 17 of whom had no Type II antibodies and the others who did have antibodies. Ultimately, feeding of the virus induced antibodies in all 17 children (Table 2). The first report of this trial was made at the closed meeting called by the National Foundation for Infantile Paralysis in March 1951 in Hershey, Pennsylvania (Immunization in Poliomyelitis, Conference, 1951). I had been asked to speak about attenuation of rabies virus, but after my talk, I told the Chairman that I had some data on the attenuation of polio virus. I was asked to present these data after lunch, probably because some of the audience, including Sabin, Salk, and other notables in polio research, were in a stage of postprandial somnolence. Dr. T. Francis, who was snoozing quietly, opened one eye and asked his neighbors whether my slides showed monkeys. "They are not monkeys, they are children," was the answer. Francis woke up quickly, and the entire audience soon expressed incredulity, wonder, and suspicion. Only the late great immunologist, Jules Freund, approached me and said: "You are right; this is the only way to develop a vaccine against polio." At this meeting, Salk presented his data on immunogenic properties of various strains of polio for rhesus monkeys. Sabin, on the other hand, participated mostly in the discussion of other presentations and was rather skeptical about the attenuated virus for oral immunization.

Within a short time, we attenuated the Type I polio virus, again through passages in mice and cotton rats (Koprowski et al 1954c). A more complicated strategy was used to attenuate the Type III virus (Koprowski et al 1954b). With all three types of viruses in hand, we set about a series of vaccination

TABLE 2. FIRST ORAL VACCINATION with attenuated live polio virus

| Volunteer no. | Date fed | Neutralizing antibodies | |
		Date of bleeding	Serum dilution
1	2/27/50	2/27/50	<1:2
		10/19/50	1:298
2	6/20/50	3/14/50	1:6
		7/19/50	>1:600
3	7/24/50	3/14/50	<1:2
		10/28/50	1:472
4	7/24/50	3/14/50	<1:2
		8/25/50	1:260
5	7/24/50	3/14/50	1:4
		10/28/50	>1:600
6	12/3/50	11/10/50	<1:2
		1/8/51	1:294
7	1/9/51	12/8/50	1:6
		2/12/51	1:111
8	1/9/51	12/8/50	1:3
		3/8/51	>1:600
9	1/9/51	12/8/50	<1:2
		2/12/51	1:422
10	2/18/51	2/5/51	1:7
		3/26/51	>1:600
11	12/3/50	12/10/50	1:3
		5/7/51	1:306
12	12/10/50	11/10/50	1:3
		1/8/51	1:1,414
13	1/16/51	1/15/51	<1:2
		2/16/51	1:32
14	1/16/51	2/16/51	1:2,580
15	10/4/50	8/30/50	>1:6
		10/28/50	>1:2,000
16	4/12/50	4/12/50	<1:2
		10/10/50	1:1,016
17	6/20/50	3/14/50	<1:6
		8/25/50	1:220

American Journal of Hygiene, 1952

trials in many parts of the United States in collaboration with many individuals. In the course of this work, we encountered several interesting situations and personalities. For example, we were asked to vaccinate infants born to prisoners in the only prison for women located in Clinton Farms in New Jersey (Koprowski et al 1956b). How do women get pregnant in a women's prison? The answer was simple. This was a model prison directed by the great prison system reformer in the United States, Ida Mann. There were no bars and no gates in the prison, and younger inmates often wandered out on the nearby highway, flagged down a passing truck, climbed into the cab of the truck and, after riding a mile, returned to the prison, pregnant.

An immunization trial of children in Sonoma Valley, California (Koprowski et al 1956a), was made possible through an active collaboration with Karl F. Meyer, Director of George Williams Hooper Foundation at the University of California in San Francisco. An eminent authority in infectious diseases, Meyer had a Veterinary Degree from the University of Zurich and an honorary degree from the College of Medical Evangelists in Loma Linda, California. He liked to use his M.D. title and once after presenting the results of our polio trials in California, a not very friendly member of the audience asked him, "Is it true, Dr. Meyer, that you are a veterinarian and not an M.D.?" "It is true," answered Meyer, "are you sick?" This ended the exchange.

From 1950 to 1956, we orally vaccinated more than 500 children with live attenuated virus (Koprowski et al 1960a). In the course of these studies, we learned several facts about vaccination. The vaccine could be administered in any vehicle, such as baby formula, gelatin capsules, spray, etc. In almost all vaccinees, the virus seeded the intestinal tract and was excreted in the stool from a few days to two weeks after ingestion (Plotkin et al 1959). Infants of all ages, even less than one month old or premature, could be successfully immunized against polio as indicated by a type-specific antibody response (Pagano et al 1960a). In a major study of vaccinations with all three types of polio involving 850 children age 1.5 to 6 months in Philadelphia, 91 to 100% of vaccinees had significant antibody response to all three types. Twenty-two cases of paralytic polio were observed in the city and none occurred in vaccinees or their households (Pagano et al 1960b).

In 1954, we published comparative studies on oral administration of attenuated Type II virus to humans, chimpanzees, and monkeys, and found that the fate of polio virus fed to chimpanzees and their antibody response was similar to that in humans (Koprowski et al 1954a). In 1956, Ghislain Courtois, Director of the Laboratory in Stanleyville, Belgian Congo, proposed that we enlarge our study of chimpanzees in his chimpanzee camp in Stanleyville. As the result of this request, we spent several months in 1956 and 1957 in Stanleyville. At one point, Dr. Courtois, who knew of our experience with vaccination of humans using live polio virus, proposed oral vaccination of personnel in the camps since they were handling wild polio virus strains used for challenge of immunized chimpanzees. These personnel were vaccinated in 1952 and shortly thereafter, the Belgian Congo government, with Courtois as intermediary, requested mass vaccination of young children in rural areas of Zaire and Burundi, since they feared a polio outbreak. The mass vaccination was organized in 1958 (Courtois et al 1958). The natives were called by drum beats to certain localities, and Drs. Flack and Jervis administered the vaccine in a slightly salty suspension by spray. Within six weeks of the campaign, 244,000 persons were vaccinated (Table 3). The Government of Congo soon requested immunization of a population in the Congo capitol, Leopoldville, in the face of an epidemic. It was thus possible to evaluate the efficacy of the vaccination. The estimated protection by vaccination was 68% (Table 4) (Plotkin et al 1961). This first mass vaccination with a viral vaccine was followed by numerous large population vaccinations in Poland, Switzerland, and Croatia.

In 1960, the United States Public Health Service called for the establishment of a committee to decide which groups of attenuated strains to license. The committee consisted mostly of scientists whose work was supported by the National Foundation for Infantile Paralysis. On August 24, 1960, the decision was announced by the Surgeon General that attenuated strains developed by Albert Sabin were recommended for licensing by the authorities in the United States. John Paul, in his *History of Poliomyelitis* (Paul 1971), refers to my reaction to this decision: "Koprowski remained one of the leaders; he was later to lament the fact that the vaccine against poliomyelitis which he

TABLE 3. **FIRST MASS VACCINATION (1958) TRIAL with attenuated live oral polio vaccine* in the Belgian Congo and Ruanda-Urundi**

| | No. of vaccinees | | | |
| | Adults | | Children under 15 years | Total |
Town	Males	Females		
Stanleyville	1,651	1,005	1,572	4,228
Aketi	—	—	1,978	1,978
Gombari	1,371	1,028	1,083	3,482
Watsa	3,136	4,753	4,900	12,789
Bambesa and Kula-Ponge	219	946	1,268	2,433
Ruzizi Valley	62,140	62,200	91,164	215,504
Banalia	1,267	1,343	1,572	4,182
Total	69,784	71,275	103,537	244,596

*CHAT strain type 1 poliovirus In addition, 2,511 persons were vaccinated with type III poliovirus, Fox III strain: 134 men, 130 women and 2,247 children.

TABLE 4. ESTIMATED PROTECTION BY CHAT VIRUS against paralytic poliomyelitis in Léopoldville African children age six months through two years

District	Period[a]	Group	Cases	Person-weeks	Rate per 10^5 person-weeks	Estimated protection (%)
Ancienne Cité	1	Vaccinated	4	64 800	6.2	53
		Non-Vaccinated	27	205 200	13.2	
	2	Vaccinated	13	774 800	1.7	52
		Non-Vaccinated	5	142 480	3.5	
Nouvelle Cité	1	Vaccinated	2	47 800	4.2	71
		Non-Vaccinated	41	284 200	14.4	
	2	Vaccinated	8	677 560	1.2	80
		Non-Vaccinated	18	295 880	6.1	
"Five Districts"	1	Vaccinated	4	130 400	3.0	68
		Non-Vaccinated	13	133 600	9.7	
	2	Vaccinated	5	686 920	0.7	88
		Non-Vaccinated	18	305 240	5.9	
All[b]		Vaccinated	34		2.6[c]	67
		Non-vaccinated	81		7.7[c]	

[a] period 1 = period of type 1 epidemic, October 1958 through March 1959, period 2 = end of April 1959 through April 1960; [b] excluding period 1 in Nouvelle Cité; [c] average of rates given above, excluding period 1 in Nouvelle Cité. Source: Plotkin et al (1961)

had discovered should have been named the Sabin vaccine." This is not a correct description of my reaction. First of all, I was not in the habit of "lamenting" anything. Secondly, I realized that the decision was not based on scientific evidence showing an advantage of one group of strains over another, but simply one designed to support a member of the polio "coterie" rather than an outsider. My conclusion was confirmed later in that same year when Smodel, a member of the Committee, told one of my friends that the Committee knew that there was no difference between the strains of all investigators but that Sabin was an "old boy" and was therefore granted licensure for his set of attenuated strains. "And," he added, "what I am surprised at is that Hilary Koprowski, after what we have done to him, is still on speaking terms with me."

Over the years, I came to realize that I was rather fortunate to remain a "private person," so to speak, free to pursue research in various fields of my interest instead of becoming a public figure, meeting with statesmen and heads of governments, and obliged to visit many countries in order to advise and often supervise vaccine production.

But my greatest satisfaction is in the 46th anniversary of the first successful vaccination with attenuated virus of a group of 20 children. The incidence of paralytic poliomyelitis in the Western hemisphere in the years 1931–1953 is striking (Table 5) (Payne 1954). Today, no case of paralytic poliomyelitis in the Western hemisphere has been recorded since 1992 (Table 6) (Robbins 1995), and a marked worldwide decrease in polio cases has been observed for the first time (less than 10,000) (Figure 1). This is the greatest satisfaction a scientist can expect in a lifetime.

FIGURE 1. REPORTED CASES OF POLIOMYELITIS due to wild virus in 1993. The countries of the former Soviet Union reported less than 10 polio cases during 1993. The whole of the American continent and several countries in each of the other continents are moving toward polio eradication.

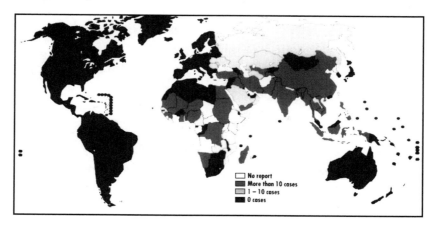

TABLE 5. POLIOMYELITIS as a world problem, 1946–1953

Region	Country	Years	Median infant mortality rate per 1,000 live births	Median poliomyelitis incidence per 100,000 population	Averages
The Americas	Argentina	1946-50	70	2.70	4.30
		1951-53	65	5.90	
	Brazil	1946-50	107	0.20	0.20
	Canada	1926-30	93	6.60	13.30
		1931-35	75	6.40	
		1936-40	63	9.90	
		1941-45	55	6.70	
		1946-50	44	14.30	
		1951-53	38	35.90	
	Chili	1946-50	149	0.30	0.30
	Colombia	1946-50	86	1.00	1.00
	Mexico	1946-50	100	1.10	1.10
	Salvador	1946-50	78	1.80	1.80
	United States	1921-25	74	4.20	11.65
		1926-30	68	4.90	
		1931-35	59	6.90	
		1941-45	41	8.80	
		1946-50	32	19.10	
		1951-53	28	26.00	
	Uruguay	1926-30	98	1.30	2.17
		1946-50	60	2.80	
		1951-53	43	2.40	
Europe	Austria	1926-30	117	1.30	1.30
	Bulgaria	1926-30	147	0.30	0.30
	Czecho-	1926-30	144	0.40	1.45
	slovakia	1936-40	98	2.50	
	Denmark	1921-25	82	2.80	21.16
		1926-30	82	2.40	
		1931-35	71	30.40	
		1936-40	60	10.70	
		1941-45	49	25.00	
		1946-50	37	16.90	
		1951-53	28	59.90	
	Eire	1946-50	51	4.20	4.40

TABLE 6. INCIDENCE OF POLIOMYELITIS in the United States and Canada

Year	Country	Median poliomyelitis incidence per 100,000 population
1951–1953	United States	26.0
	Canada	35.0
1992–1994	United States	0
	Canada	0
	(and the entire western hemisphere)	0

REFERENCES

Armstrong C (1939). Successful transfer of the Lansing strain of poliomyelitis virus from the cotton rat to the white mouse. Public Health Rep 54: 230-205.

Badham J (1834–1835). Paralysis in childhood: four remarkable cases of suddenly induced paralysis in the extremities, occurring in children, without any apparent cerebral or cerebrospinal lesion. London Med Gaz, n.s. 17: 215.

Bell C (1844). The nervous system of the human body as explained in a series of papers read before the royal society of London. Ed 3. London: H. Renshaw, 434-435.

Brodie M, Park WH (1935). Active immunization against poliomyelitis. NY State J Med 35: 815.

Burnet FM, MacNamara J (1931). Immunologic differences between strains of poliomyelitic virus. Brit J Exp Pathol 12: 57.

Committee of Typing of the National Foundation for Infantile Paralysis: Immunological classification of poliomyelitis viruses. I. A cooperative program for the typing of 100 strains (1951) Am J Hyg 54: 191-204.

Courtois G, Flack A, Jervis GA, Koprowski H, Ninane G (1958). Preliminary report on mass vaccination of man with live attenuated poliomyelitis virus in the Belgian Congo and Ruanda-Urundi. Brit Med J 2(5090): 187-190.

Immunization in Poliomyelitis. Conference. (March 15, 1951). Hershey, Pennsylvania.

Koprowski H (1960a). Historical aspects of the development of live virus vaccines in poliomyelitis. Trans Stud Coll Physicians Phila 27: 95-106.

Koprowski H (1960b). Viruses 1959. Trans NY Acad Sci 22: 176.

Koprowski H (1995). Visit to an ancient curse. Sci and Med 2: 48-57.

Koprowski H, Jervis GA, Norton TW (1952). Immune responses in human volunteers upon oral administration of a rodent-adapted strain of poliomyelitis virus. Am J Hyg 55: 108-126.

Koprowski H, Jervis GA, Norton TW (1954a). Oral administration of a rodent-adapted strain of poliomyelitis virus to chimpanzees. Archiv Fuer Die Gesamte Virusforschung 5: 413-424.

Koprowski H, Jervis GA, Norton TW (1954b). Administration of an attenuated type I poliomyelitis virus to human subjects. Proc Soc Exp Biol Med 86: 244-247.

Koprowski H, Jervis GA, Norton TW, Nelson TL, Chadwick DL, Nelsen DJ, Meyer KF (1956a). Clinical investigations on attenuated strains of poliomyelitis virus, use as method of immunization of children with living virus. JAMA 60: 954-966.

Koprowski H, Jervis GA, Norton TW, Pfeister K (1954c). Adaptation of type I strain of poliomyelitis virus to mice and cotton rats. Proc Soc Exp Biol Med 86: 238-244.

Koprowski H, Norton TW, Hummeler K, Stokes J Jr., Hunt AD, Jr., Flack A, Jervis GA (1956b). Immunization of infants with living attenuated poliomyelitis virus. Laboratory investigations of alimentary infection and antibody response in infants under six months of age with congenitally acquired antibodies. JAMA 162: 1281-1288.

Lansteiner K, Popper E (1909). Ubertragung der poliomyelitis acuta auf affen. Z Immun Forsch (orig.) 2: 377.

Leake JP (1935). Poliomyelitis following vaccination against this disease. Am J Med Assn 105: 2152.

Monteggia GB (1813). Instituzione chirurgicale. 8 Vols, Ed 2. Milano: Guiseppe Maspero.

Pagano JS, Plotkin SA, Koprowski H (1960a). Variation in response in early life to vaccination with living attenuated poliovirus and lack of immunologic tolerance. Lancet 1: 1224-1226.

Pagano JS, Plotkin SA, Koprowski H, Richardson S, Stokes J (1960b). Vaccination of families against poliomyelitis by feeding and by contact spread of living attenuated virus: including studies of virus properties after human intestinal passage. Acta Paediatrica 49: 551-571.

Paul JR (1971). A history of poliomyelitis. New Haven, London: Yale University Press, 486.

Payne AMM (1954). Poliomyelitis as a world problem. In: International poliomyelitis conference. Vol 3. Philadelphia: Lippincott, 393-400.

Plotkin SA, Koprowski H, Stokes J, Jr. (1959). Clinical trials in infants of orally administered attenuated poliomyelitis viruses. Pediatrics 23: 1041-1062.

Plotkin SA, Lebrun A, Courtois G, Koprowski H (1961). Vaccination with CHAT strain of type 1 attenuated poliomyelitis virus in Leopoldville, Congo. III. Safety and efficacy during the first 21 months of study. Bull WHO 24: 785-792.

Robbins F (1995). The quest for a vaccine. World Health 1: 14-17.

Underwood A (1789). A treatise on the diseases of children with general directions for the management of infants from the birth. Ed 2. London: Mathews.

10

Current Concepts in Poliomyelitis

Akio Nomoto

Poliomyelitis is an acute disease of the central nervous system (CNS) caused by lytic replication of poliovirus, a human enterovirus that belongs to the Picornaviridae family. The genome of poliovirus is a single-stranded RNA with positive polarity, to which a small protein, VPg, is attached at the 5' terminus. Poly(A) exists at the 3' terminus of the RNA. This genome is enclosed in an icosahedral capsid composed of 60 copies of each of four capsid proteins, VP1–4. Chemical (Kitamura et al 1981) and three-dimensional (Hogle et al 1985) structures of the virion particle have been elucidated, and poliovirus is now one of the best-characterized viruses.

A model for poliovirus pathogenesis was first described in the 1950s (Sabin 1956; Bodian 1955) (Figure 1). Poliovirus infection is initiated by oral ingestion of virus, followed by primary multiplication in the alimentary mucosa and then in tonsils and Peyer's patches. After viral multiplication in these tissues, virus moves into the blood. Finally, the virus invades the CNS and paralytic poliomyelitis occurs as a result of the destruction of neurons in the CNS, especially the motor neurons in the anterior horn of the spinal cord and neurons of the motor cortex and brain stem. During viremia, many tissues must be exposed to the virus. However, most tissues do not support poliovirus replication. Thus, poliovirus seems to have a distinct tissue tropism.

Humans are the only natural host of poliovirus. However, monkeys are also very susceptible to poliovirus when inoculated directly in the CNS, that is, the final target of the virus. Other animal species are not susceptible to most poliovirus strains. Because of this characteristic species specificity of poliovirus, inoculation of virus into the CNS of monkeys has been the only method of assessing neurovirulence levels of poliovirus strains and has been used to evaluate vaccine candidates and oral live poliovirus vaccine preparations (Sabin and Boulger 1973).

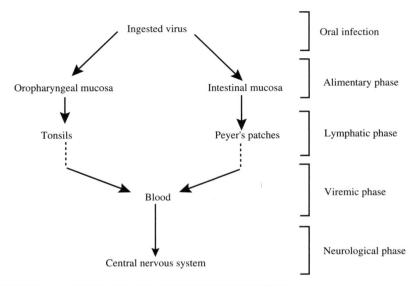

FIGURE 1. A MODEL FOR POLIOVIRUS PATHOGENESIS

To control severe paralytic disease, attenuated poliovirus strains of all three serotypes have been developed and used effectively as oral vaccines, that is, Sabin 1, 2, and 3 (Sabin and Boulger 1973). In the human alimentary tract, the attenuated Sabin strains can replicate to a sufficiently high level to elicit neutralizing antibodies. However, they seldom move into the blood and have a very poor replicating capacity in the CNS. Thus attenuated poliovirus strains have a different tissue tropism from that of virulent strains.

Although eradication of poliomyelitis in the world appears to be occurring, little is known about the mechanisms by which the poliovirus causes paralysis as well as the mechanisms responsible for attenuation. For example, it is not known how the virus moves into the blood from the primary multiplication site, how the virus invades the CNS, or how secretory IgA is elicited by the virus infection. Thus poliomyelitis is not truly understood. Humans are simply lucky that the polio vaccines worked.

VIRULENT AND ATTENUATED STRAINS OF POLIOVIRUS

Attenuated poliovirus strains now in use as vaccines were progeny virus of adapted subpopulations following a number of passages of virulent strains in simian extraneural cells and tissues. This is known as the attenuation process. However, the Sabin 2 strain is different in that the parental strain was already avirulent in the CNS of monkeys.

The attenuated Sabin 1 strain was derived from the virulent Mahoney strain of type 1 poliovirus. These two strains differ strikingly in the potential for causing the disease. The neurovirulence phenotype of poliovirus can be determined in monkeys by monitoring paralysis and development of histological lesions in the CNS after intracerebral or intraspinal injection. The different biological characteristics of the two strains must be due to the differences in genome structures. Comparative sequence studies performed on the genomes of the Mahoney and Sabin 1 strains have identified and mapped 56 nucleotide substitutions scattered over the genome (Nomoto et al 1982). These nucleotide changes were found to result in 21 amino acid replacements within the viral polyprotein. The availability of infectious cDNA clones (Racaniello and Baltimore 1981; Omata et al 1984), together with knowledge of the total nucleotide sequences of the genomes, have allowed a molecular genetics approach to analyses of the structure-function relationship of the viral genome with the use of recombinant DNA technology.

To identify the genome regions that influence the neurovirulence or attenuation phenotype of poliovirus type 1, a number of recombinants of the two strains were constructed using infectious cDNA clones of the viruses. Monkey neurovirulence tests on these recombinants revealed that the region of nucleotide positions 1 to 1122, including mainly the 5' noncoding region of the RNA, harbored a relatively strong determinant(s) influencing the neurovirulence or attenuation phenotype, and that determinants weakly influencing the phenotype were spread over several areas of the entire viral genome (Omata et al 1986; Kohara et al 1988) (Figure 2). The region encoding the viral capsid precursor protein (P1) showed only a weak influence on the neurotropic phenotype of the viruses. These results led to speculations that

FIGURE 2. RELATIVE INFLUENCE OF POLIOVIRUS genomic regions on neurovirulence. Monkey neurovirulence tests were performed on recombinant viruses between the Mahoney and Sabin 1 strains. Extent of influence was characterized as "strong" or "weak." A, K, and B represent cleavage sites of the restriction enzymes AatII, KpnI, and BglIII on the cDNA, respectively. Numbers in parentheses indicate nucleotide positions. Length from the 5' end of the genome is shown at the top of the figure in kilobases (kb). Genome organization is indicated at the bottom of the figure. P1 represents the region of the viral capsid proteins. P2 and P3 represent regions of the viral RNA replication proteins.

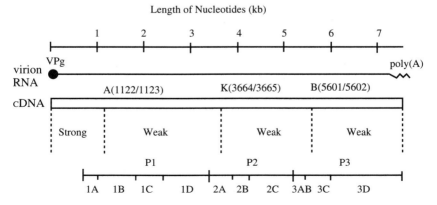

the rate of viral multiplication in neurons of the CNS is an important determinant of the neurotropic phenotype, and that there is no significant difference in recognition of the cell surface receptor between the virulent and attenuated strains. Viral RNA replication proteins (P2 and P3) also appear not to contribute significantly to the neurotropism of the virus.

To identify a relatively strong neurovirulence (neurotropism) determinant(s) in the genome region upstream of nucleotide position 1122, recombinant viruses including this region were constructed and examined for monkey neurovirulence. Those studies identified an adenine residue (Mahoney type) at nucleotide position 480 as a relatively strong determinant of the neurovirulence phenotype expressed by the 5' proximal 1122 nucleotides of poliovirus type 1, although other mutations were suggested to have some influences on the phenotype (Kawamura et al 1989). Relatively strong neurovirulence determinants at nucleotide positions 481 and 472 have also been identified for poliovirus type 2 and type 3, respectively (Evans et al 1985; Macadam et al 1991; Ren et al 1991). These nucleotides exist in an IRES (internal ribosomal entry site) within the 5' untranslated region of the RNA (Pelletier and Sonenberg 1988). Since an IRES has an important role in the initiation of virus-specific protein synthesis, it is possible that the neurovirulence phenotype is determined primarily by efficiency of translation of the virus in the CNS.

In the alimentary tract, both virulent and attenuated polioviruses replicate well, perhaps due to differences in distribution of host factors involved in the IRES function in different tissues. Large amounts of host factors may exist at sufficiently high levels in the alimentary tract, and even an inefficient IRES of the attenuated poliovirus may work well. If so, the host factors are limiting in the CNS, and only an efficient IRES of the virulent strain is highly functional. This may explain why the attenuated strain can grow well in the alimentary tract but not in the CNS. Thus, further studies of IRES activities of the virulent and attenuated polioviruses are necessary to understand the mechanisms underlying the neurotropic phenotype of poliovirus. Such experiments are not possible in monkeys, and new animal models are needed.

TRANSGENIC MICE SUSCEPTIBLE TO POLIOVIRUS

The host specificity of poliovirus is considered to be determined by a specific cell surface molecule that acts as a poliovirus receptor (PVR). Transgenic mice that express the human PVR (hPVR) gene have been produced that are susceptible to all three poliovirus serotypes (Ren et al 1990; Koike et al 1991).

The gene for hPVR is about 20 kb long and is located as a single copy at band q13.1–13.2 of human chromosome 19 (Koike et al 1990). Nucleotide sequence analysis has revealed that hPVR belongs to the immunoglobulin (Ig) superfamily and has three linked extracellular Ig-like domains, V-C2-C2, followed by a membrane-spanning domain and a cytoplasmic domain

(Mendelsohn et al 1989; Koike et al 1990). Binding of poliovirus to the PVR destabilizes the virion particle and leads to the formation of "A" particles which do not contain the capsid protein VP4 (De Sena and Mandel 1976; 1977; Kaplan et al 1990; Koike et al 1992) and which are considered to be intermediates in the process of virus uncoating. Thus, the PVR appears to have a dual function, that is, binding to poliovirus and initiation of uncoating.

The hPVR transgenic mice inoculated with poliovirus show a flaccid paralysis in the legs (Koike et al 1991), reminiscent of the clinical signs observed in human poliomyelitis. These mice served as proof that PVR determines species specificity of poliovirus infection. In our research group, four lines of poliovirus-sensitive transgenic mice have been developed so far (Koike et al 1991), of which ICR-PVRTg1 mice were the most sensitive (Koike et al 1994b). These mice showed a differential sensitivity depending on the route of inoculation, with the greatest sensitivity observed when the virus was inoculated directly into the CNS, especially into the spinal cord (Koike et al 1994a) (Table 1).

With any route of inoculation, much higher doses of virus were required for the attenuated Sabin 1 strain to cause paralysis as compared with those of the virulent Mahoney strain. Thus, the virulent and attenuated phenotypes of poliovirus in primates appear to be preserved in the transgenic mouse model, where different replication efficiencies may underlie the phenotype in the CNS, as in monkeys.

To compare the replication abilities of the Mahoney and Sabin 1 strains in the CNS, nontransgenic mice and transgenic mice were inoculated intracerebrally with 10^3 plaque-forming unites (PFU) or 10^6 PFU of the viruses (Koike et al 1991) (Table 2). One week after inoculation, virus recovered from the brain and spinal cord was quantitated. The amounts of both the Mahoney and Sabin 1 strains dramatically decreased in the CNS of nontransgenic mice during the seven days after inoculation. By contrast, approximately 10^7 PFU of the virus was recovered from both the brain and spinal cord of the transgenic mice inoculated with 10^3 PFU of the Mahoney strain, indicating that the Mahoney strain replicates well in the CNS of these

TABLE 1. SENSITIVITY OF ICR-PVRTG1 MICE to polioviruses inoculated by various routes

Inoculation route	Mahoney strain	Sabin 1 strain
Oral	>6~7[a]	>8
Intravenous	>4~5	>7
Intracerebral	>1~2	>6
Intraspinal	NT[b]	>1

[a] Logarithmic value of PFU required to cause paralysis; [b] not tested.

TABLE 2. POLIOVIRUS MULTIPLICATION in the CNS of mice

Mice	Poliovirus strain	Amount of virus inoculated	Amount of virus recovered	
			Brain	Spinal cord
Nontransgenic				
	Mahoney	3[a]	ND[b]	ND
	Sabin 1	6	1.0	ND
Transgenic				
(ICR-PVRTg1)	Mahoney	3	6.8	7.1
	Sabin 1	6	4.9	5.3

[a] Logarithmic value of PFU; [b] not detected.

mice. For transgenic mice inoculated with the Sabin 1 strain, the amount of recovered virus was less than that inoculated, although comparison of virus recovery from transgenic *vs* nontransgenic mice clearly indicates some replication of the Sabin 1 strain in the CNS of the transgenic mice. These results strongly suggest that different capacities for viral replication in the CNS underlie the different mouse neurovirulence levels of the Mahoney and Sabin 1 strains.

To determine whether the clinical similarity between the transgenic mice and primates after poliovirus inoculation reflected damage by the lytic infection primarily in the motor neurons of the spinal cord, immunocytochemical techniques were used to detect viral antigens in the CNS of the transgenic mice (Koike et al 1994a; Horie et al 1994). Poliovirus antigens were detected mainly in motor neurons in any transverse section of the lumber, thoracic, and cervical regions, but not in glial cells. *In situ* hybridization revealed hPVR mRNA in the same types of cells in which viral antigen was detected (Koike et al 1994a). Thus, the replication sites of poliovirus in the CNS of the transgenic mice are the same as those of humans and monkeys. Furthermore, the data strongly suggested that specific expression of hPVR provides cell specificity of poliovirus in the CNS. Indeed, PVRs are detected by specific antibodies only in neurons of the CNS of humans and monkeys (J Aoki, unpublished results). Thus both clinical and histopathological similarities between primates and the transgenic mice point to the suitability of the latter as a new animal model for the study of poliovirus pathogenesis.

Relative Neurovirulence Levels in the Two Animal Models

Recombinant viruses between the virulent Mahoney and attenuated Sabin 1 strains of type 1 poliovirus, whose monkey lesion scores had been determined, were subjected to neurovirulence tests using ICR-PVRTg1 transgenic mice inoculated intracerebrally (Horie et al 1994). LD_{50} values (dose at which 50% of the mice die) obtained from the mouse neurovirulence test correlated well with mean monkey lesion scores, suggesting that neurovirulence tests using the transgenic mouse model can provide a reliable estimate of relative neurovirulence levels of poliovirus. Further mouse neurovirulence tests on recombinant viruses with regard to the 5'-proximal portion of the genome revealed that every replacement involving position 480, an important nucleotide influencing monkey neurovirulence of type 1 polio-virus, resulted in significant changes in mouse LD_{50} values. This suggests an important contribution of nucleotide 480 to the expression of neurovirulence or attenuation phenotype, although one point substitution at position 480 was not sufficient for full expression of the attenuation phenotype encoded by the 5'-proximal 1122 nucleotides. These data suggest that interactions of neurovirulence determinants spread throughout the viral genome with mouse host factors are not significantly influenced by the species differences between monkeys and mice, and that the distribution of host factors involved in IRES function is the same in the CNS of the two animal models. Tissue-specific IRES function of the attenuated Sabin 1 strain may be clarified by using the mouse model.

Three poliovirus-sensitive transgenic mouse lines, ICR-PVRTg1, ICR-PVRTg5, and ICR-PVRTg21, have been examined for their susceptibilities to intracerebrally inoculated poliovirus (Koike et al 1994b). ICR-PVRTg1 and ICR-PVRTg5 showed the highest and lowest sensitivities to poliovirus among them, respectively. As mentioned above, neurovirulence tests using intra-cerebral inoculation in the most sensitive ICR-PVRTg1 mice can distinguish polioviruses with a wide range of neurovirulence levels, from the virulent strain to the attenuated strain (Horie et al 1994). However, this method is not sufficiently sensitive to distinguish polioviruses of very low neurovirulence levels, such as oral poliovirus vaccine preparations.

The susceptibilities of the transgenic mice to poliovirus differ as a function of the inoculation route, and intraspinal inoculation of the virus requires the least amount of virus to induce paralysis in 50% of the mice (Koike et al 1994a). Therefore, intraspinal inoculation may be more suitable for safety testing of oral live poliovirus vaccines. Indeed, neurovirulence tests of oral poliovirus vaccines were established using the transgenic mouse line ICR-PVRTg21 with intraspinal inoculation (Abe et al 1995a; 1995b).

hPVR EXPRESSION AND TISSUE TROPISM OF POLIOVIRUS

Susceptibility of animal tissue to poliovirus must be governed by many factors, such as physical accessibility of virus to cells, virus entry ability into cells, and virus replication ability in the subsequent steps. All these steps of virus replication also involve host immunological and physiological conditions. Thus, mechanisms for the development of poliomyelitis are very complicated and therefore difficult to elucidate completely. Nevertheless, determination of the distribution of hPVR in humans and transgenic mice appears to be a promising approach to solving this problem.

Northern blot hybridization to detect hPVR mRNA in various human tissues demonstrated expression of the message in all tissues tested. Relatively high level expression was detected in the liver, heart, and lung, which are not replication sites of poliovirus *in vivo*. The expression pattern of hPVR in transgenic mice is different from that in humans, although the hPVR molecule and hPVR mRNA were detected in all mouse tissues tested. Thus, it is clear that PVR distribution is not the only determinant of tissue-specific replication of poliovirus, and host factors other than PVR must be considered in elucidating the mechanisms that determine the tissue tropism and dissemination route of poliovirus.

ACKNOWLEDGMENTS

This work was supported in part by research grants from The Ministry of Education, Science, Sports, and Culture of Japan and The Ministry of Health and Welfare of Japan.

REFERENCES

Abe S, Ota Y, Koike S, Kurata T, Horie H, Nomura T, Hashizume S, Nomoto A (1995a). Neurovirulence test for oral live poliovaccines using poliovirus-sensitive transgenic mice. Virology 206: 1075-1083.

Abe S, Ota Y, Doi Y, Nomoto A, Nomura T, Chumakov KM, Hashizume S (1995b). Studies on neurovirulence in poliovirus-sensitive transgenic mice and cynomolgus monkeys for the different temperature-sensitive viruses derived from the Sabin type 3 virus. Virology 210: 160-166.

Bodian D (1955). Emerging concept of poliomyelitis infection. Science 122: 105-108.

De Sena J, Mandel B (1976). Studies on the *in vitro* uncoating of poliovirus. I. Characterization of the modifying factor and the modifying reaction. Virology 70: 470-483.

De Sena J, Mandel B (1977). Studies on the *in vitro* uncoating of poliovirus. II. Characteristics of the membrane-modified particle. Virology 78: 554-566.

Evans DMA, Dunn G, Minor PD, Schild GC, Cann AJ, Stanway G, Almond JW, Currey K, Maizel JV Jr. (1985). Increased neurovirulence associated with a single nucleotide change in a noncoding region of the Sabin type 3 Poliovaccine genome. Nature (London) 314: 548-550.

Hogle JM, Chow M, Filman DJ (1985). Three-dimensional structure of poliovirus at 2.9 ångstrom resolution. Science 229: 1358-1363.

Horie H, Koike S, Kurata T, Sato-Yoshida Y, Ise I, Ota Y, Abe S, Hioki K, Kato H, Taya C, Nomura T, Hashizume S, Yonekawa H, Nomoto A (1994). Transgenic mice carrying the human poliovirus receptor: new animal model for study of poliovirus neurovirulence. J Virol 68: 681-688.

Kaplan GM, Freistadt MS, Racaniello VR (1990). Neutralization of poliovirus by cell receptors expressed in insect cells. J Virol 64: 4697-4702.

Kawamura N, Kohara M, Abe S, Komatsu T, Tago K, Arita M, Nomoto A (1989). Determinants in the 5' noncoding region of poliovirus Sabin 1 RNA that influence the attenuation phenotype. J Virol 63: 1302-1309.

Kitamura N, Semler BL, Rothberg PG, Larsen GR, Adler CJ, Dorner AJ, Emini EA, Hanecak R, Lee JJ, van der Werf S, Anderson CW, Wimmer E (1981). Primary structure, gene organization and polypeptide expression of poliovirus RNA. Nature (London) 291: 547-553.

Kohara M, Abe S, Komatsu T, Tago K, Arita M, Nomoto A (1988). A recombinant virus between the Sabin 1 and Sabin 3 vaccine strains of poliovirus as a possible candidate for a new type 3 poliovirus live vaccine strain. J Virol 62: 2828-2835.

Koike S, Horie H, Ise I, Okitsu A, Yoshida M, Iizuka N, Takeuchi K, Takegami T, Nomoto A (1990). The poliovirus receptor protein is produced both as membrane-bound and secreted forms. EMBO J 9: 3217-3224.

Koike S, Taya C, Kurata T, Abe S, Ise I, Yonekawa H, Nomoto A (1991). Transgenic mice susceptible to poliovirus. Proc Natl Acad Sci USA 88: 951-955.

Koike S, Ise I, Sato Y, Mitsui K, Horie H, Umeyama H, Nomoto A (1992). Early events of poliovirus infection. Sem Virol 3: 109-115.

Koike S, Aoki J, Nomoto A (1994a). Transgenic mouse for the study of poliovirus pathogenicity. In: Wimmer E, ed. Cellular receptors for animal viruses. Cold Harbor Spring: Cold Spring Harbor Laboratory Press, 463-480.

Koike S, Taya C, Aoki J, Matsuda Y, Ise I, Takeda H, Matsuzaki T, Amanuma H, Yonekawa H, Nomoto A (1994b). Characterization of three different transgenic mouse lines that carry human poliovirus receptor gene—Influence of the transgene expression on pathogenesis. Arch Virol 139: 351-363.

Macadam AJ, Pollard SR, Ferguson G, Dunn G, Skuce R, Almond JW, Minor PD (1991). The 5' noncoding region of the type 2 poliovirus vaccine strain contains determinants of attenuation and temperature sensitivity. Virology 181: 451-458.

Mendelsohn CL, Wimmer E, Racaniello VR (1989). Cellular receptor for poliovirus: molecular cloning, nucleotide sequence, and expression of a new member of the immunoglobulin superfamily. Cell 56: 855-865.

Nomoto A, Omata T, Toyoda H, Kuge S, Horie H, Kataoka Y, Genba Y, Nakano Y, Imura N (1982). Complete nucleotide sequence of the attenuated poliovirus Sabin 1 strain genome. Proc Natl Acad Sci USA 79: 5793-5797.

Omata T, Kohara M, Sakai Y, Kameda A, Imura N, Nomoto A (1984). Cloned infectious complementary DNA of the poliovirus Sabin 1 genome: Biochemical and biological properties of the recovered virus. Gene 32: 1-10.

Omata T, Kohara M, Kuge S, Komatsu T, Abe S, Semler BL, Kameda A, Itoh H, Arita M, Wimmer E, Nomoto A (1986). Genetic analysis of the attenuation phenotype of poliovirus type 1. J Virol 58: 348-358.

Pelletier J, Sonenberg N (1988). Internal initiation of translation of eukaryotic mRNA directed by a sequence derived from poliovirus RNA. Nature (London) 334: 320-325.

Racaniello VR, Baltimore D (1981). Cloned poliovirus complementary DNA is infectious in mammalian cells. Science 214: 916-919.

Ren R, Costantini F, Gorgacz EJ, Lee JJ, Racaniello VR (1990). Transgenic mice expressing a human poliovirus receptor: a new model for poliomyelitis. Cell 63: 353-362.

Ren R, Moss EG, Racaniello VR (1991). Identification of two determinants that attenuate vaccine-related type 2 poliovirus. J Virol 65: 1377-1382.

Sabin AB (1956). Pathogenesis of poliomyelitis: Reappraisal in the light of new data. Science 123: 1151-1157.

Sabin AB, Boulger LR (1973). History of Sabin attenuated poliovirus oral live vaccine strains. J Biol Stand 1: 115-118.

Varicella Virus

Varicella-zoster virus (VZV) causes chickenpox in children and can establish a latent infection in spinal ganglia, reactivating years later to cause shingles in adults. History detailing the initial establishment of tissue culture techniques and their use for the isolation of varicella virus is presented by Tom Weller, a major pioneer in the field of live tissue culture. The initial experimental evidence that the virus causes chickenpox in children and shingles in adults is presented, together with the initial use of the fluorescent antibody test of Albert Coons. Many scientific problems are being solved, including how varicella enters cells, its replicative strategy and packaging, and how the virus is maintained in a latent state in neurons and subsequently activated. Anne Gershon summarizes the recent events leading to the development of the varicella virus vaccine.

11

History of Varicella Virus

Thomas Weller

Following our report in 1949 of the growth of the poliomyelitis viruses in cultures of human tissues (Enders, Weller, Robbins 1949), the successful culture technique was adopted by many investigators. By 1954, when we received the Nobel Prize for our research on poliomyelitis, Gard (1955) noted that the introduction of the tissue culture technique for viruses was comparable to the discovery by Koch of media for the growth of bacteria. It is not generally known that the stimulus for this development was the desire to isolate and propagate varicella virus. Thus our studies on varicella had a major impact on the field of virology.

The tissue culture method has been most useful in the study of host-specific viral diseases of humans that cannot be studied in laboratory animals or the embryonated hen's egg. Varicella-zoster virus (VZV) proved to be such an agent. Here, we review the efforts that led to the isolation and serial cultivation of VZV. The morphologic and epidemiologic evidence for the co-identity of the etiologic agents of varicella and zoster is reviewed elsewhere (Weller 1983; Weller 1992).

In 1941, Larry Kingsland and I had preclinical internships in pathology at the Children's Hospital in Boston. While in Harvard Medical School, we had both worked on growing viruses in roller tube tissue cultures. As embryonic pediatricians, we were intrigued by varicella because the etiological agent remained elusive despite extensive attempts by many investigators at viral isolation using animals. The agent appeared highly host-specific, and we therefore prepared roller cultures of human tissues and inoculated them with materials from varicella cases. Our obligation as interns plus our primitive facilities that led to frequent bacterial contaminants combined to yield negative results. The effort was terminated when we both entered military service in 1942.

The belief that varicella virus was highly host-specific was enhanced by the classical report of Goodpasture and Anderson (1944) who inoculated human skin fragments grafted on the chorioallantoic membrane of nine-day-old chick embryos with zoster vesicle fluid; typical eosinophilic intranuclear inclusions were seen on histological examination of the skin fragments harvested four to eight days after inoculation. Not until 1947 when I joined Dr. J. Enders in setting up a laboratory at the Boston Children's Hospital was it possible to return to the varicella problem. By then, the embryonated hen's egg had become a standard virological technique, but the method had not been explored as a means of isolating varicella virus. After several months in a fruitless effort to isolate varicella virus in eggs, and in the absence of roller tube equipment, I modified the classical Maitland flask culture by maintaining the tissue fragments in culture for long periods by changing the nutrient fluids at three- to five-day intervals. Dr. Enders then suggested attempting the culture of a slow-growing virus such as mumps virus which at that time was propagated in chick embryos but had not been grown in tissue culture. In March 1948, the first mumps virus cultures were prepared. The mumps virus grew well and could be assayed by determining hemagglutinin titers in the fluid phase of the cultures or by infectivity titrations in eggs (Weller and Enders 1948).

With this result in hand, I returned to the varicella problem and prepared flask cultures of arm and shoulder tissues from a four-month-old human fetus. Penicillin and streptomycin were incorporated in the medium. While varicella was the primary objective, I was also investigating viruses that produced paralysis in mice, including Lansing poliomyelitis virus. In March 1948, four cultures were inoculated with acute phase varicella throat washings and four with Lansing virus. The varicella cultures were negative, a finding that would now be expected in view of the insensitivity of tissue cultures for the isolation of VZV from throat washings. However, mice inoculated intracerebrally with the Lansing culture fluids did develop paralysis, and intensive studies on the cultivation of the poliomyelitis viruses immediately took precedence. (See Figures 1 and 2.)

Eight months later, I returned to the varicella problem. Histological examination of human embryonic tissue fragments from flask cultures that had been inoculated with varicella vesicle fluid 4 to 14 days earlier revealed focal collections of cells with eosinophilic intranuclear inclusions. This phenomenon was observed in six experiments in which the vesicle fluid inocula derived from four different cases of varicella. However, repeated attempts to subculture an inclusion-producing agent using fluids or fragments from inclusion-containing cultures as inocula failed, a frustrating situation (Weller and Stoddard 1952).

We then turned to the use of roller tube cultures of either embryonic skin-muscle tissue or foreskin tissue obtained from boys three months to three years of age. At the time, our medium consisted of bovine amniotic fluid (90%), beef embryo extract (5%), horse serum (5%), antibiotics, soybean trypsin inhibitor, and phenol red. The nutrient fluid was changed at three- or four-day

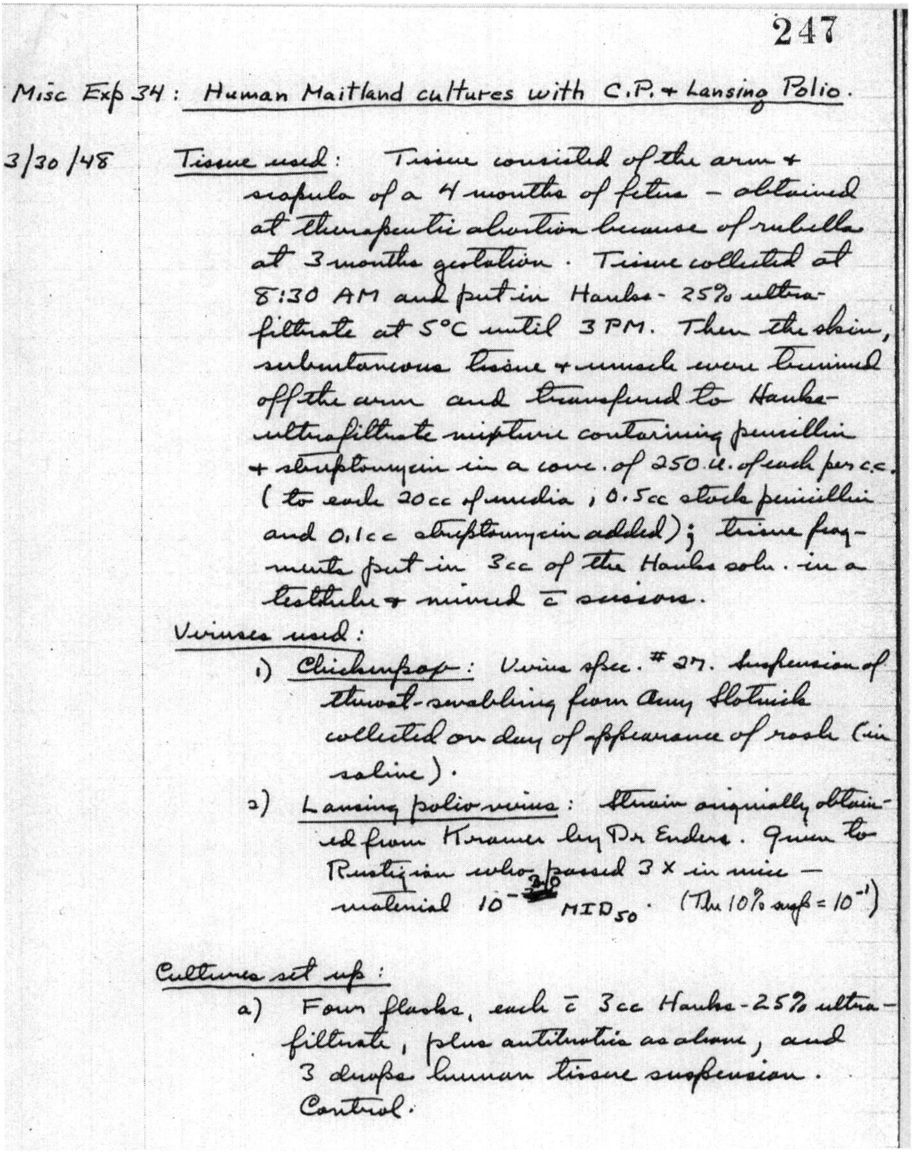

Misc Exp 34 : Human Maitland cultures with C.P. + Lansing Polio.

3/30/48 Tissue used : Tissue consisted of the arm + scapula of a 4 months of fetus — obtained at therapeutic abortion because of rubella at 3 months gestation. Tissue collected at 8:30 AM and put in Hanks- 25% ultra-filtrate at 5°C until 3 PM. Then the skin, subcutaneous tissue + muscle were trimmed off the arm and transferred to Hanks-ultrafiltrate mixture containing penicillin + streptomycin in a conc. of 250 u. of each per cc. (to each 20cc of media , 0.5cc stock penicillin and 0.1cc streptomycin added); tissue fragments put in 3cc of the Hanks soln. in a testtube + minced c̄ scissors.

Viruses used :

1) Chickenpox : Virus spec. # 27. Suspension of throat-swabbing from Amy Slotnick collected on day of appearance of rash (in saline).

2) Lansing polio virus : strain originally obtained from Kramer by Dr. Enders. Given to Rustigian who passed 3 x in mice — material $10^{-\frac{3.9}{6}}$ MID_{50}. (The 10% susp = 10^{-1})

Cultures set up :

a) Four flasks, each c̄ 3cc Hanks-25% ultra-filtrate , plus antibiotic as above, and 3 drops human tissue suspension. Control.

FIGURE 1. A PAGE FROM DR. WELLER'S RESEARCH NOTEBOOK, March 30, 1948, page 247. This was the first experiment using modified Maitland cultures of human embryonic tissues. Four cultures were inoculated with throat-washings from a varicella patient and four cultures were inoculated with Lansing virus mouse-brain material.

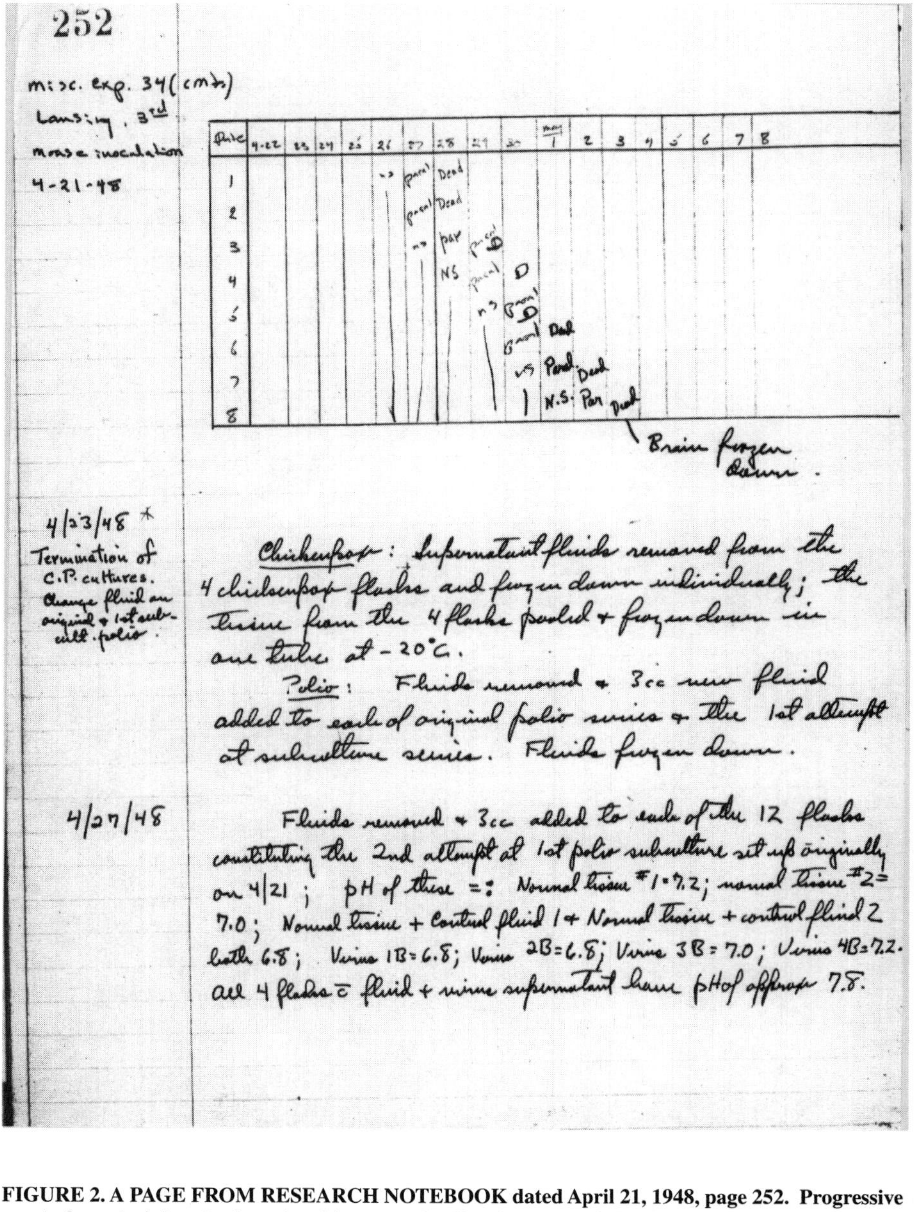

252

misc. exp. 34 (cm?s)
Lansing, 3rd mouse inoculation
4-21-48

Date	4-22	23	24	25	26	27	28	29	30	May 1	2	3	4	5	6	7	8
1					N.S.	paral	Dead										
2						paral	Dead										
3						N.S.	par	paral D									
4							NS	paral	D								
5								N.S.	paral D								
6									paral	Dead							
7									N.S.	Paral.	Dead						
8										N.S.	Par	Dead					

Brain frozen down.

4/23/48 * Termination of C.P. cultures. Change fluid on original + 1st subcult. polio.

Chickenpox: Supernatant fluids removed from the 4 chickenpox flasks and frozen down individually; the tissue from the 4 flasks pooled + frozen down in one tube at -20°C.

Polio: Fluids removed + 3cc new fluid added to each of original polio series + the 1st attempt at subculture series. Fluids frozen down.

4/27/48 Fluids removed + 3cc added to each of the 12 flasks constituting the 2nd attempt at 1st polio subculture set up originally on 4/21; pH of these =: Normal tissue #1 = 7.2; normal tissue #2 = 7.0; Normal tissue + Control fluid 1 + Normal tissue + control fluid 2 both 6.8; Virus 1B = 6.8; Virus 2B = 6.8; Virus 3B = 7.0; Virus 4B = 7.2. All 4 flasks c̄ fluid + virus supernatant have pH of approx 7.8.

FIGURE 2. A PAGE FROM RESEARCH NOTEBOOK dated April 21, 1948, page 252. Progressive onset of paralysis in mice inoculated intra-cerebrally with pooled fluids from the cultures inoculated 22 days earlier with Lansing virus; in the interim, the nutrient fluids had been changed four times at intervals of four to six days.

intervals. In our preliminary report (Weller 1953), we recorded the development of focal cytopathic changes in the cell sheet of cultures from 6 to 13 days following their inoculation with varicella vesicle fluid samples obtained from six different cases. Whereas the cytopathology observed with the polio cultures was characterized by the rapid destruction of the whole cell sheet, that produced by varicella was characterized by collections of rounded swollen cells that slowly increased in size. The swollen cells had intranuclear inclusions and the collections contrasted sharply with adjacent sheets of uninfected fibroblasts or epithelial cells. The slow increase in the size of the focal collections appeared to reflect the contact infection of adjacent normal cells. Most important was the ease with which subculture could be accomplished when suspensions of infected cells, but not the medium from the infected cultures, were used as the inoculum. We carried strains of varicella virus through ten passages. This was the first evidence of the strong cell association of varicella virus when cultivated *in vitro*, a characteristic that has frustrated many investigators.

The next pressing question involved zoster. Would zoster vesicle fluids produce similar cytopathic changes? Material from two cases of zoster produced apparently identical cytopathic changes, and the two viral strains were carried through four tissue culture passages.

While the evidence suggested that we had isolated the etiological agents of varicella and of zoster, immunologic supportive evidence was lacking and there was no information on the antigenic identity or lack of identity of the viruses from the two clinical entities. The first evidence that patients convalescing from varicella or from zoster had a significant antibody response to our viral isolates was obtained in collaboration with Albert Coons in an early application of his remarkable fluorescent antibody technique (Weller and Coons 1954). Infected focus-containing coverslip preparations of cells were used as antigen. These represented three isolates from varicella cases, two isolates from zoster cases, and the cellular alterations produced by one strain of herpes simplex virus. When these preparations were exposed to dilutions of acute and of convalescent phase sera from varicella, zoster, and herpes simplex cases, fixation was detected by the use of a fluorescent anti-human gammaglobulin conjugate. Antibody reacting to an almost identical degree appeared during convalescence from varicella and from zoster. Antibody reacting with the herpes simplex preparations was demonstrated uniformly only in a group of sera obtained from cases of recurrent herpes simplex infection. These findings provided the first immunologic evidence that the etiologic agents of varicella and of herpes zoster had been isolated and propagated *in vitro*. The findings also suggested the co-identity or close antigenic relationship of the varicella and zoster viruses.

No differences were noted between the cytopathic and histologic changes induced on cultivation by the viruses isolated from the two clinical entities (Weller et al 1958). Both agents produced apparently identical focal areas of swollen cells, with foci gradually increasing in size and number. Prolonged serial passage did not produce any adaptive change in the cytopathic behavior of the agents. The slowly progressive nature of the involvement of the cell sheet was apparent when varicella strains were kept in the culture tubes for a prolonged period; after 30 days, between 45 and 85% of the cell sheet was involved. Various human tissues supported the growth of the viruses *in vitro,* including fibroblasts, myometrial, testicular, and amnion cells, and brain and kidney tissue. Specific lesions appeared in inoculated monkey testis and kidney cultures, but the cytopathic changes progressed less rapidly than in cultures of human origin. Specific cytopathology was noted irregularly in inoculated cultures of rabbit and bovine tissues. Even after foci had become numerous in infected cultures, repeated attempts to recover virus from the cell-free medium usually failed.

In an effort to obtain additional immunologic evidence of the etiological nature of our viral isolates, we used fluids from cultures with extensive specific degenerative changes as a complement-fixing (CF) antigen; low and variable levels of complement-fixing activity were demonstrated. Attempts to enhance activity by ultrafiltration to concentrate the antigen failed because the product was anti-complementary. We therefore grew the viruses in bottle cultures of human tissues until the cytopathic changes were extensive. At that stage, we eliminated serum and embryo extract from the medium and maintained the cultures on a straight bovine amniotic fluid medium. When these fluids were concentrated 20-fold by ultrafiltration, a satisfactory CF antigen was obtained. On examination of sera from both clinical entities by CF using two varicella antigens and one zoster antigen, similar rises in titer were obtained with each of the three antigens. It was also shown that a satisfactory CF antigen could be prepared from infected monkey kidney cell cultures.

However, because of the known limited specificity of some CF antigens, we sought *in vitro* neutralization data. Classical neutralization procedures involving a preliminary incubation of serum and of viral dilutions could not be used because of the cell-associated nature of the viruses. A preliminary study involving the incubation of infected cell suspensions and serum dilutions prior to their use as inocula resulted in little inhibition of the focal process. We then turned to including the serum under examination as a continuing constituent of the culture medium. When convalescent serum was incorporated in the medium in a 10% concentration, both varicella viruses and zoster viruses were inhibited by the homologous or heterologous serum. Thus, we had evidence by another procedure that we had isolated the etiologic agents of varicella and zoster (Weller and Witton 1958). The validity of this conclusion was firmly documented years later by endonuclease analysis which showed that the genomic structure of the DNA of a virus isolated from a patient with varicella was identical with that of a virus isolated from the same patient during a subsequent attack of zoster (Strauss et al 1984).

As a physician trained in pediatric infectious disease, I have derived great satisfaction that our concept of VZV as a highly host-specific organism best studied in a human cell tissue culture system, first investigated in 1941, was finally proven to be valid seventeen years later. The work spanned an era when I isolated viruses such as the cytomegaloviruses and rubella virus, agents of more social significance than varicella-zoster, but the result of efforts not as personally satisfying. Representing the "then" category of microbe hunters, I can only look with admiration at the "now" category of investigators.

It is clear that the age of discovery of new viruses of pediatric importance persists as it did half a century ago. What is different is the vast spectrum of molecular tools available to the modern microbe hunter. The list of the hepatitis viruses, the herpes-related viruses, and many others continues to expand. Virology has grown fantastically in the past 50 years, and there is every reason to believe that achievements in the next 50 years will surpass the most optimistic of current predictions.

REFERENCES

Gard, S (1955). Presentation: The Nobel Prize for Physiology and Medicine, 1954. In: Les Prix Nobel en 1954. Stockholm: Imprimerie Royale P. A. Norstedt and Soner, 38-42.

Goodpasture EW, Anderson K (1944). Infection of human skin, grafted on the chorioallantois of chick embryos, with the virus of herpes zoster. Am J Pathol 20: 447-455.

Enders JF, Weller TH, Robbins FC (1949). Cultivation of the Lansing strain of poliomyelitis virus in cultures of various human embryonic tissues. Science 109: 85-87.

Strauss SE et al (1984). Endonuclease analysis of viral DNA from varicella and subsequent zoster infections in the same patient. N Engl J Med 311: 1362-1364.

Weller TH (1953). Serial propagation in vitro of agents producing inclusion bodies derived from varicella and herpes zoster. Proc Soc Exp Biol Med 83: 340-346.

Weller TH (1983). Medical progress. Varicella and herpes zoster. Changing concepts of the natural history, control, and importance of a not-so-benign virus. N Engl J Med 309: 1362-1368, 1434-1440.

Weller TH (1992). Varicella and herpes zoster; a perspective and overview. J Infect Dis 166: (suppl. 1) S1-6.

Weller TH et al (1958). The etiologic agents of varicella and herpes zoster. Isolation, propagation, and cultural characteristics in vitro. J Exp Med 108: 843-868.

Weller TH, Coons AH (1954). Fluorescent antibody studies with the agents of varicella and herpes zoster propagated in vitro. Proc Soc Exp Biol Med 86: 789-794.

Weller TH, Enders JF (1948). Production of hemagglutinin by mumps and influenza viruses in suspended cell tissue cultures. Proc Soc Exp Biol Med 69: 124-128.

Weller TH, Stoddard MB (1952). Intranuclear inclusion bodies in cultures of human tissue inoculated with varicella vesicle fluid. J Immunol 68: 311-319.

Weller TH, Witton HM (1958). The etiologic agents of varicella and herpes zoster. Serologic studies with the viruses as propagated in vitro. J Exp Med 108: 869-890.

12

Live Attenuated Varicella Vaccine

ANNE GERSHON, PHILIP LARUSSA, AND SHARON STEINBERG

Live attenuated varicella vaccine (Oka strain) was developed in the early 1970s in Japan and was licensed in the United States for routine use by the Food and Drug Administration (FDA) for immunization of varicella-susceptible children and adults in March 1995. It has had the longest gestational period, over 21 years, of any previously licensed vaccine in the United States. Until this licensure, the FDA had not approved any previous new live virus vaccine since live attenuated rubella vaccine was licensed in 1969.

Several aspects of varicella vaccine set it apart from other live virus vaccines, in part accounting for its long pre-licensure period. First, varicella-zoster virus (VZV) is a member of the herpesvirus group; these viruses characteristically become latent following primary infection. The potential effect of vaccination on latent infection due to VZV was initially unknown, especially since there was no useful animal model for study of the virus some 20 years ago when the vaccine was developed. Thus when studies were initiated, there was no way to know whether vaccination might predispose to or offer protection against zoster as well as chickenpox. Second, the vaccine was first tested in the United States in immunocompromised children with underlying leukemia who were at high risk to manifest severe and even fatal varicella, rather than in healthy children which is the usual approach for investigating a vaccine. A controversy developed which continued for many years over whether healthy children, who usually have only mild or moderate symptoms from chickenpox, should be vaccinated. Third, VZV is a labile virus that is difficult to propagate *in vitro* and therefore also difficult to quantitate, impeding production and standardization of a vaccine.

Although several investigators were attempting to develop a varicella vaccine for many years, a satisfactory product that was relatively nonreactogenic and immunogenic was only successfully achieved by Takahashi and colleagues, who developed the Oka strain in the early 1970s (Neff et al 1981; Takahashi et al 1974). All of the current varicella vaccines, products manufactured by Merck and Co. in the United States, by SmithKline Beecham in Europe, and by Biken Institute in Japan, originate from the Oka strain and differ only in passage of virus used and in stabilizers.

FIRST JAPANESE STUDIES

The Oka strain was isolated from an otherwise healthy three-year-old Japanese boy with acute varicella; the virus was passaged 11 times in human embryonic lung fibroblasts at 34°C, 12 times in guinea pig fibroblast cells at 37°C, and 7 to 8 times in MRC-5 cells (human lung fibroblasts) at 37°C. The resultant virus was administered to children in a hospital setting to abort nosocomial spread of varicella. The vaccine was immunogenic and appeared to be safe in these children (Takahashi et al 1974). Children with a variety of underlying medical problems were immunized in various Japanese studies (reviewed in Gershon 1980). For example, children who were immuno-suppressed due to steroid therapy for a variety of medical conditions, were safely immunized against varicella without incident (Gershon 1980). In addition, an efficacy study conducted in families involving vaccination of one healthy child within three days of onset of varicella in a sibling showed complete protection in 18 vaccinees; in a similar number of children in similar families, the attack rate for varicella in exposed unvaccinated siblings was 100% (Gershon 1980). Children with underlying leukemia in remission who were at particular risk to develop severe or fatal varicella were also immunized in Japan. It was soon determined that in order to prevent some of these vaccinees from developing a vaccine-associated vesicular rash, chemotherapy had to be withheld briefly before and after vaccination (Gershon 1980). While this strategy caused vaccinees to develop cellular and humoral immune responses to VZV, the protective efficacy of the vaccine remained unknown in high-risk children.

INITIAL STUDIES IN THE UNITED STATES

In the mid-1970s, most American investigators studying VZV were not very enthusiastic about the use of varicella vaccine due to concerns about vaccine safety. However, by the late 1970s, there was a change in attitude brought on by a report that 30% of leukemic children receiving steroids, especially during the incubation period of varicella, developed disseminated chickenpox and 7% died from the infection (Feldman et al 1975). While successful passive immunization against varicella was available (Gershon et al 1974), more children were at risk from unknown exposures to VZV than from recognized exposures, and effective antiviral chemotherapy had not yet been developed. Therefore, active immunization against varicella was clearly the best approach to prevention of severe chickenpox in this population.

A decision was made at the National Institute of Allergy and Infectious Diseases (NIAID) to sponsor a large collaborative study of varicella vaccine in leukemic children and in adults who were at high risk to develop severe varicella. Fortunately, at about that time, the first reliable test to determine immune status to varicella, the fluorescent antibody to membrane antigen assay, was developed (Williams et al 1974). Moreover, technology making it possible to distinguish between wild-type and vaccine-type virus by restriction length polymorphism analysis of DNA from VZV-positive cultures also became available (Martin et al 1982; Gershon et al 1984). Thus, it became possible to test leukemic children for susceptibility to varicella, vaccinate them, and follow them for the safety, immunogenicity, and protective efficacy of the vaccine over time. This longitudinal NIAID study involving a medically well-followed cohort of children at high risk to develop severe natural varicella due to a chronic medical condition also allowed analysis of the incidence of zoster after vaccination. Adults, also at increased risk to develop severe varicella compared to healthy children, were similarly vaccinated and followed longitudinally.

This collaborative study was begun in 1979, and many of the subjects are still being followed today. The overall fatality rate (mostly due to complications of leukemia) among the 575 leukemic children followed for at least 1 year after immunization and as long as 14 years, was approximately 12% (69 children); none died from any illness related to VZV and none had disseminated VZV infections. Theoretical expectations over a time period of about 10 years would predict the loss of 25 to 50 of these children to varicella and 75 to 100 cases of disseminated varicella. In addition, 374 adults were also vaccinated without serious incident in the NIAID study; no cases of severe varicella occurred in this cohort either.

LESSONS LEARNED FROM VARICELLA VACCINE

These clinical trials in the United States and the long period between their initiation and the licensure of the vaccine spawned a large number of analyses that provided a great deal of information about the vaccine and also about VZV.

INDICATIONS OF ATTENUATION

Since there is no practical animal model for VZV, indications of attenuation have come from studies in humans. That the vaccine rarely induces skin lesions following injection into a healthy host suggests attenuation. In addition, administration of vaccine by inhalation did not induce disease in susceptible children in one small study (Bogger-Goren et al 1982).

An apparent impaired ability for transmission of vaccine-type virus compared to wild-type virus is suggestive of attenuation. Transmissibility of the vaccine virus from skin lesions that occur in 5 to 10% of healthy hosts one to four weeks after vaccination has rarely been documented (Weibel et al 1984). Household transmission of vaccine-type VZV to varicella-susceptible siblings has, however, been reported for leukemic vaccinees, although transmission occurred in only 17% and only if the vaccinee developed a rash after immunization (Tsolia et al 1990). In contrast, there is an 80 to 90% transmission rate seen in the wild-type virus infection in households (Ross et al 1962).

An additional indication that the Oka strain is attenuated is that in reported contact cases, 27% were subclinical (in contrast to the usual 5% with wild-type infection), and the average number of skin lesions was only 38, about 10 times fewer than predicted with wild-type varicella (Tsolia et al 1990). The molecular basis for the attenuation of VZV remains unknown.

SAFETY, IMMUNOGENICITY, AND PROTECTION IN HIGH-RISK HOSTS

Table 1 summarizes selected results from the collaborative study. Both leukemic children and healthy adult vaccinees unexpectedly required two doses of vaccine to achieve a seroconversion rate over 90%. Leukemic children were at particularly high risk of developing a vaccine-associated rash after the first dose of vaccine but had a higher degree of subsequent protection against varicella than that achieved by adults. Adults were less likely to be completely protected against varicella after intimate exposure to VZV. By contrast,

TABLE 1. OVERVIEW of adverse effects, seroconversion rates, estimated efficacy, and occurrence of zoster in various populations immunized with live attenuated varicella vaccine

| | Number | Percent with vaccine-associated rash | | Percent seroconversion | | Estimated efficacy | |
		Dose 1	Dose 2	Dose 1	Dose 2	%	Zoster
Children	>8800	3.8	-	97[*]	-	85-90	18 per 100,000[t]
Adolescents, adults	>1550	5.5	0.9	75[*]	99[*]	70	
Leukemic children	575[#]	50	7	82[**]	95[**]	85	800 per 100,000[tt]

[*] glycoprotein ELISA assay; [**] fluorescent antibody to membrane assay; [#] 511 vaccinees had chemotherapy withheld for one week before and one week after immunization; [t] person-years; expected incidence for this age group 20-59 per 100,000 person-years (Guess et al 1985); [tt] person-years; comparable incidence for leukemic children after natural varicella, 2460 per 100,000 person-years (Hardy et al 1991). Source of data: Merck and Co. package insert for Live Attenuated Varicella Vaccine and NIAID Collaborative Varicella Vaccine Study.

vaccinated healthy children had only about a 5% chance of developing a mild vaccine-associated rash and, after only one dose of vaccine, achieved a level of complete protection similar to that of leukemic children after two doses (Gershon 1995).

In studies of leukemic children and adults, protective efficacy was determined by comparing the attack rate of varicella following household exposure to chickenpox in vaccinees with the known attack rate of 80 to 90% in susceptible children after this kind of intimate exposure (Ross et al 1962). Persistence of immunity has been assessed by longitudinal testing of VZV antibody titers and the overall incidence and severity of breakthrough varicella cases occurring with time after immunization. In vaccinated leukemic children, the attack rate of varicella after household exposure was 17 of 123 (14%), and in healthy adults, it was 15 of 58 (26%). Those with breakthrough varicella had modified disease with a mean of 96 and 43 skin lesions, respectively. After 11 to 13 years of follow-up, 87% of leukemic children and 82% of adults were seropositive. Neither the incidence nor severity of varicella in leukemic children or adults has increased over time (Gershon 1995).

PARTIAL IMMUNITY: BREAKTHROUGH VARICELLA AFTER VACCINATION

The phenomenon of partial immunity was not anticipated when vaccine studies were initiated. Previously, protection against illness by vaccination was regarded as an all-or-none phenomenon, and a "no take" from the immunization was assumed if a vaccine recipient was not protected. Partial immunity to VZV was first demonstrated in a leukemic vaccinee who, nine months after vaccination and two weeks after antibodies to VZV were detected, developed a mild case of varicella from which wild-type VZV was isolated. Two weeks later, her brother developed full-blown natural varicella (Gershon et al 1984). It is now recognized that breakthrough varicella occurs in a minority of vaccinees. Careful studies performed in healthy children have shown that varicella occurring after immunization is a modified illness in healthy children (Watson et al 1993). There is much current interest in the degree of transmissibility of VZV from partially immune, otherwise healthy individuals with mild breakthrough varicella. It is known that transmission is frequent from leukemic children with breakthrough varicella and that transmission *can* occur from otherwise healthy individuals with breakthrough varicella. However, transmission *rates* from individuals with breakthrough varicella, particularly normal hosts in whom there may be very few nonvesicular skin lesions, are not yet known.

STUDIES IN HEALTHY CHILDREN IN THE UNITED STATES

In the United States, the Oka vaccine produced by Merck and Co. was licensed for general use in March 1995. Approximately 9000 healthy children have been vaccinated in pre-licensure studies (Table 1). Only minor adverse effects have been reported in these vaccinees: 24% had temporary discomfort at the injection site, 4% had a rash at the injection site, 4% had a mild varicella-like rash, and 15% had a brief oral temperature over 102°F. The latter symptom was not specific for the immunization and was also seen at a similar rate in children who received placebo injections. Most of the children who developed a rash had fewer than 10 lesions, most of which were not vesicular. Children under 12 years of age had a 97% seroconversion rate, measured by an ELISA assay utilizing glycoprotein (gp) of VZV as the antigen (Provost et al 1991). Adolescents (above age 12) had an 82% seroconversion rate after one dose and a 99% seroconversion rate after two doses given three months apart (White et al 1991). Antibody titers remained positive for up to four years in over 95% of children. Breakthrough cases of varicella occurred in 1 to 2% of a cohort during each year of a seven-year follow-up, but these were clearly a modified form of illness (Kuter et al 1991; White et al 1991).

EFFICACY IN HEALTHY CHILDREN

The overall efficacy of varicella vaccine in healthy children approaches 90%, depending upon the immunizing dose of virus and the definition of breakthrough varicella that is used. For example, in early clinical trials involving 4142 children given 1000 to 1625 plaque-forming units (PFU) of vaccine-type VZV, the annual attack rate of breakthrough varicella was 2.1 to 3.6%, a vaccine efficacy of 67%. In later trials in 1164 children given 2900 to 9000 PFU, the annual attack rate of breakthrough varicella was 0.2 to 1%, an efficacy rate over three years of 93% (Krause et al 1995 and data obtained from package insert of varicella vaccine produced by Merck and Co.). This latter formulation is the licensed vaccine in the United States.

Only one double-blind controlled study of the efficacy of varicella vaccine in healthy children was reported (Weibel et al 1984). In that study, 491 children were vaccinated and 465 received placebo. During the first year of follow-up, the incidence of varicella was 8.5% in the placebo group and zero in the vaccinated group, an efficacy of 100%. In the second year of follow-up, involving less than half the original cohorts, the efficacy was 96%. In this study, the vaccine dose was higher than the currently licensed vaccine, 17,000 PFU (Kuter et al 1991).

As in high-risk immunized populations, breakthrough varicella is a modified illness, often with very few skin lesions, many of which are not vesicular. If vaccine efficacy is calculated on the basis of having as few as one nonvesicular skin lesion following household exposure, the vaccine is about 75% protective. If, on the other hand, breakthrough illness is defined as more than 10 vesicles, the efficacy is obviously greater, approaching 90% even with lower potency vaccines.

PERSISTENCE OF IMMUNITY IN HEALTHY VACCINATED CHILDREN

At present, it is clear that boosting after re-exposure to VZV is common, but whether this is important in long-term maintenance of immunity to VZV is unknown.

Studies in over 4000 American children, using the gp ELISA method to test serum from children vaccinated four years previously, have shown persistence of VZV antibodies in close to 100% (data of Merck and Co.). A 20-year follow-up study from Japan of 96 vaccinees revealed an overall rate of breakthrough varicella of 2%. Positive VZV antibody titers were detectable in 25 of 26 individuals, all of whom had a positive test for cell-mediated immunity to VZV (Asano et al 1994).

COST-BENEFIT ANALYSIS AND POTENTIAL EFFECTS ON EPIDEMIOLOGY OF VARICELLA

A financial analysis for vaccine use in the United States revealed that routine use of varicella vaccine is cost effective if benefits to society and to the individual are both taken into account, saving $5 for every $1 spent (Lieu et al 1994). A mathematical model suggests that with more than 90% vaccine coverage in the population, the numbers of cases of varicella in adults will decrease, even if immunity wanes in as many as 15% of vaccinees, assuming almost complete immunization of the population (Halloran et al 1994). This calculation strongly suggests that with widespread use of the vaccine, there will be little risk of increasing the incidence of disease in adults. This has certainly been the case for measles vaccine.

Current Status of Varicella Vaccine

The vaccine is licensed for healthy children in Japan and North Korea, where over two million children have been immunized since 1989. In several European countries, it is licensed for immunocompromised children and is being considered for licensure for healthy children.

RECOMMENDATIONS FOR USE OF VARICELLA VACCINE IN THE UNITED STATES

It is recommended that healthy children over the age of 12 months who have not had varicella be immunized with one dose of vaccine. Adolescents over age 13 years and adults are recommended to receive two doses of vaccine given one to two months apart. Two doses to achieve a seroconversion greater than 90% in adolescents and adults is probably necessary due to the lower cell-mediated immune response to VZV in these groups compared to that in children (Gershon 1995; Nader et al 1995). The vaccine should not be administered to pregnant women or children with intercurrent illnesses, nor can it be relied upon for post-exposure prophylaxis.

FUTURE STUDIES

While use of varicella vaccine has engendered great controversy in the United States for many years, it is predicted to be well received now that it is licensed. Continued surveillance is planned to examine safety and protection as well as to monitor waning immunity and incidence of zoster.

Responses of children with chronic diseases other than leukemia need to be assessed in studies to determine whether these children can be safely immunized. Such children include those with asthma, diabetes, human immunodeficiency virus infection, rheumatoid arthritis receiving aspirin therapy, and children with various other forms of cancer.

ZOSTER

It is now known that zoster is related to the presence of skin lesions during primary infection, the presence of latent VZV in sensory ganglia, and a low cell-mediated immune response to VZV at some point after primary infection (often many years later). There is a high probability of all of these occurrences in leukemic vaccinees, and, indeed, zoster has occurred in this population (Hardy et al 1991; Williams and Gershon et al 1985). Vaccine-type VZV can cause zoster in leukemic vaccinees, as determined in restriction fragment length polymorphism and polymerase chain reaction (PCR) analysis (LaRussa et al 1992; Williams et al 1985). However, the frequency with which zoster occurs is significantly lower than that seen in similar children who have had the natural infection (3% vs 15% over several years, respectively) (Hardy et al 1991). Thus vaccination offers potential protection against zoster as well as against varicella. That zoster due to vaccine-type VZV has occurred in a few vaccinees is proof that zoster is due to reactivation of latent VZV (Hayakawa et al 1984; Williams et al 1985).

Zoster caused by vaccine virus has not been documented in any adult vaccinees. Interestingly, although vaccinated adults develop lower cell-mediated immune responses to VZV than vaccinated healthy children, it may be that latency after immunization is unusual in adult vaccinees since vaccine-associated rashes are unusual (Gershon 1995). Several immunized children have developed zoster, but none of the cases have been identified as due to vaccine-type VZV. All indications thus far in healthy vaccinated populations are that the incidence of zoster is lower after vaccination than after natural infection (Gershon 1995). It is hoped that this will eventually hold true for elderly patients who were immunized as children, but obviously it will be many years before this can be determined. Since additional doses of vaccine can increase cell-mediated immunity to VZV (Berger et al 1984), it may be that vaccination of varicella-immune individuals at high risk to develop zoster can be protected by administration of varicella vaccine.

Conclusions

It is anticipated that the incidence of varicella will decrease sharply and that full-blown cases will become unusual or rare after the vaccine is used routinely in a population. Today it is possible to demonstrate the presence of VZV DNA by PCR and thus to detect VZV DNA without propagating the virus *in vitro* (LaRussa et al 1992). This powerful technique should become increasingly useful as we proceed into the vaccine era in the United States. Zoster will continue to occur in about 15% of the overall population for some additional years and then it too may become rare. Due to a loss of natural boosting effects that are currently operative, routine boosters of varicella vaccine may be necessary in the population at some point to maintain a low rate of disease caused by VZV. This might be conveniently done in adolescence and in middle age, along with booster doses of other routinely used vaccines.

Acknowledgments

This work was supported by National Institutes of Health grant A124021 and Merck and Co.

References

Asano Y, Suga S, Yoshikawa T, Kobayashi H, Yazaki T, Shibata M, Tsuzuki K, Ito S (1994). Experience and reason: twenty year follow up of protective immunity of the Oka live varicella vaccine. Pediatrics 94: 524-526.

Berger R, Luescher D, Just M (1984). Enhancement of varicella-zoster-specific immune responses in the elderly by boosting with varicella vaccine. J Infect Dis 149: 647.

Bogger-Goren S, Baba K, Hurley P, Yabuuchi H, Takahashi M, Ogra P (1982). Antibody response to varicella-zoster virus after natural or vaccine-induced infection. J Infect Dis 146: 260-265.

Feldman S, Hughes W, Daniel C (1975). Varicella in children with cancer: 77 cases. Pediatrics 80: 388-397.

Gershon A (1980). Live attenuated varicella-zoster vaccine. Rev Infect Dis 2: 393.

Gershon A (1995). Varicella-zoster virus: prospects for control. Adv Ped Infect Dis 10: 93-124.

Gershon A, Steinberg S, Brunell P (1974). Zoster immune globulin: a further assessment. N Engl J Med 290: 243-245.

Gershon AA, Steinberg S, Gelb L, NIAID-Collaborative-Varicella-Vaccine-Study-Group (1984). Live attenuated varicella vaccine: efficacy for children with leukemia in remission. JAMA 252: 355-362.

Guess H, Broughton D, Melton L, Kurland L (1985). Epidemiology of herpes-zoster in children and adolescents: a population-based study. Pediatrics 76: 512-517.

Halloran E, Cochi S, Lieu T, Wharton M, Fehrs L (1994). Theoretical epidemiological and morbidity effects of routine immunization of preschool children with varicella vaccine in the United States. Am J Epidemiol 140: 81-104.

Hardy IB, Gershon A, Steinberg S, LaRussa P, the NIAID Collaborative Varicella Vaccine Study Group (1991). The incidence of zoster after immunization with live attenuated varicella vaccine. A study in children with leukemia. N Engl J Med 325: 1545-1550.

Hayakawa Y, Torigoe S, Shiraki K, Yamanishi K, Takahashi M (1984). Biologic and biophysical markers of a live varicella vaccine strain (Oka): identification of clinical isolates from vaccine recipients. J Infect Dis 149: 956-963.

Krause P, Klinman DM (1995). Efficacy, immunogenicity, safety, and use of live attenuated chickenpox vaccine. J Pediatr 127: 518-525.

Kuter BJ, Weibel RE, Guess HA, Matthews H, Morton DH, Neff BJ, Provost PJ, Watson BA, Starr S, Plotkin S (1991). Oka/Merck varicella vaccine in healthy children: final report of a 2-year efficacy study and 7-year follow-up studies. Vaccine 9: 643-647.

LaRussa P, Lungu O, Hardy I, Gershon A, Steinberg S, Silverstein S (1992). Restriction fragment length polymorphism of polymerase chain reaction products from vaccine and wild-type varicella-zoster virus isolates. J Virol 66: 1016-1020.

Lieu T, Cochi S, Black S, Halloran ME, Shinefield HR, Holmes SJ, Wharton M, Washington E (1994). Cost-effectiveness of a routine varicella vaccination program for U.S. children. JAMA 271: 375-381.

Martin JH, Dohner D, Wellinghoff WJ, Gelb LD (1982). Restriction endonuclease analysis of varicella-zoster vaccine virus and wild type DNAs. J Med Virol 9: 69-76.

Nader S, Bergen R, Sharp M, Arvin A (1995). Comparison of cell-mediated immunity (CMI) to varicella-zoster virus (VZV) in children and adults immunized with live attenuated varicella vaccine. J Infect Dis 171: 13-17.

Neff BJ, Weibel RE, Villerajos VM, Buynak E, McLean A, Morton D, Wolanski B, Hilleman M (1981). Clinical and laboratory studies of KMcC strain of live attenuated varicella virus. Proc Soc Exp Biol Med 166: 339-347.

Provost PJ, Krah DL, Kuter BJ, Morton DH, Schofield TL, Wasmuth EH, White J, Miller W, Ellis RW (1991). Antibody assays suitable for assessing immune responses to live varicella vaccine. Vaccine 9: 111-116.

Ross AH, Lencher E, Reitman G (1962). Modification of chickenpox in family contacts by administration of gamma globulin. N Engl J Med 267: 369-376.

Takahashi M, Otsuka T, Okuno Y, Asano Y, Yazaki T, Isomura S (1974). Live vaccine used to prevent the spread of varicella in children in hospital. Lancet 2: 1288-1290.

Tsolia M, Gershon A, Steinberg S, Gelb L (1990). Live attenuated varicella vaccine: evidence that the virus is attenuated and the importance of skin lesions in transmission of varicella-zoster virus. J Pediatr 116: 184-189.

Watson BM, Piercy SA, Plotkin SA, Starr SE (1993). Modified chickenpox in children immunized with the Oka/Merck varicella vaccine. Pediatrics 91: 17-22.

Weibel R, Neff BJ, Kuter BJ, Guess HA, Rothenberger CA, Fitzgerald AJ, Connor KA, McLean AA, Hilleman MR, Buynak EB, Scolnick EM (1984). Live attenuated varicella virus vaccine: efficacy trial in healthy children. N Engl J Med 310: 1409-1415.

White CJ, Kuter BJ, Hildebrand CS, Isganitis KL, Matthews H, Miller WJ, Provost PJ, Ellis RW, Gerety RJ, Calandra GB (1991). Varicella vaccine (VARIVAX) in healthy children and adolescents: results from clinical trials, 1987 to 1989. Pediatrics 87: 604-610.

Williams DL, Gershon A, Gelb LD, Spraker MK, Steinberg S, Ragab AH (1985). Herpes zoster following varicella vaccine in a child with acute lymphocytic leukemia. J Pediatr 106: 259-261.

Williams V, Gershon A, Brunell P (1974). Serologic response to varicella-zoster membrane antigens measured by indirect immunofluorescence. J Infect Dis 130: 669-672.

Vignette

Influenza

Up through the nineteenth century, respiratory diseases were frequently characterized as tubercular (tuberculosis) and nontubercular (influenza/ pneumonia). Influenza was then primarily an epidemiologic concept, with characterization based on general clinical and epidemiological resemblance because the specific causative organism had not yet been identified in any epidemic. With the outbreak of the pandemic in 1918–1919 that killed over 20 million people, the question arose of whether this was the same disease described earlier in 1802–1803, 1805–1806, 1830–1833, 1836–1837, 1847–1848, and 1889–1890.

Edward Kilbourne, a major figure and contributor to the field of influenza virus research, provides an overview of the discoveries that helped to identify and define the influenza virus and to develop the field of virology in general. John Skehel, current director of the National Institute for Medical Research (NIMR) in Mill Hill, England, recounts the major initial and continuing fundamental contributions toward understanding influenza virus infection made at the NIMR, which was established in direct response to the influenza pandemic of 1918–1919.

13

A History of Influenza Virology

EDWIN D. KILBOURNE

The term "virology" in the title of this chapter was chosen rather than "virus," which would have been the history of a pool of polymorphic and promiscuous genes of unknown origin and presumptive phylogeny, and rather than simply "influenza," the disease, which would have implied an account only slightly less speculative and extending back to credible observations of epidemics in the fifteenth century. Instead, "a history of influenza virology" is finite, contemporary, and consistent with the overall emphasis of this volume on past and present microbe hunters.

There were, of course, more microbe hunters then than now. Rather than "discovering" the causes of disease, modern efforts are focused increasingly on linking known agents to unknown diseases. Thus, the emphasis is now, quite properly, on the virus and what it does.

Influenza viruses are unique, not only in their speed of evolution and antigenic change and in their potential for genetic interchange in nature, but also in the contribution their study has provided to virology and to other fields of biology and chemistry. The somewhat untidy packages of RNA that we call influenza viruses may have been hunted down, but they dissemble even as we study them. Today's hunters find that the chase is still on as they pursue the protean proteins of an ephemeral quarry.

THE MICROBE HUNTERS

The isolation by Centanni and Savonucci of a filter-passing agent from chickens with fowl plague (geflügelpeste) in 1900 marked the beginning of orthomyxovirology (Table 1). From a preserved specimen taken in that year, the virus was isolated in 1902 as A/Brescia/1902, and thereafter preserved (cited by Klimov et al 1992). However, it was not until 1955, that fowl plague, in fact, was recognized as a type A influenza virus based on chemical studies of the virus (Schafer 1955).

TABLE 1. THE MICROBE HUNTERS

Hunter	Virus	Isolation date	Host
Centanni and Savonucci	fowl plague (influenza A)	1900	chicken
Shope	swine (influenza A)	1930	swine
Smith, Andrewes and Laidlaw	influenza A (human)	1932	ferret
Francis; Magill	influenza B (human)	1940; 1940	ferrets and mice
Taylor; Francis et al	influenza C (human)	1947; 1950	chick embryo

The year 1930 marked the first recovery of an influenza virus from a mammalian host when virus was isolated from swine with respiratory tract symptoms (Shope 1931). Shortly thereafter, Smith, Andrewes and Laidlaw (1933) infected an unusual experimental host, the ferret, with human throat washings and thereby isolated the prototypical WS (Wilson Smith) strain of influenza A virus that is still a workhorse in laboratories interested in influenza virus genetics. A word about serendipity, at this point. Ferrets were available because of their susceptibility to the canine distemper virus (later to be classified as a *para*myxovirus) which induced symptoms similar to those of influenza and which was then under study in Laidlaw's laboratory.

Still relying on ferrets, but also using mice, Francis (1940), and independently Magill (1940), recovered the antigenically distinct influenza B virus. Taylor, in 1947, completed the identification of human influenza virus types with his isolation of a virus not identifiable as type A or B from a subject with mild respiratory disease (Taylor 1949). Francis et al (1950) serologically linked the virus to epidemics of mild influenza in children and thus established the epidemic potential of the newly named influenza C virus. Other candidates have since sought the position of "influenza D," including the Sendai virus, now known to be a parainfluenza virus of mice, and more recently, the tick-borne Thogoto virus, which, although technically classified as an orthomyxovirus, does not cause influenza.

THE VIRUS *IN VITRO*

Early virology was, of course, an exercise in experimental pathology because the detection and measurement of the virus depended on its propagation in animal hosts, specifically, in the lungs of mice and ferrets. The introduction of the chick embryo for influenza virus propagation by Hoyle and Fairbrother (1937) and Lush and Burnet (1937) actually *followed* the use of minced chick embryo tissue culture by Smith (1936), but was a long step towards providing a less complicated, bacterially sterile host cell system for growth and study of virus (Table 2). The use of the allantoic sac (Burnet 1941) provided enormous quantities of relatively pure virus in tissue-free allantoic fluid and is still used today—50 years later—for the production of influenza vaccine. There is little doubt that this advance led George Hirst to make two observations that were of critical importance, not only to the study of influenza virus, but to the study of viruses in general. In harvested allantoic fluid tinged with red blood cells from the torn chorioallantoic membrane, Hirst (1941) noted the agglutination, then the disagglutination of these cells and postulated 1) the existence of viral receptors on the cells and 2) based on later work (Hirst 1943), the presence of an enzyme on the virus which destroyed these receptors. This simple bench observation (a Pasteurian example of opportunity favoring the prepared mind) unleashed animal virology from the fetters of *in vivo* study and, indeed, facilitated viral purification and physicochemical characterization in a way not available to the then prevailing aristocrats of virology—those working with bacteriophage. Uniquely, influenza viruses could be studied, in the absence of their replication, through the quantitation or inhibition of two surface proteins, hemagglutinin (HA) and neuraminidase (NA), both biologically active *in vitro*.

Although it seemed obvious from the beginning that HA and NA activities involved binding to the same receptor, it was not clear that the binding and

TABLE 2. THE INFLUENZA VIRUS *in vitro*

A. Host systems

   minced chick embryo tissue culture—Smith (1936).

   chick embryo—Hoyle and Fairbrother (1937); Lush and Burnet (1937).

   chick embryo allantoic sac—Burnet (1941).

Blood-tinged allantoic fluid and the discovery of hemagglutination and a virion enzyme—

With Pasteurian sagacity, George Hirst (1941) noted agglutination of erythrocytes in allantoic fluid from infected chick embryos and postulated:

   1) the existence of cell receptors for virus, and

   2) the presence of an elution-inducing (receptor-destroying) enzyme.

Animal virologists were freed thereby from the fetters of *in vivo* study and gained a leg up on the bacteriophage aristocracy who could quantitate virus only by titration of infectivity.

Receptor binding and receptor destruction were dissociable activities (Briody 1948) carried by two different proteins, HA and NA (Laver 1964; Laver and Kilbourne 1966).

B. Plaquing systems

   chick embryo fibroblasts—Granoff (1955); Ledinko (1955).

   continuous (aneuploid) cell line—Sugiura and Kilbourne (1965).

eluting properties were contained in the same or separate proteins. However, Briody (1948) demonstrated that the capacities of virus to agglutinate red blood cells and to destroy receptors could be dissociated by heating virus at 55°C. Later, Laver (1964) achieved separation of the two structural proteins following detergent disruption of the viral membrane and electrophoresis on cellulose acetate strips, a finding soon confirmed by segregation of the proteins by genetic reassortment (Laver and Kilbourne 1966).

Granoff (1955) and Ledinko (1955) pioneered in the production of plaques in primary chick embryo monolayer cell cultures, but paradoxically, influenza virologists were the last aboard with respect to viral plaquing in continuous cell lines—essential for precise and reproducible titration of infective virus. This was first accomplished by Sugiura and Kilbourne (1965) in an aneuploid human conjunctival line.

RECOGNITION AND CHARACTERIZATION
OF VIRAL ANTIGENIC VARIATION

Only three years after the first isolation of influenza virus from humans in 1933, Magill and Francis (1936) described antigenic differences among influenza virus strains (Table 3). These observations were soon confirmed by Burnet (1937), Smith and Andrewes (1938), and Stuart-Harris et al (1938). It is of interest that an early attempt at vaccination of humans using the originally isolated (1933) WS strain was ineffective against strains epidemic in 1936–1937 (Smith 1937). Both Hirst and Francis separately noted and stressed the homogeneity of strains isolated within the same epidemic or epidemic period, although minor changes were noted—sometimes ascribed to differences in laboratory passage history. Earlier assessment of the mechanism and significance of antigenic variation was confounded by the limited methods available for measurement of virus neutralization (assay in intact animals), and a limited understanding of viral genetics and immunology. However, hypotheses were ingenious, including Smith and Andrewes' postulation of four antigenic sites (1938) (prophetic, in view of the later identification of sites A, B, C and D deduced from quaternary hemagglutinin structure and monoclonal antibody definition). Magill and Francis (1936) envisioned a rotating "mosaic" of antigens on the surface of the virus which revealed hidden antigens with each strain change. Interestingly, the first use of the *in vitro* hemagglutination-inhibition (HI) reaction failed to show significant differences among the strains studied (Hirst 1943). Note, however, that the method used most widely still in the demonstration of small antigenic differences among strains is the HI test of sera from infected ferrets.

It was the epidemic of 1946–1947 that demonstrated beyond doubt the degree and significance of antigenic variation of the virus, because preceding that epidemic there had been widespread vaccination of the United States military, yet the vaccine essentially failed to protect (Davenport 1967). So different were the strains of 1946 and 1947 that initial subtyping of human

TABLE 3. ANTIGENIC VARIATION

Antigenic differences among influenza A strains noted only four years after the first isolation of human viruses in 1932—Magill and Francis (1936).

Early vaccine failure with original WS strain in 1936–1937.

Strain homogeneity within epidemics—Hirst, Francis.

Postulation of four antigenic sites on basis of neutralization tests in animals—Smith and Andrewes (1938).

Rotating "mosaic" of antigens—Magill and Francis (1936).

Antigenic "drift"—progressive accumulation of point mutations affecting epitopic sites of HA and NA (reviewed in Webster and Laver 1983).

Antigenic "shift"—major, nonsequential change in HA or NA protein, a major contribution by Laver and Webster (1972) based on peptide mapping.

influenza A viruses designated them as "A prime," and somewhat later as "H1" in contrast to the "H0" designation of all earlier strains. Currently, H0 and H1 strains are lumped together in the H1 category, so that this significant bit of history (and its categorical and epidemiological implications) has been lost.

In both the degree and frequency of antigenic change, the influenza viruses appeared to be unique—an assumption buttressed by the early application of analysis with monoclonal antibodies to HA variation (Yewdell et al 1979). However, similar analysis of other viruses revealed a minor degree of antigenic variation in all viruses. Nonetheless, influenza virus inferred amino acid sequence data in concert with definition of HA (and somewhat later, NA) epitopes showed that natural variation during the "drift" of interpandemic periods resulted from a progressive accumulation of point mutations in peripheral parts of the molecule in which clusters of substitutions occurred (reviewed by Webster et al 1983). The term "antigenic drift" (originally, immunological drift) appears to have been introduced by Burnet in the 1955 edition of *Principles of Animal Virology* and to have been derived from the phenomenon of *genetic drift* "in which new characters tend to become normal for a species as a result of isolation and reduction in numbers."[1]

Laver and Webster (1972) made a profoundly important contribution through their early peptide mapping studies of extracted HA proteins. Those experiments demonstrated that major, i.e., subtype-defining or pandemic-associated, antigenic change in the HA represented the evolution or acquisition of proteins too different to be ascribed to the sequential acquisition of point mutations. Therefore, change of this magnitude was described as antigenic "shift."

---

[1] I am indebted to Dr. Frank Fenner for directing me to this interesting historical footnote.

INFLUENZA VIRUS EVOLUTION

Unlike the situation with any other virus, the antecedents of human influenza viruses (and those of some animal strains) are preserved in the freezers of a number of laboratories—an archeological archive of particular importance in the case of a virus apparently capable of antigenic recycling after long periods of absence (Masurel and Marine 1973) (Table 4). Although several investigators have used amino acid sequence data to construct phylogenetic dendrograms, Peter Palese in collaboration with Walter Fitch recognized the importance of tracing changes in the more conserved internal or nonstructural proteins of the virus in looking at its evolutionary clock. Their demonstration that surface (and potentially immune-selected) glycoproteins have changed at a faster rate has provided not only a confirmation of the driving force of immunoselection in influenza virus antigenic variation, but represents one of the more persuasive recent arguments for Darwinian evolution in general (Fitch et al 1991). On the other hand, Frank Horsfall's early recognition that at least minor antigenic variation need not be immunologically driven (Horsfall 1952) has been confirmed with influenza (Kilbourne 1978) and other viruses (Dietzschold et al 1983) and may require only a single point mutation (Both et al 1983).

Although the mutation rate of influenza virus may be no more rapid than that of other RNA viruses, its rate of evolution in nature certainly is. Separately, Easterday, Scholtissek, and Webster have emphasized the slower rate of change in the genes of animal influenza viruses as a probable consequence of shorter life spans and less immunoselective pressure.

TABLE 4. INFLUENZA VIRUS EVOLUTION

Archeological archive of viruses going back to 1900.

Evidence of recycling of pandemic viruses.

Influenza viruses evolve faster in nature than do other viruses.

Selection of HA and NA is driven by host antibody (Darwinian selection).

Non-immunoselected antigenic change—bio-selection for mutants may pleiotropically select for proteins changed in antigenicity—first emphasized by Horsfall (1952) and later by Kilbourne (1978) and Schild et al (1983).

Virion proteins evolve at different rates.

INFLUENZA VIRUS GENETICS

Mutants are the basis of microbial and viral genetics, and their existence was demonstrated early in the history of influenza virology (Table 5). However, early definitions of phenotype relied on imprecise markers manifest by *in vivo* replication. Much later, stable *in vitro* markers were described (Choppin and Tamm 1960). These "plus" and "minus" inhibitor sensitive mutants were found in varying ratios in wild-type virus and were an early indication of the dimorphic, "quasispecies" nature of influenza viruses. Other instances of influenza viral dimorphism were subsequently described (Kilbourne and Murphy 1960; Mowshowitz and Kilbourne 1975; Kilbourne 1978.)

The first and most fundamental discovery concerning the genetics of influenza viruses occurred early on when Burnet and Lind (1949), and later Hirst and Gotlieb (1953), presented evidence (not well received by those schooled in classical genetics) for high frequency "recombination" (intratypically) of strains of different phenotype that was most readily explained as the result of co-replication and exchange of discrete, noncontiguous genome segments. Early markers for genetic interchange were ambiguous, being dependent upon such polygenic characters as "virulence." Transfer of filamentous morphology between strains of different subtype provided striking evidence of reassortment of genes encoding other than external proteins (Kilbourne and Murphy 1960). However, with the confirmation that hemagglutinating and enzymatic functions were segregable and identified with two different surface proteins (Laver 1964; Laver and Kilbourne 1966), unequivocal evidence for reassortment was provided by the ready production of "antigenic hybrids" of the HA and NA molecules (Kilbourne et al 1967).

The application of the formal techniques of microbial genetics awaited the development of adequate plaquing systems. Simpson and Hirst (1968) pioneered the use of chick embryo fibroblasts and conditional lethal (temperature-sensitive) mutants. This approach was facilitated by the use of a continuous cell line in which reproducible results were more easily obtained (Sugiura and Kilbourne 1972; 1975). These and other studies of temperature-sensitive (ts) mutants (summarized in detail by Mahy 1983) established the existence of seven or eight recombination-complementation groups. Together with physicochemical evidence that virion RNA did exist in segments (Pons 1976; Bean and Simpson 1976; McGeoch et al 1976; Scholtissek 1976; Duesberg 1968; Pons and Hirst 1968), the assumption was made that each gene was the source of a monocistronic message. With two exceptions (RNAs 7 and 8 of influenza A and the NA and NS genes of influenza B virus), this proved to be the case. A critical discovery of Lamb and Choppin (1979) demonstrated first that both RNA 7 and RNA 8 of influenza A viruses contained overlapping coding regions using different reading frames in the transcription of M1 and M2 and of NS1 and NS2 proteins, respectively, involving splicing of their mRNAs, and later, that RNA 6 of influenza B virus encoded not only the NA protein, but also a smaller (NB) protein, functionally equivalent to M2 (Shaw et al 1983).

A significant advance in influenza virus genetics was the demonstration (Palese and Schulman 1976; Ritchey et al 1977) that influenza viruses differed sufficiently in the electrophoretic migration of their genes and gene products to be distinguished thereby and their genomes mapped—according to functions established previously by complementation groups and antigenic analysis. This technique received early practical application in the recognition of genetic reassortment in nature (Young and Palese 1979) and in the construction and genotyping of vaccine strains (Baez et al 1980).

An early and important contribution to analysis of influenza virus genes and to assessment of their relative conservation was made by Scholtissek and Rott (1969) using the technique of RNA-RNA hybridization. This technique was also a primary factor in the establishment of a new nomenclature for influenza A viruses in which the antigenically discrete H0, H1 and Hsw1 viruses were grouped together in the H1 subtype (WHO 1980). Similarly, human H3, equine Heq2, and Hav7 subtypes were found to have sufficient sequence homology to warrant their reassignment to a single H3 subtype.

Finally, recombinant DNA technology applied to this negative-stranded RNA virus enabled the production of pure proteins for analysis and for manipulation of their structure by directed alteration of their coding sequences (reviewed in Gething and Sambrook 1983). Most recently, "reverse genetics" has provided an approach to indirect manipulation of the virion RNA by altering or introducing new sequences into cDNA copies (Enami et al 1990).

TABLE 5. INFLUENZA VIRUS GENETICS

Influenza viruses were the first to be shown to have segmented genomes

Genetic reassortment occurs in nature.

Six genes are monocistronic—two are spliced during transcription.

Variation in electrophoretic migration rates of RNAs allows physical mapping of genome.

DNA cloning of genes permits isolation and study of individual gene products.

"Reverse genetics" facilitates gene manipulation.

INFLUENZA VIRAL REPLICATION

In this brief review of the history of influenza, it is possible to mention here only those "microbe hunters" who have made major contributions to the study of replication of the virus, bearing in mind that replication is not easily dissociated from studies of viral structure and genetics.

Influenza virus replication was initially defined by its anomalies (Table 6). In early attempts to quantify viral replication in non-plaquing systems, Horsfall (1955) recognized that most virus particles were noninfectious, and the Henles (1943) and others (reviewed in Schlesinger 1959) described viral "autointerference" and multiplicity reactivation. The first defective interfering (DI) particles were described by von Magnus (1951), and the mechanism of their evolution bears his name. Many years later, Nayak et al (1985) showed that DI particles have reduced amounts of viral RNA due to internal deletions chiefly involving the P (polymerase) genes. Other studies showed that viruses infective in one host system may prove defective in another. A notable early example is the abortive replication of influenza A viruses in mouse brain (Werner and Schlesinger 1954).

The essential role of post-translational proteolytic cleavage of the HA in the entry of the virus into cells was demonstrated by Lazarowitz and Choppin (1971). This cleavage was carried out by host cell proteases, and activation of infectivity was found to involve cleavage of a peptide bond with arginine or lysine in carboxyl linkage (Klenk et al 1977). The importance of an acid environment in the fusion of the HA during viral entry subsequent to exposure of a specific amino acid sequence at the amino terminus of HA2 was also demonstrated (Skehel et al 1982; Garten and Klenk 1983). The discovery that the virus uses a unique mechanism for mRNA synthesis involving the capture of host cell-capped RNA fragments as primers for virus transcription (Plotch et al 1981) was also a major contribution to the understanding of primary viral RNA transcription.

At the end stage of viral replication—the release of completed particles from the cell—Palese and co-authors (1974) demonstrated a clear role for the NA in cleavage of HA-binding sialic acid residues, consonant with earlier and later evidence that the NA has no role in viral entry.

TABLE 6. INFLUENZA VIRUS REPLICATION

Anomalies of replication were first evident

    Majority of particles noninfectious—Horsfall (1955).

    Viral interference—Henle and Henle (1943); Schlesinger (1959).

    von Magnus phenomenon (first DI particles)—1951; Nayak et al (1985).

HA role in replication

    HA cleavage for viral entry—Lazarowitz and Choppin (1971).

    Acid environment needed for fusion—Skehel et al (1982).

Transcription

    Host cell-capped primers required for transcription—Plotch et al (1981).

Viral exit and release—Palese et al (1974).

VIRAL STRUCTURE

Mosely and Wykoff (1946), and later Murphy and Bang (1952), visualized influenza virus particles by electron microscopy, while Morgan et al (1956) demonstrated the budding mechanism of viral detachment later found to characterize many enveloped viruses (Table 7).

Laver and Valentine (1969) first clearly demonstrated that the spike-like projections of the virion surface were morphologically distinguishable as two different proteins carrying either hemagglutinating or neuraminidase activity. The purification of these proteins in large quantity led to their availability for the important crystallographic studies described below.

Influenza viruses stand unique, again, in having twice appeared on the cover of *Nature* with representations of their major external glycoproteins. The importance of Wilson, Skehel, and Wiley's paper on the structure of the influenza virus hemagglutinin (1981) cannot be overestimated. Not only did this x-ray diffraction study of crystallized protein confirm previous suspicions about the location of the viral receptor site and evidence for the trimeric structure, but, correlated with amino acid sequence data from naturally arising variants, it also demonstrated four principal antigenically reactive sites on the peripheral globular portion of the molecule (Wiley et al 1981), also verified by study of mutants selected with monoclonal antibodies (Webster et al 1983).

Later (Varghese et al 1983), studies of the crystallized neuraminidase have been no less important, defining a tetrameric protein with an unusual "head" structure comprising six beta-sheets arranged in propeller-like formation.

Recently, location of a third glycoprotein (M2) in the viral membrane (Lamb et al 1985), which serves as an ion channel (Hay 1989), completes the structural definition of the epidemiologically important surface of the virus.

TABLE 7. VIRAL STRUCTURE

First visualization by electron microscopy—Mosely and Wykoff (1946); Murphy and Bang (1952).

Budding mechanism—Morgan et al (1956).

Distinction of HA and NA spikes—Laver and Valentine (1969).

Quaternary structure of HA—Wilson et al (1981).

Structure of crystallized NA—Varghese et al (1983).

IMMUNOLOGY

The immunology of influenza is complex and offers many lessons in natural immunoselection and host response (reviewed in Kilbourne 1978). Of special interest in the present review, however, is the concept of "original antigenic sin" as espoused by Francis et al (1953) (Table 8). This doctrine, more traditionally termed the "anamnestic response," refers to the initial priming of the immune system by the virus strain first encountered in life and the rapid secondary response to epitopes shared by the original and challenge viruses, but a typical (less effective) *primary* response to the novel epitopes of later variants. The anamnestic response is unusually important with influenza viruses because of the repeated infections that occur throughout life. Another legacy of recurrent viral challenges is the antigenic competition which occurs intravirionically between the HA and NA antigens (Kilbourne 1976; Johansson and Kilbourne 1987), resulting in an unbalanced immunological response in which HA is immunodominant.

TABLE 8. IMMUNOLOGY

Concept of "original antigenic sin"—Francis et al (1953). (The anamnestic response is especially important with recurrent virus).

Intravirionic antigen competition (HA immunodominant over NA)— Kilbourne (1976); Johansson et al (1987).

Virulence

"Virulence genes" have been sought since the first isolation of the virus, particularly since its reassortment potential was appreciated. Only a few notable advances are cited here (Table 9). Pursuant to the early studies of Burnet with the mouse neurovirulence marker, Fraser (1959) demonstrated its polygenic nature, soon confirmed by Mayer et al (1973) and Rott et al (1976). More definitive studies of neurovirulence showed the contributions of three genes in the mediation of maximum neurovirulence (Nakajima and Sugiura 1980). Although Kilbourne (1969) predicted the likelihood of reduced virulence (attenuation) of reassortants of virulent and avirulent parental viruses, Scholtissek et al (1979) found reassortants which exceeded either parent in virulence—emphasizing the importance of gene combinations or "constellations" (Rott et al 1979) in virulence. This is not to say, however, that even single nucleotide changes in single genes cannot have critical importance (Almond et al 1977). Rott et al (1980) repeatedly demonstrated the importance of mutations affecting HA cleavage in determining the virulence of avian viruses and also demonstrated the potential of the proteases of co-infecting bacteria for enhancing viral invasiveness. A dramatic example of the potential of HA mutations at the cleavage site was provided by Webster and colleagues (1986) in the case of an H_5N_2 avian virus which decimated domestic chicken flocks in the USA in 1983, based on a single point mutation in the HA. History repeated itself with a similar outbreak in Mexico in 1995 (Wuethrich 1995).

TABLE 9. VIRULENCE

Virulence is polygenic—Fraser (1959); Mayer et al (1973); Rott et al (1976).

Gene "constellations" are critical—Rott et al (1979); Scholtissek et al (1979).

HA is a critical determinant of virulence

HA cleavage in avian viruses—Rott et al (1980).

HA mutation affecting cleavage site caused fatal outbreak in chickens in 1983—Webster et al (1986).

MOLECULAR EPIDEMIOLOGY

The concept of a molecular epidemiology of viruses was first introduced by Kilbourne (1973) following the demonstration (Schulman and Kilbourne 1969) that the "cross-reactivity" of the new pandemic Hong Kong virus and its antecedent, the H_2N_2 Asian virus, was mediated by an unchanged NA (N_2) antigen, to which antibody in the population probably influenced the severity of the Hong Kong pandemic (Table 10). Since then, the contributions of viral genotyping and gene sequencing to the understanding of influenza epidemics have been many. A few examples include evidence for viral reassortment in humans (Palese and Young 1983), the appearance of 1950s-like, virtually unchanged "fixed" H_1N_1 virus in 1977 (Nakajima et al 1978; Scholtissek et al 1978), and the possible origin of human pandemic viruses from animal influenza viruses. The latter possibility was suggested soon after the 1957 pandemic by Rasmussen (1964) and independently by Andrewes (1959). That reassortment of human and animal viruses could, in fact, occur was shown in the laboratory by Tumova and Pereira (1965) for avian and human strains, and by Kilbourne (1968) for viruses of equine and human origin. Webster, Laver, and others deserve great credit for the accumulation of an enormous amount of information defining the epizootiology of influenza, and contributing to the credibility of the hypothesis. Most recently, Webster and colleagues presented evidence for the transfer not only of HA and NA genes from avian viruses in 1957 and of the NA gene in 1968, but also of transfer of PB1 genes to both human pandemic strains (Kawaoka et al 1989).

Scholtissek and Naylor (1989) made an arresting conceptual contribution with their suggestion that swine may be "mixing vessels" in which recombination between mammalian (human) and avian influenza viruses can occur prior to their transfer to man.

TABLE 10. MOLECULAR EPIDEMIOLOGY

Term first proposed in 1973 (Kilbourne) to describe the shift of H_2N_2 virus to H_3N_2. (Recognized by segregation of HA and NA antigens in hybrid reassortants.)

Return of unchanged 1950s H_1N_1 virus in 1977—Nakajima et al (1978); Scholtissek et al (1978).

Reassortment in nature of human viruses—Young and Palese (1979).

Influenza virus phylogeny.

SUMMARY

Depending on viewpoint, the history of influenza virology begins almost a century ago with the isolation of geflügelpeste virus, or 31 years later with the isolation of swine influenza virus, or one year after that with isolation of the virus from man.

Influenza is unique among diseases in the importance of historical perspective, and its causative virus is probably unique in the archival frozen record of its mutations and peregrinations during this past century. Indeed, the control of influenza (to the very limited degree that it has been accomplished) depends on knowledge of the immediate past history of its transgressions.

The entire field of virology owes much to influenza virologists: for discovery of RNA reassortment genetics, antigenic variation and virus evolution, defective interfering virus, viral hemagglutination, *in vitro* viral quantitation, and correlation of viral structure and function. Influenza virus remains the paradigm of emerging viruses—indeed, the only model to which we can turn in predicting the effect of novel agents in non-immune human populations. And influenza virus will return.

REFERENCES

Almond JW (1977). A single gene determines the host range of influenza virus. Nature 270: 617-618.

Andrewes CH (1959). Asian influenza: A challenge to epidemiology. In: Pollard M, ed. Perspectives in virology. New York: Wiley, 184-196.

Baez M, Palese P, Kilbourne ED (1980). Gene composition of high-yielding influenza vaccine strains obtained by recombination. J Infect Dis 141: 362-365.

Bean WJ, Simpson RW (1976). Transcriptase activity and genome composition of defective influenza virus. J Virol 18: 365-369.

Both GW, Shi CH, Kilbourne ED (1983). Hemagglutinin of swine influenza virus: A single amino acid change pleiotropically affects viral antigenicity and replication. Proc Natl Acad Sci USA 80: 6996-7000.

Briody BA (1948). Characterization of the "enzymic" action of influenza viruses on human red cells. J Immunol 59: 115-127.

Burnet FM (1937). Influenza virus of developing egg; differentiation of 2 antigenic types of human influenza virus. Aust J Exp Biol Med Sci 15: 369-374.

Burnet FM (1941). Growth of influenza virus in the allantoic cavity of the chick embryo. Aust J Exp Biol Med Sci 19: 291-295.

Burnet FM, Lind PE (1949). Recombination of characters between two influenza virus strains. Aust J Sci 12: 109-110.

Centanni (1902). Die Vogelpest; Beitrag zu dem durch Kerzen filtrirbaren Virus. Cent f Bakt, Bd. XXXI, S. 145.

Choppin PW, Tamm I (1960). Studies of two kinds of virus particles which comprise influenza A2 virus strains. I and II. J Exp Med 112: 895-921.

Davenport FM (1967). Specific prophylaxis: Inactivated influenza virus vaccines. Am Rev Respir Dis 83: 146-150.

Dietzschold B, Wunner WH, Wiktor TJ, Lopes AD, Lafon M, Smith CL, Koprowski H (1983). Characterization of an antigenic determinant of the glycoprotein that correlates with pathogenicity of rabies virus. Proc Natl Acad Sci USA 80: 70-74.

Duesberg PH (1968). The RNAs of influenza virus. Proc Natl Acad Sci USA 59: 930-937.

Enami M, Luytjes W, Krystal M, Palese P (1990). Introduction of site-specific mutations into the genome of influenza virus. Proc Natl Acad Sci USA 87: 3802-3805.

Fitch WM, Leiter JM, Li XQ, Palese P (1991). Positive Darwinian evolution in human influenza A viruses. Proc Natl Acad Sci USA 88: 4270-4274.

Francis T Jr (1940). A new type of virus from epidemic influenza. Science 92: 405-408.

Francis T Jr, Davenport FM, Hennessy AV (1953). A serological recapitulation of human infection with different strains of influenza virus. Trans Assoc Am Physicians 66: 231-239.

Francis T Jr, Quilligan JJ Jr, Minuse E (1950). Identification of another epidemic respiratory disease. Science 112: 495-497.

Fraser KB (1959). Features of the MEL x NWS recombination systems in influenza A virus. IV. Increments of virulence during successive cycles of double infection with the strain of influenza A virus. Virology 9: 202-214.

Garten W, Klenk H-D (1983). Characterization of the carboxypeptidase involved in the proteolytic cleavage of the influenza haemagglutinin. J Gen Virol 64: 2127-2137.

Gething M-J, Sambrook J (1983). Expression of cloned influenza virus genes. In: Palese P, Kingsbury DW, eds. Genetics of influenza viruses. New York: Springer-Verlag, 169-192.

Granoff A (1955). Plaque formation with influenza strains. Virology 1: 252.

Hay AJ (1989). The mechanism of action of amantadine and rimantadine against influenza viruses. In: Notkins AL, Oldstone MBA, eds. Concepts in viral pathogenesis. Vol 3. New York: Springer-Verlag, 361-367.

Horsfall FL Jr (1955). Reproduction of influenza viruses. Quantitative investigations with particle enumeration procedures on the dynamics of influenza A and B virus reproduction. J Exp Med 102: 441-473.

Henle W, Henle G (1943). Interference of inactive virus with the propagation of virus of influenza. Science 98: 87-89.

Hirst GK (1941). The agglutination of red cells by allantoic fluid of chick embryos infected with influenza virus. Science 94: 22-23.

Hirst GK (1943a). Studies of antigenic differences among strains on influenza A by means of red cell agglutination. J Exp Med 78: 407-423.

Hirst GK, Gotlieb T (1953). The experimental production of combination forms of virus. II. A study of serial passage in the allantoic sac of agents that combine the antigens of two distinct influenza A strains. J Exp Med 98: 53-70.

Hoyle L, Fairbrother RW (1937). Further studies of complement-fixation in influenza: Antigen production in egg-membrane culture and the occurrence of a zone phenomenon. Brit J Exp Pathol 18: 425-429.

Johansson BE, Moran TM, Kilbourne ED (1987). Antigen-presenting B cells and T_H cells cooperatively mediate intravirionic antigenic competition between influenza A virus surface glycoproteins. Proc Natl Acad Sci USA 84: 6869-6873.

Kawaoka Y, Krauss S, Webster RG (1989). Avian-to-human transmission of the PB1 gene of influenza A viruses in the 1957 and 1968 pandemics. J Virol 63: 4603-4608.

Kilbourne ED (1968). Recombination of influenza A viruses of human and animal origin. Science 160: 74-76.

Kilbourne ED (1969). Future influenza vaccines and the use of genetic recombinants. Bull WHO 41: 643-645.

Kilbourne ED (1973). The molecular epidemiology of influenza. J Infect Dis 127: 478-487.

Kilbourne ED (1976). Comparative efficacy of neuraminidase-specific and conventional influenza virus vaccines in the induction of antibody to neuraminidase in humans. J Infect Dis 134: 384-394.

Kilbourne ED (1978). Genetic dimorphism in influenza viruses: Characterization of stably associated hemagglutinin mutants differing in antigenicity and biological properties. Proc Natl Acad Sci USA 75: 6258-6262.

Kilbourne ED, Lief FS, Schulman JL, Jahiel RI, Laver WG (1967). Antigenic hybrids of influenza viruses and their implications. In: Pollard M, ed. Perspectives in virology. Vol 5. New York: Academic Press, 87-106.

Kilbourne ED, Murphy JS (1960). Genetic studies of influenza viruses. I. Viral morphology and growth capacity as exchangeable genetic traits. Rapid *in ovo* adaptation of early passage Asian strain isolates by combination with PR8. J Exp Med 111: 387-406.

Klenk HD, Rott R, Orlich M (1977). Further studies on the activation of influenza virus by proteolytic cleavage of the haemagglutinin. J Gen Virol 36: 151-161.

Klimov A, Prosch S, Schafer J, Bucher D (1992). Subtype H7 influenza viruses: Comparative antigenic and molecular analysis of the HA-, M- and NS-genes. Acta Virol 122: 143-161.

Lamb RA, Choppin PW (1979). Segment 8 of the influenza virus genome is unique in coding for two polypeptides. Proc Natl Acad Sci USA 76: 4908-4912.

Lamb RA, Choppin PW (1981). Identification of a second protein (M_2) encoded by RNA segment 7 of influenza virus. Virology 112: 729-737.

Lamb RA, Zebedee SL, Richardson CD (1985). Influenza virus M_2 protein is an integral membrane protein expressed on the infected-cell surface. Cell 40: 627-633.

Laver WG (1964). Structural studies on the protein subunits from three strains of influenza virus. J Mol Biol 9: 109-124.

Laver WG, Kilbourne ED (1966). Identification in a recombinant influenza virus of structural proteins derived from both parents. Virology 30: 493-501.

Laver WG, Valentine RC (1969). Morphology of the isolated hemagglutinin and neuraminidase subunits of influenza virus. Virology 38: 105-109.

Laver WG, Webster RG (1972). Studies on the origin of pandemic influenza. II. Peptide maps of the light and heavy polypeptide chains from the hemagglutinin subunits of A2 influenza viruses isolated before and after the appearance of Hong Kong influenza. Virology 48: 445-455.

Lazarowitz SG, Choppin PW (1975). Enhancement of the infectivity of influenza A and B viruses by proteolytic cleavage of the hemagglutinin polypeptide. Virology 68: 440-454.

Ledinko N (1955). Production of plaques with influenza viruses. Nature 175: 999.

Lush D, Burnet FM (1937). Influenza virus on developing egg; complement fixation with egg membrane antigens. Aust J Exp Biol Med Sci 15: 375.

Magill TP (1940). A virus from cases of influenza-like upper-respiratory infection. Proc Soc Exp Biol Med 45: 162-164.

Magill TP, Francis T Jr (1936). Antigenic differences in strains of human influenza virus. Proc Soc Exp Biol Med 35: 463-468.

Mahy BWJ (1983). Mutants of influenza virus. In: Palese P, Kingsbury DW, eds. Genetics of influenza viruses. New York: Springer-Verlag, 192-254.

Masurel N, Marine N (1973). Recycling of Asian and Hong Kong influenza A virus hemagglutinin in man. Am J Epidemiol 97: 44-49.

Mayer V, Schulman JL, Kilbourne ED (1973). Non-linkage of neurovirulence exclusively to viral hemagglutinin or neuraminidase in genetic recombinants of A/NWS (H0N1) influenza virus. J Virol 11: 272-278.

McGeoch D, Fellner P, Newton C (1976). The influenza virus genome consists of eight distinct RNA species. Proc Natl Acad Sci USA 73: 3045-3049.

Morgan C, Rose HM, Moore DH (1956). Structure and development of viruses observed in the electron microscope. III. Influenza virus. J Exp Med 104: 171-182.

Mowshowitz S, Kilbourne ED (1975). Genetic dimorphism of the neuraminidase in recombinants of H3N2 influenza virus. In: Barry RD, Mahy BWJ, eds. Negative strand viruses. Vol 2. London: Academic Press, 765-775.

Mosley VM, Wyckoff RWG (1946). Electron micrography of virus of influenza. Nature (London) 157: 263.

Murphy JS, Bang FB (1952). Observations with the electron microscope on cells of the chick chorioallantoic membrane infected with influenza virus. J Exp Med 95: 259-268.

Nakajima S, Desselberger U, Palese P (1978). Recent human influenza A (H1N1) viruses are closely related genetically to strains isolated in 1950. Nature 274: 334-339.

Nakajima S, Sugiura A (1977). Neurovirulence of influenza virus. Virology 81: 486-489.

Nakajima S, Sugiura A (1980). Neurovirulence of influenza virus in mice. II. Mechanism of virulence as studied in a neuroblastoma cell line. Virology 101: 450-457.

Nayak DP, Chambers TM, Akkina RK (1985). Defective-interfering (DI) RNAs of influenza viruses: Origin, structure, expression, and interference. Curr Top Microbiol Immunol 114: 103-151.

Palese P, Compans RW (1976). Inhibition of influenza virus replication in tissue culture by 2-deoxy-2,3-dehydro-N-trifluoroacetylneuraminic acid: Mechanism of action. J Gen Virol 33: 159-163.

Palese P, Schulman JL (1976). Differences in RNA pattern of influenza viruses. J Virol 17: 876-884.

Palese P, Schulman JL (1976). Mapping of the influenza virus genome: Identification of the hemagglutinin and the neuraminidase genes. Proc Natl Acad Sci USA 73: 2142-2146.

Palese P, Young JF (1983). Molecular epidemiology of influenza virus. In: Palese P, Kingsbury DW, eds. Genetics of influenza viruses, New York: Springer-Verlag, 321-336.

Plotch SF, Buoloy M, Ulmanen I, et al (1981). A unique cap(m^7GpppXm)-dependent influenza viron endonuclease cleaves capped RNAs to generate the primers that initiate viral RNA transcription. Cell 23: 847-858.

Pons MW (1976). A re-examination of influenza single and double stranded RNAs by gel electrophoresis. Virology 69: 789-792.

Rasmussen AF Jr, Hanson RP (1964). Avian myxoviruses and man. In: Hanson RP, ed. Newcastle disease virus: an evolving pathogen; proceedings of an international symposium. Madison: University of Wisconsin Press, 313.

Ritchey MB, Palese P, Schulman JL (1977). Differences in protein patterns of influenza A viruses. Virology 76: 122-128.

Rott R, Orlich M, Scholtissek C (1976). Attenuation of pathogenicity of fowl plague virus by recombination with other influenza A viruses, apathogenic for fowl. Non-exclusive dependence of pathogenicity on the hemagglutinin and neuraminidase. J Virol 19: 54-60.

Rott R, Orlich M, Scholtissek C (1979). Correlation of pathogenicity and gene constellation of influenza A viruses. III. Non-pathogenic recombinants derived from
highly pathogenic parent strains. J Gen Virol 44: 471-477.

Rott R, Reinacher M, Orlich M, et al (1980). Cleavability of hemagglutinin determines spread of avian influenza viruses in the chorioallantoic membrane of chicken
embryo. Arch Virol 65: 123-133.

Schafer W (1955). Vergleichende sero-immunogische Untersuchungen neber die Viren
de Influenza und klassichen Geflügelpest. Zentrabl Naturforsh 106: 81-91.

Schlesinger RW (1959). Interference between animal viruses. In: Burnet FF, Stanley
WM, eds. The viruses. New York, London: Academic Press, 157-194.

Schild GC, Oxford JS, de Johg JC, Webster RG (1983). Evidence for host cell selection of influenza virus antigenic variants. Nature 303: 706-709.

Scholtissek C, Harms E, Rohde W, Orlich M, Rott R (1976). Correlation between
RNA fragments of fowl plague virus and their corresponding gene functions. Virology 74: 332-344.

Scholtissek C, Naylor E (1989). Fish farming and influenza pandemics. Nature 331:
215.

Scholtissek C, Rohde W, von Hoyningen V, Rott R. (1978). On the origin of the
human influenza virus subtypes H2N2 and H3N2. Virology 87: 13-20.

Scholtissek C, Rott R (1969). Hybridization studies with influenza virus RNA. Virology 39: 400-407.

Scholtissek C, Vallbracht A, Fehmig B et al (1979). Correlation of pathogenicity and
gene constellation of influenza A viruses. II. Highly neurovirulent recombinants
derived from non-neurovirulent or weakly neurovirulent parent virus strains. Virology 95: 492-500.

Scholtissek C, von Hoyningen V, Rott R (1978). Genetic relatedness between the new
1977 epidemic strains (H1N1) of influenza and human influenza strains isolated
between 1947 and 1957 (H1N1). Virology 89: 613-617.

Schulman JL, Kilbourne ED (1969). Independent variation in nature of the hemagglutinin and neuraminidase antigens of influenza virus: Distinctiveness of the hemagglutinin antigen of Hong Kong/68 virus. Proc Natl Acad Sci USA 63: 326-333.

Shaw MW, Choppin PW, Lamb RA (1993). A previously unrecognized influenza B
virus glycoprotein from a bicistronic mRNA that also encodes the viral neuraminidase. Proc Natl Acad Sci USA 80: 4879-4883.

Shope RE (1931). Swine influenza. I. Experimental transmission and pathology. J
Exp Med 54: 349-372.

Simpson RW, Hirst GK (1968). Temperature-sensitive mutants of influenza A virus:
Isolation of mutants and preliminary observations on genetic recombination and
complementation. Virology 35: 41-49.

Skehel JJ, Bayley PM, Brown EB, Martin SR, Waterfield MD, White JM, Wilson IA,
Wiley CD (1982). Changes in the confirmation of influenza virus hemagglutinin at
the pH optimum of virus-mediated membrane fusion. Proc Natl Acad Sci USA 79:
968-972.

Smith W (1936). The complement-fixation reaction in influenza. Lancet 2: 1256-
1259.

Smith W (1937). Specific prophylaxis of influenza. Lancet 1: 575.

Smith W, Andrewes CH (1938). Serological races of influenza virus. Brit J Exp Pathol
19: 293-314.

Smith W, Andrewes CH, Laidlaw PP (1933). A virus obtained from influenza patients. Lancet 2: 66-68.

Stuart-Harris CH, Andrewes CH, Smith W (1938). A study of epidemic influenza with special reference to the 1936–1937 epidemic. In: Medical Research Council, Special Report, Series No. 228, London, 107.

Sugiura A, Kilbourne ED (1966). Genetic studies of influenza viruses. III. Production of plaque type recombinants with A_0 and A_1 strains. Virology 29: 84-91.

Sugiura A, Tobita K, Kilbourne ED (1972). Isolation and preliminary characterization of temperature-sensitive mutants of influenza virus. J Gen Virol 10: 639-647.

Taylor RM (1949). Studies on survival of influenza virus between epidemics and antigenic variants of the virus. Am J Public Health 39: 171-178.

Tobita K, Kilbourne ED (1975). Structural polypeptides of antigenically distinct strains of influenza B virus. Arch Virol 47: 367-374.

Tumova B, Pereira HG (1956). Genetic interaction between influenza A viruses of human and animal origin. Virology 27: 253-261.

Varghese JN, Laver WG, Colman PM (1983). Structure of the influenza virus glycoprotein antigen neuraminidase at 2.9 Å resolution. Nature 303: 35-40.

von Magnus P (1951). Propagation of PR8 strain of influenza A virus in chick embryos. I. The influence of various experimental conditions on virus multiplication. Acta Pathol Microbiol Scand 28: 250-277.

Webster RG, Kawaoka Y, Bean WJ Jr (1986). Molecular changes in A/chicken/Pennsylvania/83 (H5N2) influenza virus associated with acquisition of virulence. Virology 149: 165-173.

Webster RG, Laver WG, Air GM (1983). Antigenic variation among type A influenza viruses. In: Palese P, Kingsbury DW, ed. Genetics of influenza viruses. New York: Springer-Verlag, 127-168.

Werner GH, Schlesinger RW (1954). Morphological and quantitative comparison between infectious and non-infectious forms of the influenza virus. J Exp Med 100: 203-215.

WHO Memorandum (1980). A revision of the system of nomenclature of influenza viruses. Bull WHO 58: 585-591.

Wiley DC, Wilson IA, Skehel JJ (1981). Structural identification of the antibody-binding sites of Hong Kong influenza hemagglutinin and their involvement in antigenic variation. Nature 289: 373-378.

Wilson IA, Skehel JJ, Wiley DC (1981). Structure of the hemagglutinin membrane glycoprotein of influenza virus at 3 Å resolution. Nature 289: 366-373.

Wuethrich B (1995). Playing chicken with an epidemic. Science 267: 1594.

Yewdell JW, Webster RG, Gerhard W (1979). Antigenic variation of three distinct determinants of an influenza type A hemagglutinin molecule. Nature 279: 246-248.

Young JF, Palese P (1979). Evolution of human influenza A viruses in nature: Recombination contributes to genetic variation of H1N1 strains. Proc Natl Acad Sci USA 76: 6547-6551.

14

Discovery of Human Influenza Virus and Subsequent Influenza Research at the National Institute for Medical Research

JOHN SKEHEL

Research on influenza has been a major interest of the National Institute for Medical Research (NIMR) in England since its establishment in the early 1920s when, because of the enormous impact of the 1918–1919 influenza pandemic, the Medical Research Council decided to develop a program of research on "filterable viruses." The initial experiments at the Institute, however, did not involve influenza directly. Dr. P.P. Laidlaw chose instead to work on canine distemper in the hope that since both distemper and influenza are acute infections of the respiratory tract, his results with the animal virus system would be relevant to the human infection. With financial support provided in response to a public appeal in *The Field* magazine, Dr. Laidlaw and Mr. G.W. Dunkin, the veterinary superintendent at the newly acquired farm at Mill Hill in north London, established isolation quarters for dogs and began their studies of immunity to distemper. They were successful and in 1928, in collaboration with Burroughs Wellcome, an effective vaccine was produced. As models for influenza, the distemper studies were also successful but in a completely different way. Dogs did not prove to be particularly suitable animals for studies of immunity since they were occasionally naturally immune. In 1926 at the suggestion of Captain S.R. Douglas, the Head of Experimental Pathology, Bacteriology and Protistology, ferrets were introduced as experimental animals at Mill Hill. This suggestion was apparently based on the experience of Douglas' gamekeeper's acquaintances that distemper was frequently transmitted from infected dogs to ferrets and indeed their value was proven in the immunity studies; they were the only domesticated animals other than dogs infectible with distemper. In 1933, on the occasion of the first influenza epidemic since 1919, attempts were made by Drs. Laidlaw, C.H. Andrewes, and W. Smith to find an animal susceptible to influenza. The ferrets in the established colony at Mill Hill were again the only animals to develop the disease and human influenza viruses were discovered.

The discovery set the scene for research on influenza in the Institute until the end of World War II in a period when many of the procedures for handling influenza viruses experimentally were established. Serological tests were devised for detecting and comparing different isolates; viruses obtained from ferrets were successfully transmitted to mice and inadvertently to humans on at least one well-documented occasion; fowl plague virus, an avian influenza, was found to replicate in eggs; and studies of immunity and vaccination were initiated. However, it was not until after the war that the overriding importance of variations in virus antigenicity for these studies was appreciated. The failure of vaccines to provide protection during epidemics in 1947 highlighted the importance of differences between virus isolates and confirmed opinion at the World Health Organization that international surveillance of influenza variants would be essential for effective control by vaccination. Dr. Andrewes, by then Head of Bacteriology and Virology, was given the responsibility of establishing the World Influenza Centre to collaborate with national influenza laboratories throughout the world in the detailed comparison of influenza isolates and in recommending annually which viruses should be included in vaccines.

The Centre opened at the NIMR in 1948 under the direction of Dr. C.M. Chu who appropriately had a special interest in serological techniques and had become well known in influenza research for characterizing a group of nonspecific inhibitors of hemagglutination, now called the Chu inhibitors. On his return to China in 1950, his position was taken by Dr. A. Isaacs. Against the background of surveillance responsibility, each successive director of the World Influenza Centre has, by his research, influenced its character. None has done this more than Dr. Isaacs. He was an expert on influenza before he came to Mill Hill and initially continued his research on the variation in virus properties which occurs when they are grown in different tissues and in cells from different animals, a topic which has considerable practical importance in current vaccine production. His major research theme, however, became the mechanism of virus interference, and in 1957, his studies of interference between influenza viruses led to the discovery of interferon. "Virus interference" is a collective term for processes by which cells infected by one virus are protected from infection by a second which may or may not be related to the first. A number of mechanisms can be involved in such phenomena depending on the particular virus-cell system, but the discovery by Dr. Isaacs and his colleague Dr. J. Lindenmann was by far the most interesting and the most generally applicable. They showed that cells infected with partially inactivated influenza virus secreted a protein which, when absorbed by uninfected cells, rendered them less able to support the replication of either influenza or unrelated viruses. The potential importance of interferon in recovery from natural infections and as an antiviral therapeutic agent was obvious, and studies of the mechanism of its induction and its action occupied the remainder of Dr. Isaacs' short career.

At about the time of the Asian influenza pandemic in 1957, the World Influenza Centre became the responsibility of Dr. H.G. Pereira, who, despite favoring adenoviruses, nevertheless participated in some of the most important influenza research, particularly in influenza virus genetics. His studies with Dr. B. Tumova on the isolation of recombinant viruses from eggs co-infected with avian and human influenza viruses are the mechanistic foundation of the proposals that "new" influenza viruses, such as those of the Asian and Hong Kong pandemics, originated from mammalian or avian reservoirs. Dr. Pereira's interests in the relationships between human and avian viruses were shared by Dr. G.C. Schild, who succeeded him in the World Influenza Centre in 1970. Combined with his research on the antigenic properties of the influenza virus membrane glycoproteins, Dr. Schild's comparative serological analyses of avian, porcine, equine, and human influenza viruses were the basis for the system of influenza A subtype classification now in use, which recognizes the distinctive and independently variable antigenic properties of the hemagglutinin and neuraminidase glycoproteins, distinguishing 14 subtypes of hemagglutinin and 9 of neuraminidase. In this system, the viruses initially isolated in the 1933 epidemic, the Asian pandemic of 1957, and the Hong Kong pandemic of 1968 are members of the H_1N_1, H_2N_2 and H_3N_2 subtypes, respectively.

In 1975, Dr. Schild joined the National Institute of Biological Standards and Control as Head of Virology and I became Director of the World Influenza Centre. Since that time, research in our laboratory has mainly concerned the hemagglutinin glycoprotein of the influenza virus membrane. As an example of our current understanding of the mechanisms involved in influenza replication, I will describe briefly recent studies with colleagues in the laboratories of our long-term collaborators Dr. D. Wiley in Harvard and Dr. M. Knossow in Gif-sur-Yvette on the structural changes in the hemagglutinin required for its role in membrane fusion.

Hemagglutinins (HA) have two functions in the initial stages of virus infection: 1) they bind viruses to cells by complexing with sialic acid residues of cell surface glycoproteins or glycolipids; and 2) following endocytosis, they mediate fusion of virus and endosomal membranes, allowing transfer of the genome-transcriptase complex into the cell. Membrane fusion activity requires changes in HA structure that are specifically induced at endosomal pH (between pH 5 and 6), and recent x-ray crystallographic analyses of fragments derived from the molecule in its fusion pH conformation have revealed the nature of these changes.

Two soluble fragments have been used to prepare crystals (Figure 1). One is released by endoproteinase LysC digestion from the membrane distal region of the molecule, whereas the other derives from the fibrous central region upon thermolytic digestion. The structure of the membrane distal fragment was determined in studies of a complex with the Fab of a monoclonal antibody (Bizebard et al 1995) and found to be very similar to the structure of the equivalent region of native HA. The conclusion from this observation is that the changes in structure in the membrane distal region of HA induced at the pH of fusion primarily involve de-trimerization of otherwise structurally unchanged subunits.

By contrast, the central fibrous domain containing the region proposed to participate directly in membrane fusion, the "fusion peptide," is extensively reorganized (Bullough et al 1994). The fusion peptide itself is displaced more than 100Å from its buried location in the native molecule, in the formation of a new, over 105Å-long trimeric coiled-coil. In addition, the membrane proximal components of the native structure are reoriented through 180° (Figure 1). Only 30 residues of more than 500 in each subunit of the trimer form the same quaternary structure in the fusion pH and native structures.

As a consequence of this extensive reorganization, it seems likely that fusion-active HA interacts through its fusion peptides with the endosomal membrane while retaining association with the virus membrane through its membrane anchor sequence. In this way, it appears to form a molecular bridge between the two membranes which is presumably an intermediate required in the process of their fusion.

The importance of the pH of various intracellular vesicles for different stages of influenza replication has also been a central research interest of Dr. A.J. Hay, Director of the Centre since 1993. His group's analyses of the mechanism of action of the anti-influenza drug amantadine led to the description of the function of the virus-specified transmembrane protein, M_2 as a proton-specific, proton-activated membrane channel. Logically, sequence analyses of drug-resistant variants are now coupled in the Centre with the routine sequence and antigenic typing of viruses recovered from outbreaks throughout the world.

REFERENCES

Bizebard T, Gigant B, Rigolet P, Rasmussen B, Diat O, Bösecke, P, Wharton SA, Skehel JJ, Knossow M (1995). Structure of influenza virus haemagglutinin complexed with a neutralizing antibody. Nature 376: 92-94.

Bullough PA, Hughson FM, Skehel JJ, Wiley DC (1994). The structure of influenza haemagglutinin at the pH of membrane fusion. Nature 371: 37-43.

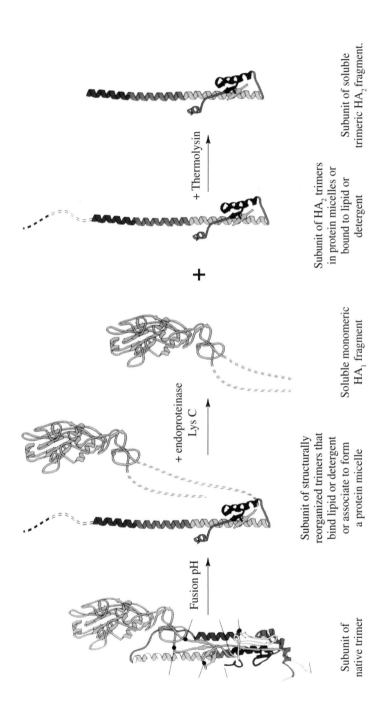

FIGURE 1. DERIVATION OF SOLUBLE FRAGMENTS OF HEMAGGLUTININ (HA) used for structural analyses and the structures of their subunits. In the diagram of the subunit of a native trimer, the locations of the N- and C-termini of the component polypeptides, HA_1 and HA_2; the site of cleavage by endoproteinase LysC, HA_1 27; the site of cleavage by thermolysin, HA_2 37, and the residues beyond which the structure of HA_1 is not available for the fusion pH form, HA_1 43 and HA_1 309, are indicated.

Subunit of native trimer

Fusion pH →

Subunit of structurally reorganized trimers that bind lipid or detergent or associate to form a protein micelle

+ endoproteinase Lys C →

Soluble monomeric HA_1 fragment

+

Subunit of HA_2 trimers in protein micelles or bound to lipid or detergent

+ Thermolysin →

Subunit of soluble trimeric HA_2 fragment.

Hepatitis B Virus

Hepatic cancer involving hepatitis viruses is likely to be the first human cancer preventable by vaccination. The hepatitis viruses can be separated clinically and biochemically, and the development of the first successful vaccine against hepatitis B virus is described by its developer, Saul Krugman. The subsequent events that led to a more defined vaccine through molecular engineering and the realization that hepatitis B virus infection was associated with hepatic cancer is presented by Maurice Hilleman, a major player in those efforts.

Many questions about hepatitis B virus and the other hepatitis viruses remain to be addressed. Mechanisms of viral persistence, host immune control or failure of control, as well as vaccine strategies against non-hepatitis B viruses are not clear and are the subject of vigorous research efforts by many investigators.

15

The History of Viral Hepatitis

SAUL KRUGMAN

The disease that is called viral hepatitis was first reported as "epidemic jaundice" in Greece during the fifth century BC, and many outbreaks were described during the nineteenth and twentieth centuries.

Epidemics of so-called "campaign jaundice" were prevalent during various wars. For example, more than 70,000 cases occurred among Union troops during the American Civil War, and many hundreds of thousands of cases occurred among American, British, and French troops during World War II. In retrospect, it is likely that these outbreaks were caused by hepatitis A virus.

The first recognized outbreak of hepatitis B occurred in Bremen, Germany, in 1883 during a smallpox immunization program. At that time, thousands of individuals were inoculated with vaccine prepared from glycerinated lymph of human origin. Of 1289 vaccinated shipyard workers, 191 (15%) developed jaundice several weeks to eight months post-inoculation. By contrast, jaundice was not observed in several hundred uninoculated employees. It is likely that the source of the human lymph was a hepatitis B carrier.

During the first half of the twentieth century, outbreaks of "long incubation period" hepatitis were observed in many countries of the world among several groups: patients who attended clinics for venereal disease, diabetes, and tuberculosis, persons who received blood transfusions, persons who were inoculated with mumps or measles convalescent serum, and military personnel who received yellow fever vaccine during World War II. These infections were caused by the use of hepatitis B-contaminated needles, syringes, blood, and blood products. At that time, the yellow fever vaccine contained human serum that was obtained from an unrecognized hepatitis B carrier.

Initially, knowledge of the pathogenesis of the disease was based on clinical and pathologic observations. In 1865 Virchow proposed that so-called "catarrhal jaundice" was caused by a plug of mucus in the ampulla of Vater, a concept that was accepted for many years. However, the presence of diffuse hepatic necrosis demonstrated by Eppinger in 1923 and by Rich in 1930 indicated that an infectious agent was the cause of the disease.

During the late 1930s and early 1940s, studies in human volunteers by various investigators in Europe and in the United States provided convincing evidence of a viral etiology. Their findings indicated that two viral agents were responsible for the epidemics of jaundice in military personnel during World War II. At that time, MacCallum proposed the terminology of hepatitis A virus (HAV) and hepatitis B virus (HBV). The evidence for two types of viral hepatitis was based on differences in the incubation period and on presumed differences in the mode of transmission (MacCallum and Bauer 1944; MacCallum and Bradley 1944).

Human volunteer studies during the 1940s revealed that feces or serum obtained during the acute phase of infectious hepatitis (hepatitis A) induced the disease after an incubation period of 15 to 33 days. In contrast, acute-phase serum obtained from patients with homologous serum jaundice (hepatitis B) induced the disease after a longer incubation period (50-160 days). Both Havens et al and Neefe et al failed to transmit HBV by oral inoculation of infectious serum that produced the disease regularly when administered parenterally. On the basis of these findings, it was assumed at that time that percutaneous inoculation was the only mode of transmission of HBV (Havens et al 1944; Havens et al 1947; Neefe et al 1945).

The observations of the epidemiology, natural history, and prevention of hepatitis A and hepatitis B during the 1940s were confirmed and extended by our Willowbrook hepatitis studies during the 1950s and 1960s (Krugman et al 1959; Ward et al 1959; Krugman et al 1962; Krugman et al 1967; Krugman and Giles 1970). The development of serum enzyme assays in 1955 provided sensitive markers of hepatitis infection with or without jaundice. The ability to detect hepatitis without jaundice enabled us to clarify further the epidemiology of the disease.

During the 1960s, we identified two types of viral hepatitis, each with distinctive epidemiological, clinical, and immunological features. One type, designated MS-1, resembled hepatitis A, and the other, designated MS-2, resembled hepatitis B. The MS-2 strain of HBV was infectious orally as well as parenterally. Therefore, contrary to the prevailing concept at that time, the findings indicated that HBV could be transmitted from person to person by intimate physical contact. These new epidemiological findings were confirmed during the late 1960s, when Blumberg's discovery of Australia antigen and its association with hepatitis B led to the development of specific tests for identification of hepatitis B infections (Blumberg et al 1969). The Willowbrook studies confirmed the existence of homologous immunity following HAV and HBV infections. However, one hepatitis agent did not confer immunity against the other virus.

The period between the mid-1960s and the mid-1980s proved to be a golden era in hepatitis A and hepatitis B research. Other types of hepatitis that were identified during that period included hepatitis D, parenterally transmitted non-A, non-B (hepatitis C) and enterically transmitted non-A, non-B hepatitis (hepatitis E).

HEPATITIS A

In 1973, Feinstone et al detected HAV particles in stools from patients who were infected with the MS-1 strain of the virus. The identification of hepatitis A antigen in stool specimens and in the liver of marmoset monkeys provided a source of hepatitis A antigen (HAAg). By the mid-1970s, highly sensitive immunoassays were developed for the detection of HAV and its antibody (anti-HAV). The development of IgM-specific anti-HAV serologic tests enabled physicians to distinguish recent HAV infection from a past infection.

The first successful cultivation of HAV was reported by Provost and Hilleman in 1979. This important contribution was confirmed by other investigators; and it provided the technology needed for the development of a hepatitis A vaccine. Studies with live attenuated and formalin-inactivated hepatitis A vaccines are currently in progress.

The RNA genome of HAV, a picornavirus, was molecularly cloned and partially sequenced during the 1980s. These clones can be used as sensitive probes for the detection of viral RNA in clinical samples and cell cultures. These new developments make it possible to prepare recombinant hepatitis A vaccines.

HEPATITIS B

The sequence of events that led to the identification of HBV began in 1965, when Blumberg, Alter, and Visnich identified an antigen in serum obtained from an Australian aborigine. The subsequent association of the Australia antigen with hepatitis B was confirmed by various investigators. Extensive studies during the 1960s and 1970s revealed the distribution of this antigen in various population groups and in patients whose diseases appeared to be unrelated to hepatitis B. Seroepidemiological surveys found Australia antigen to be present in the blood of 0.1 to 0.3% of healthy blood donors, in 10 to 20% of persons living in various African and Asian countries, in 10 to 15% of patients with leukemia or Hodgkin's disease, in 20 to 30% of institutionalized patients with Down's syndrome, and in about 20% of patients with viral hepatitis.

Convincing data to support the conclusion that Australia antigen was a hepatitis B antigen were reported by Prince (1968) and by Giles et al (1969). Thus, it was clear that the prevalence of the Australia antigen in African and Asian persons and in patients with leukemia and Down's syndrome indicated that these high-risk groups were likely to contract chronic hepatitis B infection.

Electron microscopic studies by Dane et al in 1970 demonstrated the presence of a 42-nm particle in the blood of patients with hepatitis B. The surface component of this particle was shown to be immunologically distinct from the core component. The accumulated evidence indicated that the Dane particle was HBV; its surface component was designated hepatitis B surface antigen (HBsAg). The core component contained endogenous DNA and hepatitis B core antigen (HBcAg). A third antigen related to infectivity, hepatitis B e antigen (HBeAg), was subsequently described by Magnius and Espmark in 1972.

The development and use of sensitive specific markers of hepatitis B infection enabled many investigators to clarify the natural history of hepatitis B infection. During the 1970s, it was observed that chronic hepatitis B infection was a cause of cirrhosis of the liver and primary hepatocellular carcinoma.

The development of tests to measure hepatitis B antibody provided the technology to identify units of blood containing the antibody. These antibody-positive units were used for the preparation of hepatitis B immunoglobulin (HBIG). Our preliminary studies indicated that HBIG was effective in preventing or modifying hepatitis B.

During the course of our studies on the natural history of hepatitis B in 1970, we developed, by serendipity, a crude heat-inactivated hepatitis B vaccine. This development was the result of an investigation designed to determine the effect of heat (boiling for one minute) on the infectivity of a 1:10 dilution of MS-2 serum in distilled water. A previous study had revealed the presence of HBV and HBsAg in MS-2 serum. The one-minute boil destroyed the infectivity of HBV, but the heat-inactivated MS-2 serum proved to be antigenic. Later, it was shown that the inactivated MS-2 serum was immunogenic and partially protective; it possessed the characteristics of an inactivated hepatitis B vaccine. By 1975, various investigators had developed plasma-derived subunit hepatitis B vaccines. The successful cloning of HBsAg by DNA recombinant technology led to the development and licensure of a yeast recombinant hepatitis B vaccine by 1987.

HEPATITIS D VIRUS

Before discussing non-A, non-B hepatitis, it is appropriate to comment briefly about hepatitis D virus (HDV). In 1977, Rizzetto and colleagues detected by immunofluorescence an antigen in liver cell nuclei and in serum of HBsAg-carriers who had chronic liver disease. This unique antigen (HDAg) was distinct from HBcAg. Studies in chimpanzees have revealed that it is a transmissible agent associated with, but distinct from, HBV. It is a 35-nm particle containing an RNA core and HBsAg as its surface component. Replication of this agent is initiated and maintained by the helper function provided by HBV infection. Delta infection of a chronic hepatitis B carrier may increase the risk of a complicating fulminant hepatitis B infection.

NON-A, NON-B HEPATITIS

Before the 1970s, HAV was recognized as the most common cause of enterically transmitted hepatitis and HBV was thought to be the most common cause of parenterally transmitted viral hepatitis. Despite routine screening of blood for HBsAg and the recruitment of voluntary blood donors, the problem of post-transfusion hepatitis remained unsolved. It soon became apparent that non-A, non-B agents were present in the blood of certain donors. In addition, by the 1980s an enterically transmitted non-A, non-B agent was recognized as the cause of certain water-borne epidemics of hepatitis. This agent has been designated the hepatitis E virus (HEV).

In 1989, two articles in *Science* provided convincing evidence that a parenterally transmitted non-A, non-B agent, termed hepatitis C virus (HCV), was identified by molecular cloning and presented characterization of its RNA genome (Choo et al 1989; Kuo et al 1989). In addition, a specific serologic assay was developed for the detection of antibody (anti-HCV) to a part of this virus.

CONCLUSION

Katkov and Dienstag (1991) summarized the present status of hepatitis research:

> Progress in hepatitis research has continued at a frenzied pace. Hepatitis virology has become the province of molecular biologists, molecular and immunological approaches have been joined to elucidate virus and host responses during acute and chronic infection, work on nonhuman examples of hepadnaviruses and on transgenically created models of HBV infection have taught us important lessons about HBV replication and pathogenesis, and both hepatitis C and hepatitis E viruses have been identified and characterized. Our understanding of the epidemiology of hepatitis B has become more sophisticated, prevention of hepatitis B with vaccine has become a routine part of preventive medicine, and chronic viral hepatitis is now considered a treatable disease!

The extraordinary progress in hepatitis research during the past 35 years was achieved by the contributions of many investigators. It has been an exciting and stimulating experience for me to be one of the participants in the studies of the natural history and prevention of viral hepatitis.

REFERENCES

Blumberg BS, Sutnick AI, London WT (1969). Australian antigen and hepatitis. JAMA 207: 1895-1896.

Blumberg BS, Alter HJ, Visnich S (1965). A new antigen in leukemia sera. JAMA 191:541-546.

Choo QL, Kuo G, Weiner AJ, Overby LR, Bradley DW, Houghton M (1989). Isolation of a cDNA clone derived from a blood-borne non-A, non-B viral hepatitis genome. Science 244: 359-362.

Dane DS, Cameron CH, Briggs M (1970). Virus-like particles in serum of patients with Australia-antigen-associated hepatitis. Lancet 1: 695-698.

Eppinger H (1923). Allgemeine und spezielle Pathologic des Ikterus. In: Kraus F and Brugsch T. Spezielle Pathologie und Therapie innerer Krankheiten, Vol 6. Berlin: Urban & Schwarzenberg, 97-340.

Feinstone SM, Kapikian AZ, Purcell RH (1973). Hepatitis A: detection by immune electron microscopy of a viruslike antigen associated with acute illness. Science 182: 1026-1028.

Giles JP, McCollum RW, Berndtson LW, Krugman S (1969). Viral hepatitis: relationship of Australia/SH antigen to the Willowbrook MS-2 strain. N Engl J Med 281: 119-122.

Havens WP, Jr, Ward R, Drill VA, Paul JR (1944). Experimental production of hepatitis by feeding icterogenic materials. Proc Soc Exp Biol Med 53: 206.

Havens WP, Jr, Kushlan SD, Green MR (1947). Experimentally induced infectious hepatitis; roentgenographic and gastroscopic observations. Arch Int Med 79: 457.

Katkov WN, Dienstag JL (1991). Prevention and therapy of viral hepatitis. Semin Liver Dis 11: 165-174.

Krugman S, Giles JP (1970). Viral hepatitis: new light on an old disease. JAMA 212: 1019-1029.

Krugman S, Giles JP, Hammond J (1967). Infectious hepatitis: evidence for two distinctive clinical, epidemiological and immunological types of infection. JAMA 200: 365.

Krugman S, Ward R, Giles JP (1962). The natural history of infectious hepatitis. Am J Med 32: 717-728.

Krugman S, Ward R, Giles JP, Bodansky O, Jacobs AM (1959). Infectious hepatitis: detection of virus during the incubation period and in clinically inapparent infection. N Engl J Med 261: 731-734.

Kuo G, Choo QL, Alter HJ, Gitnick GL, Redeker AG, Purcell RH, Miyamura T, Dienstag JL, Alter MJ, Stevens CE et al (1989). An assay for circulating antibodies to a major etiologic virus of human non-A, non-B hepatitis. Science 244: 362-364.

MacCallum FO, Bauer DJ (1944). Homologous serum jaundice: transmission experiments with human volunteers. Lancet 1: 622.

MacCallum FO, Bradley WH (1944). Transmission of infective hepatitis to human volunteers. Lancet 2: 228.

Magnius LO, Espmark A (1972). A new antigen complex co-occurring with Australia antigen. Acta Pathol Microbiol Scand [B] 80: 335-337.

Neefe JR, Gellis SS, Stokes J, Jr (1945) Homologous serum hepatitis and infectious (epidemic) hepatitis: Studies in volunteers bearing on immunological and other characteristics of the etiological agents. Am J Med 1: 3.

Prince AM (1968). An antigen detected in the blood during the incubation period of serum hepatitis. Proc Natl Acad Sci USA 60: 814-821.

Provost PJ, Hilleman MR (1979). Propagation of human hepatitis A virus in cell culture in vitro. Proc Soc Exp Biol Med 160: 213-221.

Rich AR (1930). The pathogenesis of the forms of jaundice. Bull Johns Hopkins Hosp 47: 338.

Rizzetto M, Canese MG, Aricò S, Crivelli O, Trepo C, Bonino F, Verme G (1977). Immunofluorescence detection of new antigen-antibody system (delta/anti-delta) associated to hepatitis B virus in liver and in serum of HBsAg carriers. Gut 18: 997-1003.

Virchow R (1865). Ueber das Vorkommen und den Nachweis des hepatogenen, insbesondere des katarrhabischen Icterus. Virchow's Arch Pathol Anat 32: 117.

Ward R, Krugman S, Giles JP, Jacobs AM, Bodansky O (1958). Infectious hepatitis: studies of its natural history and prevention. N Engl J Med 258: 407-416.

16

Immunology, Vaccinology, and Pathogenesis of Hepatitis B

Maurice R. Hilleman

Vaccinologists are ever alert to potential sources for antigens of infectious disease agents, since these antigens are the indispensable element in vaccine development. Just such an opportunity arose from the discoveries by Blumberg et al (1967) and Prince et al (1968) of 22-nm particles of hepatitis B surface antigen (Figure 1) in the blood of hepatitis B virus carriers. Early participants in hepatitis B vaccine development included Maupas et al (1978) of the Institute of Virology at Tours and our group at the Merck Institute (Hilleman et al 1978; Hilleman 1993) in Pennsylvania where vaccine development was initiated in 1968. Because acute infection with hepatitis B virus is rarely severe, the primary interest in hepatitis B vaccine is to prevent cirrhosis and cancer of the liver, which are long-term consequences of the hepatitis B carrier state.

The best evidence for a causal association of the carrier state with hepatocarcinoma is the finding of a 100-fold or greater increase in the occurrence of liver cancer in hepatitis B virus carriers compared with noncarrier matched controls (Beasley et al 1981). The appearance of cancer may be delayed for decades after the initial hepatitis B virus infection. About 10% of hepatitis B virus-infected individuals over five years of age develop a persistent carrier state, and the probability increases with diminishing age, reaching about 90% in infants born to highly infectious *e* antigen-positive maternal carriers (Figure 2).

Vaccine Development

First efforts made in our laboratories (Hilleman et al 1978; Hilleman 1993) and in those of Maupas et al (1978) were to isolate and purify the hepatitis B surface antigen from the plasma of hepatitis B virus carriers and to render it safe for use in a vaccine. This was a substantial task since blood may be infected with any of a large number of different viruses, requiring different

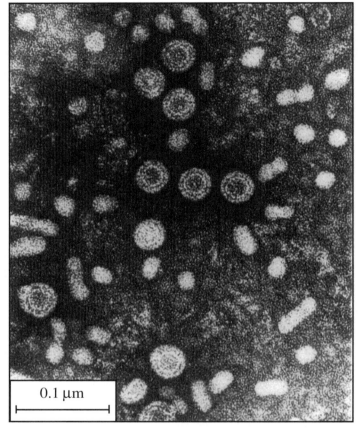

FIGURE 1. ELECTRON MICROGRAPH of partially purified human hepatitis carrier plasma showing virus and spherical and tubular surface antigen particles

procedures to assure inactivation. The Maupas vaccine and our vaccine, both in alum adjuvant, were licensed early in the 1980s.

It was evident from the start that the supply of suitable hepatitis B carrier plasma would not be sufficient to meet the needs to manufacture and that another source of antigen was needed. The new science of molecular biology was established by 1975 and collaborative arrangements were undertaken between Rutter and Hall's group (Valenzuela et al 1982) and our group (McAleer et al 1984) to develop an *in vitro* expression system for hepatitis B surface antigen in bacteria and in yeast. A recombinant yeast vaccine that contained the 226-amino acid S component of the surface antigen was licensed

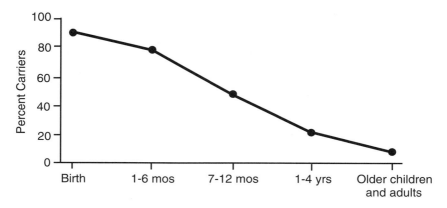

FIGURE 2. RELATIONSHIP OF AGE OF HEPATITIS B INFECTION and development of the carrier state (after Margolis, CDC, 1993)

in 1986. Michel and colleagues (see Adamowicz et al 1988) developed an expression system for hepatitis B surface antigen in mammalian Chinese hamster ovary cells at the Pasteur Institute, and a vaccine was developed by Adamowicz et al (1988) and by Girard et al (1990) at Pasteur Vaccines, now part of the Pasteur Mérieux Laboratories. This vaccine, now licensed and distributed, contains the pre-S2 component of the surface antigen as well as the S antigen.

The hepatitis B surface antigen particle (Figure 3) resembles a liposome that is made of dimeric S antigen and host cell phospholipid. There are about 50 dimers of the S antigen per 22-nm particle.

FIGURE 3. CONCEPTUAL STRUCTURE of hepatitis B surface antigen particle

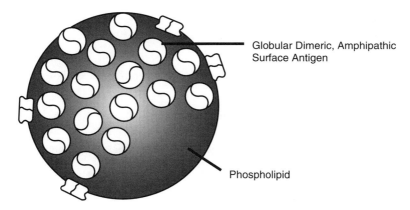

Globular Dimeric, Amphipathic
Surface Antigen

Phospholipid

VACCINE PERFORMANCE

Hepatitis B vaccine given in a regimen of 0, 1, 6 months or 0, 1, 12 months induces antibody in 95% or more of susceptible persons. All individuals who develop measurable antibody are protected on exposure to wild-type virus. Protective immunity lasts for at least 15 years, possibly a lifetime, even though the amount of antibody may decline to less than detectable levels. Immunologic memory is induced by the vaccine and a very rapid anamnestic recall follows exposure to the virus in nature or on reinjection of a single dose of vaccine (Table 1).

Remarkably, the vaccine is effective in preventing perinatal infection and the hepatitis B carrier state in 75% or more of infants born to *e* antigen-positive carrier mothers and vaccinated initially within hours of birth. Protective efficacy may be increased to at least 90% by coadministration of hepatitis B immune globulin at the time the first dose of vaccine is given in the newborn.

VACCINE APPLICATION

Vaccination programs to prevent hepatitis B worldwide were initiated nearly one and a half decades ago, giving primary emphasis to the newborn and to persons of defined high risk for infection. Hepatitis B immunization was recently added to the Expanded Program on Immunization of the World Health Organization. Hepatitis B is caused by a single virus that occurs only in the human species and is ultimately eradicable. However, eradication will be achieved only after many decades if the program is devoted to the newborn alone. For example, universal infant immunization in the United States would reduce the hepatitis B carrier state in the total population only by 30% in 15 years. Margolis et al (1991) recently proposed that routine infant

TABLE 1. HEPATITIS B VACCINE PERFORMANCE

95% or greater response after primary and secondary dose.

Antibody response means immunity.

Protective immunity lasts for 15 years or longer even if antibody is no longer detectable.

Long-term immunity is based on immunologic memory and anamnestic recall.

75% or more of infants born to *e* antigen-positive carrier mothers are protected by immediate postnatal vaccination alone.

Protective efficacy may be increased to 90% or more if hepatitis B immune globulin is also given at birth.

TABLE 2. HEPATITIS B VACCINE APPLICATION

Worldwide immunization began in 1981.

Primary interest has been in newborns and in older persons at high risk for infection.

Hepatitis B vaccine was added to the Expanded Program on Immunization of the World Health Organization.

Hepatitis B virus has but a single serotype and is eradicable.

Eradication would be facilitated by universal immunization of all susceptible individuals.

Prospective controlled studies to measure prevention of cirrhosis and cancer by vaccination were initiated about 10 years ago by the World Health Organization.

---

immunization must be complemented with vaccination of all adolescents and high-risk individuals since such a program would rapidly reduce hepatitis B occurrence in the United States. Inclusion of adolescents and high-risk individuals would mean a 70% decline in 15 years.

There is a strong basis for belief that prevention of the hepatitis B carrier state by vaccination will also prevent cirrhosis and cancer of the liver. Long-term prospective studies to measure prevention of cirrhosis and cancer were initiated in The Gambia (Fortuin et al 1993) and in Qidong County in China (Sun et al 1991; personal communication 1994) under the auspices of the World Health Organization (Table 2). These areas have a high endemic prevalence of hepatitis B infection. Vaccine was given to newborns without immune globulin and without prior testing for susceptibility. Prevention of the carrier state measured at four years after vaccination was found to be about 94% in the Gambian study (Fortuin et al 1993) and about 75% in China (Sun et al 1991; personal communication 1994) at five years after vaccination, findings which are indeed promising.

VACCINE ENIGMAS

Two findings in hepatitis B immunization have been enigmatic, one relating to early protection after vaccination, and the other to "immune carriers."

EARLY PROTECTION

It is difficult to explain protection against hepatitis B through perinatal vaccination based on an antibody response alone since humoral immunity may require weeks to months to appear. A likely explanation may be found in recent studies (Schirmbeck et al 1994a, 1994b) showing that the liposomal hepatitis B antigen, even though exogenous, can be processed by the cytosolic

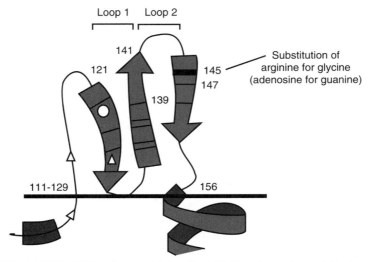

FIGURE 4. MUTATION in the second loop of hepatitis B surface antigen *a* (after Howard et al 1988)

class I pathway, which elicits a CD8[+] cytotoxic T-cell response, as well as by the class II pathway, which induces T-helper lymphocytes. Cytotoxic T-cell immunity appears within days after vaccination and it is reasonable to assume that cytotoxic T-cell clearance of virus-infected cells is the mechanism for protective immunity before a humoral response is in place.

"IMMUNE CARRIERS"

A substantial portion of individuals who were satisfactorily immunized against hepatitis B are now being found to be carriers and to circulate escape mutants of the virus of different antigenic specificities despite the presence of antibody against the native virus (Carman et al 1993). The most common change in surface antigen (Figure 4) is caused by a mutation from guanine to adenosine that brings about substitution of arginine for glycine in the group-specific *a* antigen loop against which antibody is ordinarily directed (Howard et al 1988). It is not known whether persons who are carriers of the mutant virus are contagious.

It is possible (Table 3) that the selective immune forces generated by routine use of hepatitis B vaccine will eventually give rise to transmissible viruses of altered specificity, and that vaccine of more than a single specificity may be required for immunization in the future. Selective advantage may also be given to mutant viruses by coadministration of hepatitis B immune globulin together with the vaccine.

TABLE 3. ENIGMAS OF VACCINATION against hepatitis B

Enigma 1

Early immunity—Hepatitis B vaccine prevents infection/carrier state in babies born to *e* antigen-positive carriers. It takes weeks to months for antibody to appear. Liposomal hepatitis B vaccine can be processed by the cytosolic class I pathway, which might induce a cytotoxic T-cell response within days.

Enigma 2

"Immune Carriers"—A portion of vaccinated persons may be carriers of hepatitis B virus despite circulating antibody against the surface antigen.

Antibody is directed against native virus and the circulating virus is a mutant of different antigenic specificity.

Survival of such escape mutants may be increased by coadministration of hepatitis B immune globulin together with vaccine at birth.

It is not known whether carriers of surface antigen mutant virus are infectious, but there might be need for vaccine of more than one antigenic specificity in the future.

HEPATITIS B VIRUS

The mechanics of hepatitis B virus replication and mutation were incompletely understood, mostly because of lack of a simple *in vitro* system for viral propagation. This is being changed now by the application of the tools of molecular biology to the problem.

THE VIRUS

The 60-nm enveloped hepatitis B virus bears a partially double-stranded DNA genome (Figure 5) that is circular, and the genetic flow is from DNA to RNA to DNA. There is no viral integrase, and viral replication is episomal. There are four open reading frames that include the pre-core and core regions of the nucleocapsid; the pre-S1, pre-S2, and S regions of the surface antigen; the polymerase-reverse transcriptase region; and the X transactivator region. The pre-core protein is truncated following transcription, allowing it to be secreted and circulated in the body fluids as soluble *e* antigen.

MUTATION

Hepatitis B virus is subject to a high mutation rate (Table 4) as a result of errors in reverse transcription that is reflected especially in the pre-core/core gene. Hepatitis B virus may exist in the host as a swarm of multiple

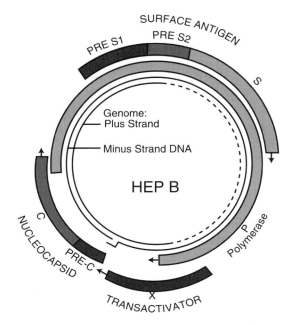

FIGURE 5. GENOME AND OPEN READING FRAMES of human hepatitis B virus

variants or quasi-species (Carman et al 1993). Viral variability, however, is restrained since so much information is packed into a small genome and there is extensive overlap in the open reading frames. These genetic overlaps provide a large chance for deleterious and lethal consequences of viral mutation and for back-selection on transmission to susceptible hosts.

PATHOGENESIS

The specifics that are responsible for the pathogenesis of hepatitis B are not established in detail, but a scenario (Table 5) can be created that is consistent with contemporary observations (Hilleman 1994): Hepatitis B virus replicates mainly in the hepatocytes. The virus is not cytopathic and the persistent

TABLE 4. MUTATION of hepatitis B virus

Errors in reverse transcription give a high mutation rate, especially in the pre-core/core gene.

Variants may exist as swarms or quasi-species.

Small size of viral genome, together with extensive overlap in open reading frames, limits viability of variants and facilitates back-selection on transmission.

TABLE 5. LIKELY SCENARIOS of hepatitis B pathogenesis

Likely scenarios:

Hepatitis B virus is not cytopathic, and clinically inapparent persistent infection may exist for prolonged periods.

Pathologic damage in hepatitis B is due to host immune response against virus antigen and not to the virus itself.

Damage is directed mainly against an antigenic specificity shared between *e* antigen and nucleocapsid core antigen that is presented on the surface of infected hepatocytes.

Cytotoxic T cells recognize and destroy infected hepatocytes.

Chronic active hepatitis fluctuates between remission and exacerbation depending on the degree of host cytotoxic T-cell activity (conditioned by percentage of hepatocytes displaying each specificity when more than a single virus is present).

Mechanisms

1. *Remission* may follow mutation of *e*/core nucleocapsid antigen to a new antigenic specificity.

    Exacerbation may follow appearance of cytotoxic T-cell activity against the new immunologic specificity.

2. *Remission* may follow induction of tolerance to the specificity of the cytotoxic T cells induced by soluble *e* antigen.

    Exacerbation may follow loss of tolerance.

3. Soluble *e antigen* may escape to the fetus *in utero* and establish tolerance that facilitates induction of the carrier state. Tolerance probably can also develop on infection in later life.

    The basis for tolerance probably resides in T-helper cells (Milich et al 1993) rather than in cytotoxic T cells themselves.

carrier state may be present for very prolonged periods without clinically apparent disease. The pathologic damage seen in chronic active hepatitis is brought about by the host response to the virus and not by the virus itself. The mechanism for damage is based mainly on presentation of hepatitis antigen of *e*/core specificity on the surface of infected liver cells that are recognized by cytotoxic T cells of corresponding specificity and effect their destruction (Carman et al 1993).

Chronic active hepatitis commonly presents a changing picture characterized by alternating clinical exacerbation and remission depending upon the degree of cytotoxic T-cell activity (Carman et al 1993). *One mechanism* for clinical remission is by mutation of the predominant virus to a new *e*/core nucleocapsid immunologic specificity that evades the activity of the resident cytotoxic T cells. Exacerbation may follow when a cytotoxic T-cell response is mounted against the specificity of the mutant virus. A *second mechanism* for exacerbation and remission may result from immunologic tolerance (Milich et al 1993) of cytotoxic T cells induced by circulating *e* antigen and directed against the shared *e*/core specificity of the nucleocapsid antigen. Loss of *e* antigen tolerance through mutation may lead to exacerbation of disease. Induction of tolerance against the new mutant specificity may again bring about clinical remission.

Tolerance to *e* antigen is of special importance in facilitating establishment of the carrier state in infants born to *e* antigen-positive mothers and probably also in persons who acquire infection later in life (Milich et al 1993). Tolerance may be established directly in the cytotoxic T cells themselves, but, more likely, indirectly in the specific T-helper cells (Milich et al 1993).

CARRIERS AND SURFACE ANTIGEN ANTIBODY

It remains unclear why hepatitis B carriers do not develop antibody against the surface antigen. Possible considerations (Table 6) include a state of tolerance to the surface antigen in carriers or that the antibody is obscured by a great excess of surface antigen. A more likely explanation (Barnaba et al 1990) is based on the fact that B cells are antigen-presenting cells, such that specific B cells recognize, process, and present exogenous hepatitis B surface antigen on the cell surface via the class I pathway, which brings about recognition and destruction by cytotoxic T lymphocytes and thereby eliminates the particular B cells that are the source of antibody against surface antigen.

TABLE 6. FAILURE OF HEPATITIS B CARRIERS to develop antibody against the surface antigen

Conventional wisdom:

Tolerance for surface antigen or presence of excess antigen obscures presence of antibody.

Present likely explanation:

B cells are antigen-presenting cells and engulf virus, exogenous hepatitis B antigen is processed by the class I pathway and presented on the B-cell surface, and cytotoxic T cells recognize and destroy B cells that otherwise would produce specific antibody.

Carriers of hepatitis B virus infection continue to produce antibody against the core antigen.

Persistent carriers of hepatitis B virus infection continue to produce anti-core antibody in contrast to surface antigen antibody.

CANCER

End stages of the pathologic changes in the hepatitis B carrier state are cirrhosis and cancer of the liver (Table 7). Both are slow to develop and may be delayed for decades after the initial infection. Cirrhosis and cancer of the liver appear to be separate entities caused by destruction of the liver parenchyma in cirrhosis and by mutative changes that lead to cancer through integration of fragments of viral DNA into the host cell genome.

The mechanisms for induction of hepatocarcinoma are not known with certainty, but there are two views. One holds that the continuing death and regeneration of liver parenchyma increases the random chance for a mutational event during hepatocyte replacement that is responsible for neoplastic transformation (Yoffe and Noonan 1992; Lau and Wright 1993). An alternative view (Koshy and Meyer 1992; Feitelson 1992) is that integration of viral

TABLE 7. HEPATITIS B CIRRHOSIS AND CANCER

End stages of pathologic changes in the hepatitis B infection are cirrhosis and/or cancer of the liver.

Cirrhosis and hepatocarcinoma appear to be separate entities:

Cirrhosis results from destruction of liver parenchyma.

Cancer may result from integration of fragments of viral DNA into the host cell genome.

Two views for generation of hepatocarcinoma:

Conventional: Continuing hepatocyte death increases random chance for neoplastic transformation on regeneration of liver cells.

Alternative: Oncogenic integration of viral DNA into the host cell genome.

DNA into the host cell chromosomes is responsible. The latter is a more satisfying explanation and is supported by the improved understanding of the molecular events that follow genetic integration of viral DNA.

As pertains to the DNA integration hypothesis, hepatitis B virus does not possess an oncogene (Lau and Wright 1993), and hepatitis B virus integration involves insertion of DNA fragments into the cell genome at random sites. A prime suspect in the induction of cancer by hepatitis B virus is the X gene (Koshy and Meyer 1992; Feitelson 1992; Balsano 1993), which is a potent transactivator. The pre-S2/S gene may also be involved in transactivation (Feitelson 1992) (Table 8).

TABLE 8. A HYPOTHESIS FOR VIRAL DNA INTEGRATION in cancer caused by hepatitis B virus

Hepatitis B virus does not possess an oncogene.

DNA fragments of virus are inserted into the host cell genome at random sites.

A prime suspect is the X gene which is a potent transactivator (hepatitis B pre-S2/S may also transactivate).

Patients may carry integrated viral DNA even though the carrier state is resolved.

Neoplasia may be a probability event most likely to occur in hosts with the largest amount of integrated viral DNA.

Possible mechanisms for oncogenesis:

1. *Cis* or *trans*-activation of protooncogenes or mutational insertion into them.

2. Mutational insertion into anti-oncogenes or binding to an anti-oncogene product (e.g., p53).

3. *Trans*-activation of viral or cellular promoters by integrated X or pre-S2/S gene.

It is possible that most persons, once infected with hepatitis B virus, carry some integrated viral DNA even though clinical recovery is complete. Neoplastic transformation is likely a probability event and is most likely to occur in hosts that have the largest amount of integrated viral DNA. Though the mechanism for viral oncogenesis in hepatitis B is unknown, there are at least three possibilities. The first is by *cis*- or *trans*-activation of protooncogenes of the host or by mutational insertion into them. The second is by mutational insertion into anti-oncogenes or by binding of a normal anti-oncogene protein, such as the p53 product, by hepatitis B virus protein. The third mechanism is by transactivation of viral and cellular promoters by integrated viral X gene or pre-S2/S gene. Clarification of the precise mechanism for oncogenic transformation will rely heavily on application of the sophisticated tools of molecular biology.

CLOSING

Hepatitis B virus is a disease of ancient origins that may cause fulminant disease or chronic persistent infection, which often terminates in death from cirrhosis or cancer of the liver. The solution to the problem of hepatitis is remarkably simple and is through application of the effective vaccine. The vaccine is being increasingly applied worldwide and may, in a future century, result in complete eradication of the virus. The host immune response in hepatitis B brings about chronic immunologic destruction of the liver in a situation in which virus and liver cells might otherwise have survived in silent harmony. Cirrhosis and cancer of the liver appear to be entirely separate entities. Cirrhosis is caused by continuing destruction of the hepatocytes. Cancer may result from integration of fragments of viral DNA into the chromosomes of the hepatocytes, causing a different set of events. It is possible that virus-induced cancer occurs in individuals who have fully recovered from the viral infection but who retain fragments of viral DNA integrated into their chromosomes.

REFERENCES

Adamowicz Ph, Tron F, Vinas R, Mevelec MN, Diaz I, Couroucé AM, Mazert MC, Lagarde D, Girard M (1988). Hepatitis B vaccine containing the S and PreS-2 antigens produced in Chinese hamster ovary cells. In: Zuckerman AJ, ed. Viral hepatitis and liver disease. New York: Alan R Liss, 1087-1090.

Balsano C, Billet O, Bennoun M, Cavard C, Zider A, Grimber G, Natoli G, Briand P, Levrero M (1993). The hepatitis B virus X gene product transactivates the HIV-LTR in vivo. Arch Virol 8 (Suppl): 63-71.

Barnaba V, Franco A, Alberti A, Benvenuto R, Balsano F (1990). Selective killing of hepatitis B envelope antigen-specific B cells by class I-restricted exogenous antigen-specific T lymphocytes. Nature 345: 258-260.

Beasley RP, Hwang LY, Lin CC, Chien CS (1981). Hepatocellular carcinoma and hepatitis B virus: A prospective study of 22,707 men in Taiwan. Lancet 2 :1129-1133.

Blumberg BS, Gerstley BJS, Hungerford DA, London WT, Sutnich AI (1967). A serum antigen (Australia antigen) in Down's syndrome, leukemia and hepatitis. Ann Int Med 66: 924-931.

Carman W, Thomas H, Domingo E (1993). Viral genetic variation: hepatitis B virus as a clinical example. Lancet 341: 349-353.

Feitelson M (1992). Hepatitis B virus infection and primary hepatocellular carcinoma. Clin Microbiol Rev 5: 275-301.

Fortuin M, Chotard J, Jack AD, Maine NP, Mendy M, Hall AJ, Inskip HM, George MO, Whittle HC (1993). Efficacy of hepatitis B vaccine in the Gambian expanded programme on immunisation. Lancet 341: 1129-1131.

Girard M, Adamowicz P, Vinas R, Mevelec MN, Tron F, Mazert MC, Fritzell B (1990). Development and clinical experience of a recombinant hepatitis B vaccine containing preS2 and S proteins. In: Chearamonte M, Floreani A, Fagiuoli S, Naccarato R, eds. Progressi clinici: Medicine. Vol 5. Padova: Piccin editore, 271-278.

Hilleman MR, Bertland AU, Buynak EB, Lampson GP, McAleer WJ, McLean AA, Roehm RR, Tytell AA (1978). Clinical and laboratory studies of HBsAg vaccine. In: Vyas GN, Cohen SN, Schmid R, eds. Viral hepatitis. Philadelphia: Franklin Institute Press, 525-537.

Hilleman MR (1993). Plasma-derived hepatitis B vaccine: A breakthrough in preventive medicine. In: Ellis R, ed. Hepatitis B vaccines in clinical practice. New York: Marcel Dekker, 17-39.

233

Hilleman MR (1994). Comparative biology and pathogenesis of AIDS and hepatitis B viruses: Related but different. AIDS Res Hum Retroviruses 10: 1409-1419.

Howard CR, Stirk HJ, Brown SE, Steward MW (1988). Towards the development of synthetic hepatitis B vaccines. In: Zuckerman AJ, ed. Viral hepatitis and liver disease. New York: Alan R. Liss, 1094-1101.

Koshy R, Meyer M (1992). Oncogenicity of hepatitis B virus. Rev Med Virol 2: 131-140.

Lau JYN, Wright TL (1993). Molecular virology and pathogenesis of hepatitis B. Lancet 342: 1335-1340.

Margolis H, Alter M, Krugman S (1991). Strategies for controlling hepatitis B in the United States. In: Hollinger FB, Lemon SM, Margolis HS, eds. Viral hepatitis and liver disease. Philadelphia: Williams & Wilkins, 720-722.

Maupas Ph, Goudeau A, Coursaget P, Drucker J, Barin F, André M (1978). Immunization against hepatitis B in man: A pilot study of two years' duration. In: Vyas GN, Cohen SN, Schmid R, eds. Viral hepatitis. Philadelphia: Franklin Institute Press, 539-556.

McAleer WJ, Buynak EB, Maigetter RZ, Wampler DE, Miller WJ, Hilleman MR (1984). Human hepatitis B vaccine from recombinant yeast. Nature 307: 178-180.

Milich DR, Jones J, Hughes J, Maruyama T (1993). Role of T-cell tolerance in the persistence of hepatitis B virus infection. J Immunotherapy 14: 226-233.

Prince AM (1968). An antigen detected in the blood during the incubation period of serum hepatitis. Proc Natl Acad Sci USA 60: 814-821.

Schirmbeck R, Melber K, Kuhröber, Janowicz ZA, Reimann J (1994a). Immunization with soluble hepatitis B virus surface protein elicits murine H-2 Class I-restricted CD8[+] cytotoxic T lymphocyte responses in vivo. J Immunol 152: 1110-1119.

Schirmbeck R, Melber K, Mertens T, Reimann, J (1994b). Selective stimulation of murine cytotoxic T cell and antibody responses by particulate or monomeric hepatitis B virus surface (S) antigen. Eur J Immunol 24: 1088-1096.

Sun Z(T), Zhu Y, Stjernsward J, Hilleman M, Collins R, Zhen Y, Hsia CC, Lu J, Huang F, Ni Z, Ni T, Liu GT, Yu Z, Liu Y, Chen JM, Peto R (1991). Design and compliance of HBV vaccination trial on newborns to prevent hepatocellular carcinoma and 5-year results of its pilot study. Cancer Detect Prev 15: 313-318.

Valenzuela P, Medina A, Rutter WJ, Ammerer G, Hall BD (1982). Synthesis and assembly of hepatitis B virus surface antigen particles in yeast. Nature 298: 347-350.

Yoffe B, Noonan CA (1992). Hepatitis B virus. New and evolving issues. Dig Dis Sci 37: 1-9.

Vignette

Retroviruses

More than 7 million people worldwide have contracted acquired immune deficiency syndrome (AIDS) in the last two decades, and more than 21 million are living with the disease or the retrovirus that leads to it—human immunodeficiency virus (HIV). Some 8,500 people are newly infected each day, or more than 5 people every minute, as AIDS expands faster than ever.

Retroviruses, unlike most other viruses and all cellular organisms, carry their genetic blueprint in the form of RNA. By means of a special enzyme called reverse transcriptase, they use RNA to synthesize DNA, a reversal of the usual cellular processes of transcription of DNA into RNA, making it possible for genetic material from a retrovirus to become permanently incorporated into the DNA genome of an infected cell.

The ever-expanding spread of AIDS mandates the study of retroviruses in humans and animals in an attempt to curb this present-day plague. Unlike the vaccines developed for smallpox, poliomyelitis, measles, yellow fever, and rubella, which are based on the use of attenuated viruses that establish an infection without disease, an effective AIDS vaccine cannot permit virus replication. Moreover, the use of killed virus, which generally provides only a limited group of antigens in the face of constantly mutating virus, will most likely not prove to be protective against all the emerging strains of HIV. Thus, new approaches to vaccine development are required. John Moloney, a major contributor in the history of retroviruses, details the numerous retroviruses of animals and humans discovered since the beginning of this century. Robert Gallo, the codiscoverer of the first retrovirus in man, describes the struggle that may lead to the vaccination and control of all retrovirus-induced human diseases.

17

The History of Retroviruses

J.B. MOLONEY

The history of retroviruses could well describe the history of the oncornaviruses, that is, the RNA tumor viruses. About 220 years ago, the French researcher Peyrilhe of the Royal College of Surgeons in Paris presented a paper on the cause of cancer to the Academy of Sciences in Lyon. His dissertation proposed that cancer was caused by a contagious agent. Peyrilhe had attempted to transmit human cancer to his dog by aspirating fluid from a mammary tumor of one of his female patients and inoculating the fluid subcutaneously into the dog. Several days later, Peyrilhe's manservant, feeling very sorry for the animal, drowned the suffering dog which had developed a massive local infection. This experiment was the first to show the early interest in the so-called "cell-free transmission" and infectious nature of cancer.

In 1908, Ellermann and Bang in Copenhagen described the first true "cell-free" transmission of cancer, that is, the transfer of avian erythromyeloblastosis. Tumor induction was by an agent "smaller than a bacterium," an agent which was postulated to be "an ultra-visible" virus and later described as a true retrovirus.

Three years later in 1911, Peyton Rous from the Rockefeller Institute reported that a sarcoma of chickens could be transmitted to closely related animals of the same line and that acellular filtrates of the tumor, which was a spindle cell sarcoma, induced lesions in the recipients with the same phenotypic properties as the original sarcoma.

The reports of Rous indicated that he had isolated not just one virus or retrovirus, but many similar agents which were derived from different primary tumors of chickens brought to him directly by local farmers. Rous emphasized that the agents alone were competent and capable of inducing tumors. It has been suggested that this observation was the first to reflect the activity of "Src" as one portion of the cellular genome that can be picked up or transduced by the RNA tumor viruses. The fact that Peyton Rous was awarded the

Nobel Prize in 1966 was a reflection of the time needed to fully appreciate the impact of his discovery in defining the etiology of certain cancers.

Shope in 1932 and 1933 found that the fibromatosis and papillomatosis lesions of rabbits were transmissible as cell-free filtrates. These discoveries, although not in the retrovirus domain, were the earliest precursors of studies in the DNA oncogenic virus area and uncovered many important principles relevant to mammalian carcinogenesis.

The development of inbred strains of mice (for example, the C_3H, Balb/c, C57 black, and the DBA lines) made possible the discovery of the non-transforming but oncogenic viruses, the mammary tumor virus of Bittner, and the murine leukemia virus of Ludwig Gross.

In 1936, John J. Bittner reported on the unusually high incidence of breast carcinomas in female mice of the C_3H strain. It was noted that the female offspring also developed a high incidence of spontaneous mammary carcinomas. Foster nursing of the offspring with females of low-incidence lines resulted in mice with a relatively low incidence of breast tumors. Thus it was accepted that the development of breast cancer was due to an "extra-chromosomal" factor transmitted from mother to offspring via the milk of the lactating female. In the early 1940s, Bittner, and later Andervont and Bryan, confirmed that there was a "milk factor," "an agent" recoverable from females of the C_3H strain of mice. For many years, the mammary tumor virus remained "a factor, an agent" before it was recognized and accepted as a true retrovirus.

In 1937, a committee of leading scientists was formed to advise the newly created United States National Cancer Institute (NCI) on "various lines of work which merit investigation." The consensus of that committee regarding tumor virus research was as follows: "It has been definitely shown that the animal parasites and bacteria which may incite malignancy in other organisms play no part in the continuation of the process. The present evidence tends to indicate that the same may be true of viruses. As causes of the continuation of the malignant process, the many microorganisms which have been described as specific etiological agents may be disregarded."

However, in 1940, Dr. W. Ray Bryan, one of the first Fellows of the NCI, arrived in Bethesda, Maryland. Bryan had been working with Joseph Beard at Duke University and had great expectations of developing a comprehensive program in retrovirology at the NCI. These interests, of course, had to be tempered in the research environment created in part by the report of the advisory committee just cited. Thus, for more than two years, Bryan collaborated with Dr. Michael Shimkin in the development of quantitative methods for the assay of chemicals as carcinogenic agents.

In 1943, Bryan did receive approval of the National Cancer Advisory Board and the new NCI director, Dr. Rosco R. Spencer. He received permission to initiate studies with the Rous sarcoma virus. A sample of the agent, then called chicken tumor I, was received in lyophilized form from Dr. Albert Claude, then at the Rockefeller Institute. Bryan's quantitative statistical procedures incorporated the principals of virus-dose-host response and shifted attention from the chance distribution of virus particles in graded virus dilutions to the influence of variations in the host. This was the basis for the reproducible *in vivo* assay of various retroviruses including several avian leukosis viruses, which later complemented the biological studies of Beard and Burmester. In 1956, Manaker and Groupé at Rutgers University described the first quantitative *in vitro* or focus-forming assay for the Rous sarcoma retrovirus. Studies with other lines of the Rous virus were undertaken in 1950s and early 1960s. Transmission of the Carr-Zilber line and the Praeue strain in mammals was reported. The Schmidt-Ruppin and the Bryan strain were also shown to be pathogenic for mammals. The latter virus induced malignant brain tumors in dogs.

Joseph Beard, who was Bryan's mentor, was a strong advocate of and pioneer in the science of tumor virology. Beard worked with Peyton Rous and Richard Shope in the 1930s and was one of the first scientists to study the role of host susceptibility in the induction of cancer by retroviruses. His research efforts were primarily with the avian leukemia viruses, myeloblastosis and erythroblastosis. Chickens inoculated with the myeloblastosis virus developed a severe viremia. The virus titer was measured by the unusual activity of the agent to dephosphorylate ATP. This, when correlated with actual particle counts, served as a rapid *in vivo* and *in vitro* assay for the retrovirus. In 1946, Burmester and colleagues in East Lansing, Michigan, showed that lymphomatosis, an infectious disease of chickens, was induced by a retrovirus. This virus is a member of the family of the avian oncornaviruses.

Leon Dmochowski, who joined the faculty of Baylor University in 1954 and later the MD Anderson Hospital and Tumor Institute in Houston, conducted some of the more intense systematic explorations of tumor viruses by electron microscopy. His investigations included morphological studies on the avian tumor viruses, the ultrastructure of mouse mammary tumors, and associated retroviruses. He extended his investigations to human neoplasms and was the first to describe particles morphologically similar to retroviruses present in the milk of human cancer patients and in concentrated blood plasmas derived from human leukemia patients. To complement the studies of Dmochowski, Bernhard and Guerin in 1958 classified the retroviruses as visualized in the electron microscope. The RNA tumor viruses were grouped according to their morphology and biological activities into types A, B, and C and, somewhat later, type D.

Bryan, Beard, Burmester, and Dmochowski were indeed the early pioneers of tumor virology and retrovirology and may be considered the fathers of the exact science as we know it today. These scientists with but few others had the knowledge, the foresight, and indeed the persistence not to disregard viruses as etiologic agents of the malignant process. Certainly, from the late

1940s to the early 1950s, it did require great conviction to exist in a research environment where it was still believed that there was very little to learn from the study of chicken cancers; after all, chemicals and not retroviruses were the cause of cancer.

A major milestone in the history of retroviruses was the discovery of the murine leukemia virus by Gross in 1951. This finding did change the climate of opinion. The direct transmission of murine leukemia with acellular filtrates in a mammalian species was surely as dramatic as the discovery of the avian leukosis virus made by Ellermann and Bang some 40 years earlier. It is believed that the difference in the impact of the Gross and Bittner discoveries in mammals was the result of design and presentation of the studies rather than any basic difference between the murine leukemia virus and the mammary tumor retrovirus systems. In both cases, host factors contribute to the likelihood of tumor development, just as genetic factors influence retroviral replication. The studies of Gross contributed to a rapid expansion of the field of retrovirology, giving impetus to investigations of the viral etiology of neoplasms of murine, feline, and bovine species as well as those of nonhuman and human primates.

In addition to the spontaneous murine leukemias, retroviruses were recovered from neoplasms initially induced by physical agents. For example, the Kaplan virus and the Passage X of Gross were both recovered from x-ray-induced neoplasms. Several leukemia agents, including the Friend virus and the Graffi virus, were recovered from solid tumors of mice. These retroviruses were carried as passengers and not etiologically related to the tumor of origin. The virus of Gross, the virus of Abelson, and the viruses of the FMR (Friend, Moloney, and Rauscher) group are today the most widely used as models for the study of human leukemia.

In 1964, Jennifer Harvey in London described "an unidentified virus which causes rapid production of tumors in mice." This virus was recovered from rats which had been inoculated with Moloney leukemia virus. The retrovirus, later designated Harvey sarcoma virus, when inoculated into susceptible mice, rats, and hamsters, induced sarcomas, angiomas, and leukemias similar to the Rauscher and Friend diseases.

Within the same time period, the Moloney sarcoma virus, which induces rhabdomyosarcomas in mice, was described. This agent was derived from sarcomas which had been induced in Balb/c mice following the inoculation of high doses of mouse-derived Moloney leukemia virus. The Kirsten murine sarcoma virus was isolated by Kirsten and Mayer in 1967 by passage of the Kirsten leukemia virus in rats. Cell-free filtrates from the plasma or spleen of rats induced sarcomas in mice and rats. As with the Harvey sarcoma virus, erythroblastosis was common.

Interestingly, the sarcoma viruses appear to be recombinants, generated through the recombination of portions of the genomes of helper-independent type C viruses and additional genetic information from the host cells from which they were originally isolated. Identification of the transforming proteins of these retrovirus has made possible studies on the molecular mechanisms of transformation of normal human cells to cancer.

In 1964, Jarret et al showed that feline leukemia is a retrovirus-induced neoplasm. Their observations were extended by Kawakami and Theilen as well as Rickard who described the isolation of additional strains of feline leukemia virus (FeLV). Three strains of feline sarcoma virus have been identified. All of these retroviruses are replication-defective transforming viruses and all are found with an excess of helper feline leukemia virus.

Bovine leukemia has been recognized as a neoplasm of infectious origin for more than 60 years. The causative agent, a retrovirus, was first described by Miller et al in 1969. The scientific importance of the bovine virus system was emphasized by the recognition by Gallo of the human T-cell retroviruses, which share several biological and biochemical properties with the bovine leukemia virus.

Since Gross first described the virus induction of murine leukemia, regular attempts have been made to identify similar agents in subhuman primates and in man. However, the first primate retroviruses were not identified until 1970 when the Mason Pfizer monkey virus (MPMV), originally thought to be a type B virus but later classified as a D type, was recovered from a mammary carcinoma of a rhesus monkey. The virus did not induce tumors, although the monkeys developed neutralizing antibodies. At approximately the same time, a type C retrovirus was observed by electron microscopy in multiple fibrosarcomas of a woolly monkey. The virus with its associated helper was designated "simian sarcoma virus and simian sarcoma-associated virus." A type C virus (GALV) was also recovered from a spontaneous lymphoma of a gibbon ape. Somewhat later the simian immunodeficiency virus (SIV) was described. This retrovirus is approximately 40% homologous to the human immunodeficiency virus (HIV) and produces an AIDS-like syndrome in Asian monkeys (macaques).

Transforming and oncogenic activity have not been associated with the primate endogenous viruses, whereas exogenous primate retroviruses transform cells *in vitro* and are oncogenic *in vivo*.

The directly transforming RNA tumor viruses or retroviruses represented by the avian and murine retroviruses contain different kinds of sequences derived from the cellular genome. These viruses were the first to call into question a major teaching of molecular biology, namely that "DNA can make RNA but the reverse is not possible." The discovery in 1970 of reverse transcriptase by Howard Temin and David Baltimore led to the complete collapse of this dictum. It was later recognized that the nontransforming but oncogenic leukemia and mammary tumor viruses are equally capable of transcribing their RNA into DNA and of integrating into the host cell genome without a direct transforming effect. With the reports of Baltimore and Temin, the field of retrovirology entered into the new era of molecular virology.

It was quickly realized that all of the RNA viruses (and the "slow" virus visna) contained RNA-dependent DNA polymerase. The enzyme became an additional marker protein of the retroviruses. In 1974, the entire family of oncornaviruses was classified as "retraviruses," a term used interchangeably with "retroviruses"; the latter term persists today.

Spieglman from Columbia University reported that human milk from leukemic individuals contained particulates banding at densities characteristic for the known RNA tumor viruses. These particulates also contained reverse transcriptase activity when tested in response to synthetic RNA templates. Similar assays revealed enzymatic activity in white blood cells of leukemia patients but not in nonleukemic controls. In the 1970s, there were reports of reverse transcriptase-containing particles recovered from human tumors and normal tissue. Some of these "viruses" could not be propagated in cell culture, and those that could were considered contaminants of other animal retroviruses and their true nature debatable.

In the early 1970s, Michael Bishop and Harold Varmus recognized the retroviral oncogene "Src." They detected copies of the Rous sarcoma virus gene, the prototype of the family of the protein kinases, in healthy chickens and in mammals, including humans. It was later shown that retroviral oncogenes are normal cellular genes, that is, they arise from cellular genes or proto-oncogenes. Homologues of viral oncogenes have been found in host DNA at sites of virus integration and of chromosome translocation.

More than two dozen retrovirus oncogenes and their products have been identified. In the late 1970s, transfer experiments revealed the presence of transforming genes in a significant fraction of human cancers. Many of the human transforming genes are members of the *ras* gene family which was first identified as the transforming principle of several strains of the murine sarcoma viruses. The *ras* oncogenes, particularly H-ras, and K-ras have been detected in human bladder, lung, and liver carcinomas. Additionally, it has been shown that the human transforming *ras* genes are immediately involved in the progression of relatively benign lesions in the large bowel of humans to colon adenocarcinomas.

The first evidence of the retroviral etiology of human disease was the isolation and characterization of the human T-cell leukemia virus, HTLV-1, by Robert Gallo in 1980. The identification of T-cell growth factor or interleukin 2, described by Gallo in 1978, made possible the development of several cell lines from patients with adult T-cell malignancies. In one such line, type C particles, with associated reverse transcriptase, were identified by electron microscopy. HTLV-2, a second virus isolate, was obtained from a patient with T-cell hairy cell leukemia. This virus is related to HTLV-1, sharing partial nucleic acid and structural protein homology to the virus. However, these two retroviruses appear to be distinct, since there is no immunological relatedness of the structural proteins to those of other mammalian or avian species. The causal association of both of these viruses with human cancer is well established. *In vitro* studies have repeatedly demonstrated the transformation of nominal T-lymphoid cells in culture. Further, virus can be transmitted either cell-free or by the cocultivation of HTLV-positive cells with nominal T-lymphocytes. Gallo firmly and for the first time established retroviruses as etiologic agents of certain human neoplasms. His studies contributed to our understanding of the cause of human disease.

Acquired immune deficiency syndrome (AIDS) in humans was first recognized in 1981. The etiology of the disease, a retrovirus, the human immunodeficiency virus (HIV), was first described by Luc Montagnier as lymphadenopathy-associated virus (LAV) and by Gallo as HTLV-3. The disease is characterized by profoundly diminished cell counts and functions of the immune system, major factors in host susceptibility to neoplasms such as Kaposi's sarcoma and to opportunistic infections.

The history of retrovirology from its inception in 1908 through the discovery, development, and elucidation of the RNA virus systems, their transforming genes and gene products in avian, murine, feline, and bovine species, in nonhuman primates and in humans has been, to say the least, intellectually stimulating and truly exciting. It is a history marked by a continuous flow of new discoveries, findings that have made it possible to understand the etiology, prevention, and control of human disease, particularly cancer.

REFERENCES

The following references represent some of those in which may be found discussions and citations of literature pertaining to the history of retrovirology.

Beard JW (1980). Biology of avian oncornaviruses. In: Klein G, ed. Viral oncology. New York: Raven Press, 55-87.

Bittner JJ (1936). Some possible effects of nursing on the mammary gland tumor incidence in mice. Science 84: 162.

Bryan WR (1946). Quantitative studies on the latent period of tumors induced with subcutaneous injections of the agent of chicken tumor I. I. curve relating dosage of agent and chicken response. J Natl Cancer Inst 6: 225-237.

Dalton et al (1974). The case for a family of reverse transcriptase viruses. Retraviridae Intervirol 4: 201-206.

Deinhardt F (1980). Biology of primate retroviruses. In: Klein G, ed. Viral oncology. New York: Raven Press, 357-398.

Ellermann V, Bang O (1908). Experimentelle Leukamie bei Huhnern. Zentralbl Bakt I Orig 46: pp. 595-600.

Gallo et al (1983). Association of the human type C retrovirus with a subset of adult T-cell cancers. Cancer Res 43: 3892-3899.

Gross L (1951). "Spontaneous" leukemia developing in C_3H mice following inoculation in infancy with AK-leukemic extracts, or AK-embryos. Proc Soc Exp Biol Med 76: 27-32.

Gross L (1970). Oncogenic viruses. Ed 2. London W.I: Pergamon Press Ltd.

Peyrilhe B (1773). L'Academie des Sciences, Arts & Belles Lettres de Lyon. Paris: Chez Ruault, 1-135.

Poiesz et al (1980). Detection and isolation of type C retrovirus particles from fresh and cultured lymphocytes of a patient with cutaneous T-cell lymphoma. Proc Natl Acad Sci USA 77: 7415-7419.

Rous P (1911). A sarcoma of fowl transmissible by an agent separate from the tumor cells. J Exp Med 13: 397-411.

Santos E et al (1985). Transforming *ras* genes. In: Rigby PWJ, Wilkie NM, eds. Viruses and cancer. Cambridge University Press, 291-313.

Shope RB (1932). A filterable virus causing tumor-like condition in rabbits and its relationship to virus myxomatosum. J Exp Med 56: 803-822.

Shope RB (1933). Infectious papillomatosis of rabbits. J Exp Med 58: 607-624.

18

Human Retroviruses: Current Concepts

ROBERT GALLO AND HOWARD STREICHER

Retroviruses have developed novel strategies for propagation and survival. These tactics depend on an intimate relationship between virus and host at the molecular level, as well as the integrated biological and social behavior of the whole organism. This success is measured in terms of human epidemic diseases and by the range of problems they cause. The decades of the 1950s to the 1970s saw the identification, molecular analyses, and linkage to disease of numerous animal retroviruses. The 1980s saw their discovery, analyses, and linkage in humans. The task now is to develop ways of effective treatment and prevention of infection.

During the 1970s, several mechanisms of cellular transformation were described. These included the insertion of the virus near vital cellular genes, a stoichiometric process that requires many cycles of integration and thus both time and active replication, and the presence in the virus of a new gene, often a cellular gene, whose overexpression caused cellular transformation (oncogenes). More recently, the insertion of viral elements in the form of infectious or endogenous elements that disrupt vital cellular genes has been shown to lead to loss of function. The family of retroviruses grew to include three groups: oncoviruses, lentiviruses, and foamy viruses or spumaviruses. Reverse transcriptase (RT) itself became a central tool in molecular biology that allowed further study of viral infections.

The search for human retroviruses focused primarily on hematologic malignancies, especially during the 1970s. The failure to find a *bona fide* human retrovirus by using the same methods as those used for animal viruses led to the widespread assumption that there were none. However, another assumption, that human viruses are not detected in an actively replicating form, led to the development of sensitive and specific assays to detect RT as a marker of infection and to attempt to grow target tissues. The latter led to the discovery of interleukin-1 (IL-1), which proved critical in the discovery of all human retroviruses. Serendipitously, this factor was directed to the growth of T cells (Ruscetti et al 1977).

In the late 1970s, the first human retrovirus (HTLV-I), which was also the first human tumor virus, was isolated and productively expressed from a patient with cutaneous T-cell lymphoma (Poiesz et al 1980). This virus has been causally linked to a form of T-cell leukemia/lymphoma marked by overexpression of the IL-2 receptor (IL-2R). First described in Japan as an acute adult T-cell leukemia (ATL), the disease is often clinically indistinguishable from aggressive variants of cutaneous T-cell lymphoma or mycosis fungoides (Takatsuki et al 1977). In 1986, HTLV-I was also associated with a neurologic disease, similar to some forms of multiple sclerosis, called tropical spastic paraparesis or human T-cell leukemia/lymphotropic virus (HTLV)-associated myelopathy (TSP/HAM). A second human retrovirus (HTLV-II) was identified, which has 70% DNA homology and similar biologic behavior to HTLV-I. The molecular biology, extra genes, pathogenesis, relationship to cancer and neurologic disease, and epidemiology of the HTLVs are summarized below. The HTLVs provide a model for the pathogenesis of virally associated disease. With the proof of the existence of human retroviruses finally at hand and extensive experience in detecting and growing these viruses, the stage was set for the most dramatic medical development of the late twentieth century.

In a remarkably short time, a new virus, now known as human immunodeficiency virus type 1 (HIV-1), emerged and became a worldwide epidemic. Based on work done in Paris and Bethesda, the causative agent of a new human retrovirus was identified, isolated, and grown in large quantity and a blood test was rapidly developed (Barre-Sinoussi et al 1983; Gallo et al 1984; Popovic et al 1984; Gallo and Montagnier 1987). A second human retrovirus, HIV-2, represents the fourth known human retrovirus. HIV-2, closely related to simian immunodeficiency viruses (SIVs), appears to be less pathogenic than HIV-1 and has been used extensively in vaccine studies in simian models.

This period of rapid discovery resulted in expanded knowledge of the complex molecular mechanisms of these viruses and the range of diseases they may produce. In particular, the life history of HIV-1 has been studied and described in minute detail over the past 10 years. Most recently, attention has focused on the extensive variation or adaptability of this virus as it infects humans. Despite numerous theories of pathogenesis, we have returned to the basic notion that this virus is the direct cause of the disease and that controlling its replication is essential.

These new perspectives on the human viruses have led to the sobering recognition that although the causative agents have been identified and *in vitro* models of the disease are in hand, we have had only limited success in controlling the epidemic of HIV and in treating the HTLV diseases or acquired immune deficiency syndrome (AIDS). The efforts in the next period must be directed toward prevention and therapy.

WHERE DO THESE VIRUSES COME FROM?

Retroviruses are either genetically transmitted (endogenous) or occur as infectious agents. Generally, endogenous viruses survive from generation to generation, presumably because they provide some advantage to the host and are not known to cause disease. The endogenous viruses presumably are the relics of ancient infections that have become integrated into the host germ line. Humans, like many animals, have both forms of virus. Multiple copies of several different DNA proviruses are often found in a single species, producing viral proteins or incomplete virions from defective proviruses. The phenomenon of viral rescue has not been observed in humans. The human endogenous viruses have limited homology to the infectious viruses and are generally defective, but exist in multiple copies that comprise about 1% of the human genome. Although there is no evidence that they cause disease, one virus called HERV-K (human endogenous virus with a Lys primer) can encode all the typical retroviral genes *(gag, pol, env)* and form particles (Lower et al 1993). These elements may expand throughout the genome, presumably by reverse transcription and integration. In addition, genetic elements known as lines and sines (long and short inserted sequences) are multiple repeat genetic sequences that are spread throughout the human genome (transposons and retrotransposons).

From the time of its discovery, HTLV-I was associated with transmission by sex (primarily man-to-woman), by milk from mother to infant, and by blood exchange. HTLV-I is endemic in many parts of the world, particularly in southern Japan, parts of equatorial Africa, some Caribbean islands, and some American black populations. Although the genome has relatively little variation, molecular epidemiology has recently shown that there are several different subtypes with distinct strains in Melanesia and equatorial Africa in addition to a more widespread "cosmopolitan" strain (Franchini and Reitz 1994). HTLV-II has appeared in unexpected places—particularly in central and South American Indians that share a common language and probably ancestry. While the HTLVs cause a significant disease burden and new HTLV subtypes, strains, and disease associations may be found, it is obvious that these are old infections of mankind (hundreds or thousands of years ago) and may be on the wane.

HIV also has an increased incidence and perhaps origin in central Africa and is transmitted by sexual and parenteral routes. However, the epidemic of HIV-1 is new to much of the world and is unstable, unpredictable, and potentially unrestricted. All humans are susceptible to infection following exposure. There are currently an estimated 20 million infected people worldwide, with expectations that the number will more than double by the end of the century. As additional information is accumulated, the fundamental role of the virus as the disease-causing agent remains clear. Preventing infection prevents disease; decreasing viral replication decreases transmission and progression of disease. Whether this is done by public health measures, education, antiviral drugs, or immunologic control, the target remains the virus.

Both HTLV and HIV have counterparts in Old World monkeys known as simian T-cell leukemia virus (STLV) and SIV (Franchini and Reitz 1994). Although HIV-1 has no known primate counterpart, HIV-2 is very similar to SIV strains obtained from Sooty mangabeys. Interspecies transmission appears to be an ancient and ongoing phenomenon. Like other persistent lifelong infections, they can sometimes cause serious disease determined by chance events, genetics, or unidentified cofactors.

Infection of humans by various SIVs may also have been an old and repetitive event, probably sporadic and limited in extent. Possibly major social changes, including urban migration and loss of rain forest ecology, were precursors of the current epidemics.

SIMILARITIES AND CONTRASTS

Both groups of human retroviruses infect the same subset of human CD4$^+$ T lymphocytes. There is no clear syndrome of acute infection with HTLV-I, reflecting in part the limited viral replication and immune response to the virus. *In vitro* T-cell function is impaired, while some T cells become immortalized but remain IL-2-dependent. These immortalized cell lines, although CD4$^+$, have suppressor function, in contrast to Sezary cell lines, which have helper function. Disease usually takes decades to develop. Over a lifetime, leukemia occurs in about 2% of the infected population and a similar percentage may develop neurologic disease or an inflammatory process. The CD4+ T cells and macrophages are the targets of HIV infection. *In vitro,* the effects are dramatically different from those of HTLV and were recognized in culture well before viral reagents were available.

Many strains of HIV induce massive syncytia, in contrast to the tendency of HTLV-I to form smaller giant cells. The receptor for HTLV-I is unknown. Most transmissions require cell-cell contact, in contrast to the relatively easy cell-free transmission of HIV. HTLV-I replication from the time of infection to disease is at low levels with no detectable viremia. For HIV, there is also a "clinical" latent period from infection to disease, but recent studies have made it clear that HIV replication is continuous and high level (Ho et al 1995; Wei et al 1995).

Although there is limited sequence homology between the HTLV and HIV viruses, some of the strategies used by both groups are similar. The genes of HTLV-I were the first known to contain regulatory, nonstructural genes. These genes are formed from a two-step splicing process, which was also subsequently found in bovine leukemia virus and STLV, as well as HIV.

HTLV AS A CAUSE OF LEUKEMIA AND NEUROLOGIC DISEASE

HTLV-I is found as integrated provirus in the CD4+ T cells of infected individuals. The detection of a monoclonal population of T cells is associated with a high risk for the development of leukemia. The events leading to T-cell immortalization *in vitro*, including loss of IL-2 dependence, are just beginning to be understood. Following infection, the viral DNA form integrates and virus production remains at a low level, but several viral regulatory proteins may be expressed as HTLV immortalizes some cells *in vitro*. The viral protein *tax* activates a number of cellular genes, including IL-2 and IL-2R-alpha chain. Another protein of interest in the process of growth promotion is a newly discovered HTLV-I regulatory protein, p12, which binds to several cellular proteins, including the IL-2R-beta chain, and might be involved in intracellular signaling (Franchini et al 1993). HTLV rapidly transforms T cells *in vitro* but does not, unlike Rous sarcoma virus for example, have an oncogene (Koralnik et al 1992). We speculate that the strategy of the virus is to escape immune detection while stimulating the expansion of infected T cells.

Since virus is not actively replicating, antiviral therapy appears to be of little benefit in treating ATL. Chemotherapy has had limited success. Immune therapy utilizing toxin coupled to IL-2R antibody attacks a prominent feature of infected cells and has achieved some dramatic remissions but frequent relapses (Waldmann 1993). Recently, the use of the combination of the antiviral azidothymidine (AZT) with α-interferon has had an unpredicted and surprising *in vivo* effect, inducing remissions and dramatically clearing leukemic cells, probably by inducing apoptosis (Gill et al 1995).

In contrast, ATL patients with TSP/HAM have an increased spontaneous T-cell activation, increased viral expression and vigorous immune responses to the virus. Presumably, this is a demyelinating disease with damage caused by cytotoxic T cells, possibly directed against the *tax* protein. This hypothesis suggests a therapeutic approach based on attenuating the immune attack, possibly by inducing oral tolerance to the virus, similar to use of myelin basic protein in experimental models of multiple sclerosis.

Prevention of infection by identifying carriers and avoiding exposure through breast feeding, sexual contact, or blood products seems to be the most effective way of controlling disease. Efforts such as blood testing prior to transfusion and avoiding breast feeding may already be having an effect in endemic areas of Japan. The incidence of viral infection seems to decrease with each generation following emigration. HTLV-I infection has been induced in rats by several groups. It has been studied in the rabbit model where recombinant vaccinia vectors expressing the entire viral envelope have been shown to induce protection against all associated viral challenge (Franchini et al 1995).

HIV PATHOGENESIS AND THERAPY

HIV follows the same general genomic organization and life cycle as all retroviruses. In addition to *gag, pol,* and *env* genes, HIV contains several nonstructural genes. Genes with similar functions in HTLV-I such as *tax* and *rex* are an example of convergent evolution, yet the effects of each infection are diametrically opposite. Knowledge of the biologic effects of many of these genes is rapidly expanding. They appear to be far more versatile than expected. For example, the transactivating factor tat is a viral regulatory protein that has potent extracellular effects, stimulating endothelial cells and perhaps contributing to Kaposi's sarcoma (KS) (Ensoli et al 1994). Vpr may be essential to macrophage tropism, forming a complex that directs virus towards the nucleus and facilitates integration in resting cells. Each of these proteins may provide additional therapeutic targets.

Extensive and detailed molecular analysis of many HIV isolates has become available over the past 10 years. Passage of virus results in the introduction of a limited number of related sequences. Variation occurs within an individual during the course of an infection, among individuals, and within geographic areas. We still do not have a clear understanding of the basis of cellular tropism, virulence, and replication rates, or escape from immune control despite the enormous amount of molecular analyses. One area in which considerable progress has been made is in identifying specific variations that are selected by exposure to anti-HIV drug therapy.

Recent analyses by sensitive polymerase chain reaction techniques indicate that viral replication is far more extensive than has been supposed. Suppression of viral production with antiviral drugs results in the temporary restoration of T-cell numbers. While there is no doubt that the virus remains latent in many cells and can be activated by immune stimulation, the pathogenic process has been characterized as a dynamic equilibrium between viral destruction of T cells and T-cell replacement. The rate of viral production is directly proportional to the rate of T-cell loss. A great deal remains to be learned about the ability of the immune system to replace lost cells in adult life. It appears that chronic T-cell activation and the loss of memory cell precursors are associated with disease activity. Extensive infection of lymph nodes and trapping of viral complexes by lymph node dendritic cells progresses to the collapse of germinal centers in association with progressive viral burden (Pantaleo et al 1995). Recent studies of long-term survivors without progressive immune destruction show little lymph node damage, although the protective element(s) in these individuals is not known.

Despite the advances, the basic tenet that the pathogenesis of the disease is dependent on direct and indirect effects of viral replication has remained unchanged. While many mechanisms (including direct killing, cytokine-mediated killing, defective signaling, gp120-induced apoptosis, and syncytia formation) are involved, the overall "viral burden" is the single factor determining outcome. Ultimately, pathogenic effects involve four areas: immune destruction associated with immune impairment progressing to failure; chronic immune activation; neurologic damage; and proliferation of some cells leading to neoplasms such as KS and B-cell lymphoma.

KS was a unique AIDS-identifying lesion even before the identification of HIV. The tumor focuses attention on the crossroads of the areas of viral infections, immunodeficiency, angiogenesis, and tumorigenesis. The indirect role of HIV, cytokine production, and HIV regulatory products such as tat have been described and represent an emerging concept in tumorigenesis. Such indirect effects may be occurring in other infections such as the recently described MALT lymphomas of the gastrointestinal tract stimulated by *Helicobacter pylori*(Roggero et al 1995). A new putative herpesvirus detected in KS tissue has been associated with HIV and classical KS, as well as some Epstein-Barr virus-positive B-cell lymphomas (Moore and Chang 1995). As in KS, HIV is not found in the tumor cells of the B-cell lymphomas. The observation that pregnant mice were resistant to experimental KS-like viruses has led to the discovery that human chorionic gonadotropin may cause regression of KS strains (Lunardi-Iskandar et al 1995).

Treatment of HIV infection has gradually improved over the past decade. Perhaps the most significant advances have been in the prophylaxis and treatment of opportunistic infections. Direct anti-retroviral therapy has advanced beyond the most obvious target of RT to include protease inhibitors and potentially other viral proteins. Recent studies with IL-2 early in the infection have produced some sustained increases in CD4+ cell numbers, and IL-2 therapy has been proposed to reverse the loss of T helper type-1 cells (Kovacs et al 1995).

Immunotherapy in infected individuals has not proven effective so far, but experimental protocols continue to test new vectors and target epitopes. Since HIV infection involves integration into the host genome, eradication of the virus seems unlikely. Moreover, the virus may adapt to and eventually escape suppressive drugs. At times, resistant mutants have emerged with remarkable swiftness. A number of findings give some clues to possible approaches to better therapy. One clinical trial has shown that transmission from mother-to-child greatly diminished when AZT was given in the final weeks before delivery and in a large amount at delivery. Therefore, some combinations of RT drugs may prove more effective than expected. For example, AZT combined with other drugs to avoid resistance and new approaches to chemotherapy, such as the recently described effects of a ribonucleoside inhibitor, hydroxyurea, or blocking of the p7 gag protein, may lead to unexpected success.

HIV AND AIDS ARE DIFFERENT FROM OTHER VIRAL DISEASES BUT NOT UNIQUE

RNA viruses are single-stranded and their genetic material can vary rapidly, with frequencies of genetic change ranging from 10^4 to 10^6 per nucleotide replication cycle, resulting in a heterologous mixture. Retrovirus vectors mutate at a similar rate and the rate of base pair substitutions for HIV is probably not very different from other single-stranded RNA viruses (Coffin 1995). The variation seen in viral sequences represents the high mutation rate, the ability to fix mutations with selective pressure, and the large number of replication cycles in the infectious process. Many viruses are immunosuppressive, at least transiently. The distinctive strategy of HIV seems to be able to expand the remarkable duality of the retrovirus in its ability to be latent while at other times to be highly replicative. This may allow the virus to simultaneously adapt to drugs and to avoid immune elimination. The virus may both activate and destroy the immune system at the same time, perhaps to expand itself and to eliminate effective immune responses. AIDS may simply be the end game in this dynamic struggle between host and virus.

CONCLUSION

The past decade has been a period of expansion in retrovirology which has paralleled a remarkable era of molecular and cellular biology. The discovery of HTLV-I ended a 70-year search for a human tumor virus. This experience brought early success in the identification, isolation, and the ability to grow HIV. In *Microbe Hunters,* Paul DeKruif tells the stories of 14 pioneers of "indomitable courage, driven by a ceaseless urge to conquer the deadly enemies of mankind (bacterial diseases) with super human perseverance" (DeKruif 1926). Working together, we will need the same resolve to conquer these retroviruses.

REFERENCES

Barre-Sinoussi F, Carmine JC, Rey F, Nugeyre MT, Chamaret S, Greust J, Dauguet C, Alexer-Blin C, Veinet-Brun F, Rouzious C, Rosenbaum W, Montagnier L (1983). Isolation of a T-lymphotropic retrovirus from a patient at risk for acquired immune deficiency syndrome. Science 220: 868-870.

Coffin JM (1995). HIV population dynamics in vivo: implications for genetic variation, pathogenesis, and therapy. Science 267: 483-489.

DeKruif P (1926). Microbe hunters. New York: Pocket Books, Inc.

Ensoli B, Gendelman R, Markham P, Fiorelli V, Colombini S, Raffeld M, Cafaro A, Chang HK, Brady JN, Gallo RC (1994). Synergy between basic fibroblast growth factor and HIV-1 Tat protein in induction of Kaposi's sarcoma. Nature 371: 674-680.

Franchini G, Reitz Jr MS (1994). Phylogenesis and genetic complexity of the nonhuman primate Retroviridae. AIDS Res Hum Retroviruses 10: 1047-1060.

Franchini G, Mulloy JC, Koralnik IJ, LoMonico A, Sparkowski JJ, Andresson T, Goldstein DJ, Schlegel R (1993). The human T-cell leukemia/lymphotropic virus type I p12^I protein cooperates with the E5 oncoprotein and binds the 16-kilodalton subunit of the vacuolar H$^+$ ATPase. J Virol 67: 7701-7704.

Franchini G, Tartaglia J, Markham P, Benson J, Fullen J, Wills M, Arp J, Dekaban G, Paoletti E, Gallo RC (1995). Highly attenuated HTLV type I$_{env}$ poxvirus vaccines induce protection against a cell-associated HTLV type I challenge in rabbits. AIDS Res Hum Retroviruses 11: 305-311.

Gallo RC, Montagnier L (1987). The chronology of AIDS research. Nature 326: 435-436.

Gallo RC, Salahuddin SZ, Popovic M, Shearer GM, Kaplan M, Haynes BF, Palker TJ, Redfield R, Oleske J, Safair B, White G, Foster P, Markham PD (1984). Frequent detection and isolation of cytopathic retroviruses (HTLV-III) from patients with AIDS and at risk for AIDS. Science 224: 500-503.

Gill PS, Harrington Jr W, Kaplan MH, Ribeiro RC, Leibman HA, Berstein-Singer M, Espina BM, Cabral L, Allen S, et al (1995). Treatment of adult T-cell leukemia-lymphoma with a combination of interferon alpha and zidovudine. N Engl J Med 332: 1744-1748.

Ho DH, Neumann AU, Perelson AS, Chen W, Leonard JM, Markowitz M (1995). Rapid turnover of plasma virions and CD4 lymphocytes in HIV-1 infection. Nature 373: 123-126.

Koralnik IJ, Gessain A, Klotman ME, LoMonico A, Berneman ZN, Franchini G (1992). Protein isoforms encoded by the pX region of human T-cell leukemia/lymphotropic virus type I. Proc Natl Acad Sci USA 89: 8813-8817.

Kovacs JA, Baseler M, Dewar RJ, Vogel S, Davey Jr RT, Falloon J, Polis MA, Walker RE, Stevens R, Salzman NP, Metcalf JA, Masur H, Lane HC (1995). Increases in CD4 T lymphocytes with intermittent courses of interleukin-2 in patients with human immunodeficiency virus infection—a preliminary study. N Engl J Med 332: 567-575.

Lower R, Boller K, Hasenmaier B, Korbmacher C, Muller-Lantzsch N, Lower J, Kurth R (1993). Identification of human endogenous retroviruses with complex mRNA expression and particle formation. Proc Natl Acad Sci USA 90: 4480-4484.

Lunardi-Iskandar Y, Bryant JL, Zeman RA, Lam VH, Samaniego F, Besnier JM, Hermans P, Thierry AR, Gill P, Gallo RC (1995). Tumorigenesis and metastasis of neoplastic Kaposi's sarcoma cell line in immunodeficient mice blocked by a human pregnancy hormone. Nature 375: 64-68.

Moore PS, Chang Y (1995). Detection of herpesvirus-like DNA sequences in Kaposi's sarcoma in patients with and those without HIV infection. N Engl J Med 332: 1181-1185.

Pantaleo G, Menzo S, Vaccarezza M, Graziosi C, Cohen OJ, DeMarest JF, Montefiori D, Orenstein JM, Fox C, Schrager LK, Margolick JB, Buchbinder S, Giorgi JV, Fauci AF (1995). Studies in subjects with long-term nonprogressive human immunodeficiency virus infection. N Engl J Med 332: 209-216.

Poiesz BJ, Ruscetti FW, Gazdar AF, Bunn PA, Minna JD, Gallo RC (1980). Detection and isolation of type C retrovirus particles from fresh and cultured lymphocytes of a patient with cutaneous T-cell lymphoma. Proc Natl Acad Sci USA 77: 7415-7419.

Popovic M, Sarngadharan MG, Read E, Gallo RC (1984). Detection, isolation, and continuous production of cytopathic retroviruses (HTLV-III) from patients with AIDS and pre-AIDS. Science 224: 497-500.

Roggero E, Zucca E, Pinotti G, Pascarella A, Capella C, Savio A, Pedrinis E, Paterlini A, Venco A, Cavalli F (1995). Eradication of *Helicobacter pylori* infection in primary low-grade gastric lymphoma of mucosa-associated lymphoid tissue. Ann Intern Med 122: 767-769.

Ruscetti FW, Morgan DA, Gallo RC (1977). Functional and morphologic characterization of human T cells continuously grown *in vitro*. J Immunol 119: 131-138.

Takatsuki K, Uchiayama T, Sagawa K, Yodoi J (1977). Adult T-cell leukemia in Japan. In: Seno S, Takaku F, Irino S, eds. Topics in hematology. Amsterdam: Excerpta Medicia, 73-76.

Waldmann TA (1993). The IL-2/IL-2 receptor system: a target for rational immune intervention. Immunol Today 14: 264-270.

Wei X, Ghosh SK, Taylor ME, Johnson VA, Emini EA, Deutsch P, Lifson JD, Bonhoeffer S, Nowak MA, Hahn BH, Saag MS, Shaw GM (1995). Viral dynamics in human immunodeficiency virus type 1 infection. Nature 373: 117-122.

Vignette

Emerging Viruses

Indiscriminantly causing devastating illness and death, the newly emerging viruses such as Ebola, Lassa fever, and hantaviruses strike randomly and without warning in many parts of the world. Public awareness of these viral marauders and the resulting hemorrhagic fevers is heightened due to mass media and instantaneous news reporting.

The story of these viruses, their methods of emerging, and expectations for these and other viruses predicted to emerge is presented in this chapter by Brian Mahy, Head of Virology, Centers for Disease Control and Prevention, Atlanta.

Several factors contribute to the emergence of these viruses. 1) The evolution of new viral genetic material through alterations in viral nucleic acids primarily by mutation, but also possibly by recombination or reassortment of genetic materials and the selective pressures for survival of newly formed viruses. 2) The modification in human behavior, including changes in drug use, sexual conduct, and international travel. 3) The enhanced human contact with virus vectors as humans encroach on rain forests, explore previously uninhabited areas, and utilize farmland that previously housed rodents or other virus carriers. 4) The technological revolution that enables identification and discovery of new viruses or newly altered pre-existing viruses. These issues and others are elucidated in this chapter.

19

Current Problems
with Viral Hemorrhagic Fevers

B.W.J. Mahy and C.J. Peters

There are several human virus infections that may be associated with hemorrhagic fever. This chapter concentrates on problems we have encountered with viral hemorrhagic fevers at the Centers for Disease Control and Prevention (CDC) since May 1993. These problems have included the recognition of a new North American hantavirus, several outbreaks of hemorrhagic fever caused by arenaviruses, and most recently the resurgence of filovirus infections in Central and West Africa. In each instance, remarkable discoveries have been made, and our knowledge of the virus has expanded considerably. However, we remain woefully unable to prevent or treat most viral hemorrhagic fevers. As a consequence, they are among the most feared diseases of mankind, causing sudden and frightening morbidity and mortality, without prior warning, in many parts of the world. A vaccine has been developed against Argentine hemorrhagic fever (Junin virus) at the United States Army Research Institute for Infectious Diseases, and the antiviral drug ribavirin holds promise in treatment of arenavirus diseases, but other viral hemorrhagic fevers remain without any reasonable prospect of successful use of vaccines or antiviral drugs. In light of the severity of these diseases, appropriate resources will hopefully be brought to bear so that control or prevention therapies can be developed.

Hantavirus Pulmonary Syndrome (HPS)

RECOGNITION OF A NEW DISEASE SYNDROME

In early May 1993, a physician working in New Mexico recognized a possible association between the deaths of a young woman on the Navajo Indian Reservation and a young man with whom she had been living who died five days after her. Both had experienced sudden onset of acute fever, myalgia, headache, and cough before the development of bilateral pulmonary interstitial infiltration and death from respiratory failure. The New Mexico Department of Health was notified of the deaths on May 14, and public inquiries soon led to the recognition of more than 20 similar cases diagnosed as adult respiratory distress syndrome (ARDS) in the Four Corners area. Half of these patients had died, although all were previously healthy and most were young adults, median age 34 years. Extensive laboratory studies by state health departments, local hospitals, and university researchers excluded many of the possible causes of acute respiratory failure, such as influenza, pneumonic plague, pneumonic tularemia, or anthrax, but failed to identify the causative agent. It was noticed, however, that more than one case had occurred in association with a single household and so an infectious agent or perhaps an environmental toxin was suspected.

At the end of May, serum samples from a number of patients were sent to CDC to begin a broad-based laboratory investigation. Within a few days, serologic tests using hantavirus antigens, especially Puumula virus and Prospect Hill virus, proved positive (Ksiazek et al 1995). The hantavirus genus of the family Bunyaviridae was known to contain at least four distinct viruses, Hantaan, Seoul, Puumula and Prospect Hill, each having a different rodent species as the natural reservoir (Table 1). Of the four, only Prospect Hill virus and Seoul virus were known to exist naturally in the Western hemisphere, and none of these viruses had been associated with severe pulmonary disease (Yanagihara 1990). Nevertheless, from these and subsequent findings, it became clear that a new virus causing a unique disease had emerged, and the disease was termed hantavirus pulmonary syndrome (HPS) (Hughes et al 1993), in contrast to hemorrhagic fever with renal syndrome (HFRS), the predominant disease associated with the Hantaan, Seoul and Puumula viruses. Despite differences in the overt disease manifestation, the underlying process leading to HPS, like HFRS, appears to be immunopathologic. The prodromal illness, fever, headache, and myalgia, is similar to that observed in many virus diseases. The blood picture in hospitalized patients includes thrombocytopenia, hemoconcentration, and neutrophilic leukocytosis with a left shift. Lymphocytic cells are found in the lung interstitium and are probably involved in causing the increased vascular permeability which results in seeping of protein-rich edema fluid into the interstitial and alveolar spaces and causing frequently fatal respiratory distress. The reason for the differing tropism for lung in HPS or kidney in HFRS is presently unknown. Studies of hantaviral antigens in post-mortem examination of HPS cases revealed widespread distribution in endothelial cells throughout the microvascular system, especially in the lung, but also in lower amounts in the medulla and glomeruli of the kidney (Zaki et al 1995). There was also an accumulation of HPS viral antigen in follicular dendritic cells of the spleen as well as in occasional macrophages and lymphocytes. The presence of viral antigens in endothelial cells was not associated with evidence of endothelial cytopathic effects as observed by light microscopy (Zaki et al 1995). Although thrombocytopenia is a significant

TABLE 1. HANTAVIRUSES

Virus	Disease	Reservoir
Hantaan	HFRS	*Apodemus agrarius* (striped field mouse)
Seoul	HFRS	*Rattus norvegicus* (Norway rat)
Puumula	Nephropathia epidemica	*Clethrionomys glariolus* (bank vole)
Thai	None known	*Bandicota indica* (bandicoot rat)
Prospect Hill	None known	*Microtus pennsylvanicus* (meadow vole)
Sin Nombre	HPS	*Peromyscus maniculatus* (deer mouse)
Black Creek Canal	HPS	*Sigmodon hispidus* (cotton rat)
Bayou	HPS	*Oryzomys palustris* (marsh rice rat)
RM-97	None known	*Reithrodontomys megalotis* (western harvest mouse)
Shelter Island	HPS	*Peromyscus leucopus* (white-footed mouse)
Thottapalayam	None known	*Suncus murinus*

feature of patients with HPS, hemorrhage is not common. The reasons for these different manifestations of hantavirus infection provide a fascinating field for future studies.

SIN NOMBRE VIRUS AND ITS RELATIVES

Serological tests confirmed that HPS was caused by a hantavirus, but the pattern of cross-reactivity against the known hantavirus antigens suggested that a new virus was involved (Ksiazek et al 1995). Using published sequence information on the known hantaviruses, consensus polymerase chain reaction (PCR) primers were designed and synthesized, then used to amplify reverse transcription products of total RNA extracted from the tissues of eight clinically and serologically suspected HPS patients. Seven of these RNA preparations gave PCR bands with primers designed to detect Puumula and Prospect Hill-like viruses. The bands were then sequenced, and all seven yielded the same sequence for which specific PCR primers were then designed and synthesized (Nichol et al 1993). It immediately became clear that the virus responsible for HPS was new, and this led to an investigation to find the rodent reservoir for the new hantavirus. Trapping of several hundred rodents in peridomestic settings in areas where HPS cases had occurred indicated that the deer mouse *(Peromyscus maniculatus)* was the prevalent rodent species, and 30% of these had serum antibodies against hantaviruses (Childs et al 1994). When tissues from these seropositive rodents were RNA-extracted and amplified by reverse transcriptase-PCR, hantavirus sequences were obtained which closely matched those determined in lung autopsy tissues of HPS patients who had resided close to where the rodents were trapped (Nichol et al 1993).

These studies established that a new hantavirus was present in the rodent vector, the deer mouse *(Peromyscus maniculatus)*. Serial passage of infected lung material from a deer mouse in uninfected colonized deer mice followed by several passages in Vero E6 cell cultures yielded a virus with the identical genetic sequence to that contained in the original lung material (Elliott et al 1994). Although this virus was originally named Muerto Canyon virus, there were objections to the name from local residents and from the Navajo Nation. Accordingly, a new name acceptable to all parties was agreed upon, and the virus is now called Sin Nombre virus (Zaki et al 1995).

As further cases of HPS and infected deer mice were identified outside the Four Corners region of the southwestern United States, a considerable genetic sequence variation (up to 15%) was noted. This provided a powerful epidemiologic tool, since specific sequence changes were found to be related to the geographic locales from which the mice were caught (Nichol et al 1993). In addition, it became clear that Sin Nombre virus was not new, but had been established in the rodent reservoir for many generations (Spiropoulou et al 1994).

The distribution of *Peromyscus maniculatus* is widespread in the North American continent, from Mexico to northern Canada. Nevertheless, HPS cases have recently been detected in regions outside the normal habitat of this rodent species (Figure 1). Investigation of an HPS case in Florida led to the isolation of a new hantavirus present in the cotton rat, *Sigmodon hispidus* (Rollin et al 1995). A fatal case of HPS in Rhode Island yielded a distinct virus sequence (RI-1) and a reservoir in *Peromyscus leucopus* (Hjelle et al 1995); this virus was later renamed NY-1. New hantaviruses were also detected by sequence analysis in the harvest mouse, *Reithrodontomys megalotis* (Hjelle et al 1994), and the cotton rat, *Sigmodon alstoni* from Venezuela (S.T. Nichol and C. Fulhorst, personal communication). New hantaviruses were also detected by sequence analysis of post-mortem tissues of human HPS cases in Louisiana (Morzunov et al 1995) and Brazil (S.T. Nichol, personal communication). In the Louisiana case, the rodent reservoir has been identified as *Oryzomys palustris*, and the virus named Bayou virus. The rodent vector in Brazil has not yet been identified.

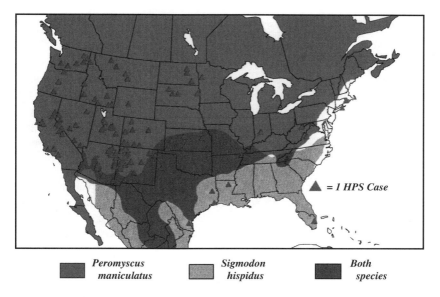

Peromyscus maniculatus   **Sigmodon hispidus**   **Both species**

FIGURE 1. DISTRIBUTION OF KNOWN RODENT HOSTS for hantavirus and location of HPS cases, as of May 3, 1996. Total cases (n=135 in 24 states). Rodent distributions from: Burt WH, Grossenheider RP. A Field Guide to the Mammals, 3rd ed. New York, New York, Houghton Mifflin Company, 1980.

Thus in the past few years, a previously unrecognized group of hanta-viruses has been revealed as the cause of extremely severe respiratory disease in the Western hemisphere. For confirmed cases of HPS, the mortality is nearly 50% (Table 2). It will be extremely important to develop antiviral drug therapies or vaccines that can control this disease in the future. In the mean-time, the control strategy is directed towards avoidance of close contact with rodents or their excreta.

TABLE 2. CHARACTERISTICS OF 107 HPS CASE PATIENTS reported to CDC as of May 22, 1996

- Age: Median 35 years; Range 11-69 years
- 79 (59%) Male
- Race: 96 (71%) White
          36 (27%) American Indian
           3 (2%) Black
- Ethnicity: 16 (12%) Hispanic
- Case Fatality Rate: 49%

Source: CDC, Division of Viral and Rickettsial Diseases

ARENAVIRUSES

BOLIVIAN HEMORRHAGIC FEVER

In August 1994, reports were received from Bolivia of seven severe hemorrhagic illnesses, including six deaths, in a family of nine. The survivor tested positive by IgM assay for Machupo virus infection, and samples from five of the six fatal cases were confirmed both by virus isolation and rapid antigen detection ELISA. Other samples were sent to a clinical diagnostic laboratory in Santa Cruz where a Bolivian scientist became exposed to an aerosol of infected blood due to a centrifuge accident and developed a fever. The scientist was treated with ribavirin and closely observed; she recovered from her mild symptoms without seroconversion. During their stay in Bolivia, the CDC team also heard of another hemorrhagic fever case in Cochabamba, but this patient died with typical Bolivian hemorrhagic symptoms before treatment could be started.

This episode emphasizes the fact that Bolivian hemorrhagic fever remains endemic in Bolivia, despite a paucity of information internationally. A growing number of South American arenaviruses is being recognized as human intervention brings increased contact with rodent vectors (Figure 2).

There is a clear need to evaluate the role of ribavirin in post-exposure prophylaxis or therapy and to develop more effective antiviral drugs and vaccines in the future. The example of Argentine hemorrhagic fever (Junin virus), for which an effective attenuated vaccine was recently developed, illustrates an approach to this problem.

BRAZILIAN HEMORRHAGIC FEVER

A new arenavirus, called Sabia, was isolated from a fatal case of hemorrhagic fever that occurred in 1990. While characterizing the virus, a laboratory technician became infected, probably by aerosol, and recovered after a severe illness lasting 15 days (Coimbra et al 1994). The vector of this virus, presumed to be a rodent, is presently unknown.

On August 8, 1994, a visiting professor working alone in the Yale Arbovirus Research Unit at Yale University was preparing a concentrate of Sabia virus from infected cell cultures when a centrifuge tube cracked, leaking its contents into the centrifuge. He cleaned out the centrifuge, but did not report the incident. On August 16, he developed myalgia and the following day, fever. He was hospitalized at the Tropical Medicine Clinic at Yale–New Haven Hospital on August 19, and the same day he was started on intravenous ribavirin therapy. Ten days later, after a brief febrile illness, he was discharged (Barry et al 1995). Infection with Sabia virus was confirmed by virus isolation.

This incident attracted national attention because of the concerns of the public about laboratory safety. The possibility that the infection might have spread was carefully investigated, but no secondary cases developed among

USA
Tamiami (1964)
(*Sigmodon hispidus*)

Tacaribe (1956)
(*Artibeus* bats)

Guanarito (1990)
(*Zygodontomys brevicauda*)

Pichindé (1965)
(*Oryzomys albigularis*)

VENEZUELA
COLOMBIA

Amaparí (1964)
(*Oryzomys capito;
Neacomys guianae*)

Machupo (1963)
(*Calomys callosus*)

BRAZIL

Flexal (1975)
(*Oryzomys* spp.)

Latino (1965)
(*Calomys callosus*)

BOLIVIA

Sabiá (1993)

Paraná (1965)
(*Oryzomys buccinatus*)

Junín (1958)
(*Calomys musculinus*)

ARGENTINA

Oliveros (1992)
(*Bolomys obscurus*)

FIGURE 2. ARENAVIRUSES isolated in South America. Source: CDC, Division of Viral and Rickettsial Diseases (DVRD)

five close contacts or 75 hospital and laboratory workers who examined the patient or handled specimens. A full inquiry was conducted by a CDC team to establish the facts and recommend improved safety procedures. The Sabia virus samples were moved to the biosafety level 4 laboratory at CDC where they can be studied in safety along with other highly pathogenic arenaviruses.

LASSA FEVER

Lassa fever virus occurs in endemic foci in the rodent reservoir, *Mastomys natalensis,* in several countries in West Africa. In August 1994, CDC was called by an Atlanta pharmacist whose brother lived in Ekpoma, Bendel State, Nigeria. The brother had developed hemorrhagic fever and was moved to Lagos for hospital care where he died of Lassa fever, exposing 15 physicians to possible infection. CDC sent ribavirin to Lagos for treatment of any who developed clinical symptoms. This incident highlights the potential for introduction of Lassa fever into the United States, closely resembling a case five years previously of a Nigerian who became infected with Lassa fever in Ekpoma while attending a family funeral and returned to his home in Chicago where he died of Lassa fever (Holmes et al 1990). There are chronic problems with Lassa fever in Nigeria as well as several other countries in West Africa, but no studies are currently in progress in these countries to monitor the disease, understand the epidemiology, or develop more effective control measures. The potential for introduction of Lassa fever into a developed industrial nation has been demonstrated and remains a continuous cause for concern.

FILOVIRUSES

EBOLA HEMORRHAGIC FEVER

After an interval of 15 years since its last occurrence in southern Sudan, a human case of Ebola hemorrhagic fever occurred in the Ivory Coast in late November 1994. Although at least four human infections occurred during the outbreak of Ebola virus in monkeys in Reston, Virginia, in 1989, no clinical disease was apparent. The case patient in the Ivory Coast was an ethologist who necropsied a chimpanzee on November 16 and developed fever, diarrhea and a pruritic rash beginning on November 24. After evacuation to Switzerland on December 1, the patient fully recovered. Investigation by physicians in the Ivory Coast and at the Institut Pasteur in Paris confirmed the infection by Ebola virus isolation from serum samples. There were no secondary cases, and no evidence of ongoing transmission in chimpanzees or humans was found (Le Guenno et al 1995).

A more detailed ecological investigation planned for the beginning of May 1995 was curtailed when CDC received reports on May 6 of a large outbreak of hemorrhagic fever centered in a hospital in Kikwit, Zaire (Figure 3). A number of nurses and doctors had died from a severe hemorrhagic illness that appeared to be Ebola to local physicians who had treated cases in the 1976 outbreak. Specimens forwarded via the Belgian Embassy to the Institute of Tropical Medicine in Antwerp could not be diagnosed there since the Institute no longer had the capability. Antwerp scientists sent the samples to CDC, where, within 6 hours, an Ebola antigen detection test showed that all but 2 of the 14 patients were actively infected (Ksiazek et al 1991). Reverse transcriptase-PCR amplification from the serum samples using Ebola-specific primers yielded a product of 528 base pairs that contained only four nucleotide sequence changes as compared to a similar product amplified from the 1976 Zaire isolate of Ebola virus (Sanchez et al 1996).

Thus Ebola virus clearly remains endemic in parts of Africa. To date, the natural reservoir has not been discovered. Although the disease in the necropsied chimpanzee from the Ivory Coast was consistent with Ebola infection, it is highly unlikely that this species is the reservoir host. Chimpanzees do, however, hunt other monkeys such as *Colobus* species and consume their meat (Stanford 1995), suggesting another possible reservoir host. We now have better reagents and molecular technologies with which to investigate this enigma. A further priority must be the development of some effective therapy for patients infected with this devastating hemorrhagic fever.

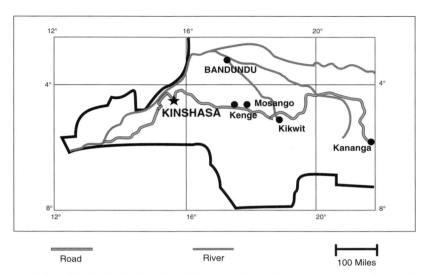

FIGURE 3. RELEVANT SITE IN ZAIRE of Ebola hemorrhagic fever outbreak, May 1995. Source: CDC, Division of Viral and Rickettsial Diseases (DVRD).

ACKNOWLEDGMENTS

In addition to the investigators cited in the references, we thank the following colleagues at CDC for providing valuable unpublished information: Lori Armstrong, Jamie Childs, James Mills, Paul Kilgore, Tom Ksiazek, Stuart Nichol, Pierre Rollin, Tony Sanchez, and Sherif Zaki.

REFERENCES

Barry M, Russi M, Armstrong L, Geller D, Tesh R, Dembry L, Gonzalez JP, Khan AS, Peters CJ (1995). Brief report: treatment of a laboratory-acquired Sabia virus infection. N Engl J Med 333: 294-296.

Childs JE, Ksiazek TG, Spiropoulou CF, Krebs JW, Morzunov S, Maupin GO, Gage KL, Rollin PE, Sarisky J, Enscore RE, Frey JK, Peters CJ, Nichol ST (1994). Serologic and genetic identification of *Peromyscus maniculatus* as the primary rodent reservoir for a new hantavirus in the southwestern United States. J Infect Dis 169: 1271-1280.

Coimbra TLM, Nassar ES, Burattini MN, Madia de Souza LT, Ferreira IB, Rocco IM (1994). New arenavirus isolated in Brazil. Lancet 343: 391-392.

Duchin JS, Koster F, Peters CJ, Simpson G, Tempest B, Zaki S, Ksiazek TG, Rollin PE, Nichol S, Umland ET, Moolenaar RL, Reef SE, Nolte KB, Gallaher MM, Butler JC, Breiman RF, and the Hantavirus Study Group (1994). Hantaviral pulmonary syndrome: a clinical description of 17 patients with a newly recognized disease. N Engl J Med 330: 949-955.

Elliott LH, Ksiazek TG, Rollin PE, Spiropoulou CF, Morzunov S, Monroe M, Gold-smith CS, Humphrey CD, Zaki SR, Krebs JW, Maupin G, Gage K, Childs JE, Nichol ST, Peters CJ (1994). Isolation of the causative agent of hantavirus pulmonary syndrome. Am J Trop Med Hyg 51: 102-108.

Hjelle B, Chavez-Giles F, Torrez-Martinez N, Yates T, Sarisky J, Webb J, Ascher M (1994). Genetic identification of a novel hantavirus of the harvest mouse *Reithrodontomys megalotis.* J Virol 68: 6751-6754.

Hjelle B, Krolikowski J, Torrez-Martinez N, Chavez-Giles F, Vanner C, Laposata E (1995). Phylogenetically distinct hantavirus implicated in a case of hantavirus pulmonary syndrome in the Northeastern United States. J Med Virol 46: 21-27.

Holmes GP, McCormick JB, Trock SC, Chase RA, Lewis SM, Mason CA, Hall PA, Brammer LS, Perez-Oronoz GI, McDonnell MK, Paulissen JP, Schonberger LB, Fisher-Hoch SP (1990). Lassa Fever in the United States, investigation of a case and new guidelines for management. N Engl J Med. 323: 1120-1123.

Hughes JM, Peters CJ, Cohen ML, Mahy BWJ (1993). Hantavirus pulmonary syndrome: an emerging infectious disease. Science 262: 850-851.

Ksiazek TG, Peters CJ, Rollin PE, Zaki S, Nichol S, Spiropoulou C, Morzunov S, Feldmann H, Sanchez A, Khan AS, Mahy BWJ, Wachsmuth K, Butler JC (1995). Identification of a new North American hantavirus that causes acute pulmonary insufficiency. Am J Trop Med Hyg 52: 117-123.

Ksiazek TG, Rollin PE, Jahrling PB, Johnson E, Dalgard DW, Peters CJ (1991). Enzyme immunosorbent assay for Ebola virus antigens in tissues of infected primates. J Clin Microbiol 30: 947-950.

LeGuenno B, Formontry P, Wyera M, Gounon P, Walker F, Boesch C, (1995). Isolation and partial characterisation of a new strain of Ebola virus. Lancet 345: 1271-1274.

Morzunov SP, Feldmann H, Spiropoulou CF, Semenova VA, Rollin PE, Ksiazek TG, Peters CJ, Nichol ST (1995). A newly recognized virus associated with a fatal case of hantavirus pulmonary syndrome in Louisiana. J Virol 69: 1980-1983.

Nichol ST, Spiropoulou C, Morzunov S, Rollin PE, Ksiazek TG, Feldmann H, Sanchez A, Childs J, Zaki S, Peters CJ (1993). Genetic identification of a hantavirus associated with an outbreak of acute respiratory illness. Science 262: 914-917.

Rollin PE, Ksiazek TG, Elliott LH, Ravkov EV, Martin ML, Morzunov S, Livingston W, Monroe M, Glass G, Ruo S, Khan AS, Childs JE, Nichol ST, Peters CJ (1995). Isolation of Black Creek Canal Virus, a new hantavirus from *Sigmodon hispidus* in Florida. J Med Virol 46: 35-39.

Ruo SL, Sanchez AS, Elliott LH, Brammer LS, McCormick JB, Fisher-Hoch SP (1991). Monoclonal antibodies to three strains of hantaviruses: Hantaan, R22, and Puumala. Arch Virol 119: 1-11.

Sanchez A, Trappier SG, Many BWJ, Peters CJ, Nichol ST (1996). The virion glycoproteins of Ebola viruses are encoded in two reading frames and are expressed through transcriptional editing. Proc Natl Acad Sci USA 93: 3602-3607.

Stanford CB (1995). Chimpanzee hunting behavior and human evolution. Am Scientist 83: 256-261.

Yanagihara R (1990). Hantavirus infection in the United States: epizootiology and epidemiology. Rev Infect Dis 12: 449-457.

Zaki SR, Greer PW, Coffield LM, Goldsmith CS, Nolte KB, Foucar K, Feddersen RM, Zumwalt RE, Miller GL, Khan AS, Rollin PE, Ksiazek TG, Nichol ST, Mahy BWJ, Peters CJ (1995). Hantavirus Pulmonary Syndrome. Pathogenesis of an emerging infectious disease. Am J Pathol 146: 3, 552-579.

Vignette

Rubella Virus Infection

Rubella (German measles) is a mild viral infection of children and young adults. However, infection of a pregnant woman and subsequent transmission of the virus to her fetus can have disastrous results. In 1941, Norman Gregg, an Australian ophthalmologist, made the profound observation of a direct association between rubella infection of pregnant women and the development of congenital cataracts in the newborn. This report was soon confirmed and extended by workers in Australia, Sweden, America, and England who also noted the association of rubella infection of the fetus with heart disease and deafness, thus completing the characteristic clinical triad of congenital rubella infection. Other clinical and epidemiologic studies confirmed this clinical picture of congenital rubella syndrome and also the risk of fetal anomalies in relation to the time in gestation when infection occurred. But it was not until 20 years after Gregg's initial report that the isolation and serologic identification of rubella virus was achieved by Weller and Neva in Boston and by Parkman, Beuscher, and Artenstein in Washington, DC. By 1969–1970, the rubella vaccine was developed for commercial use. By the late 1970s, use of the vaccine in the general population significantly reduced the incidence of rubella-associated congenital malformations, once a devastating problem in both human tragedy and expense for lifelong institutionalization of such children. The chapter by Teryl Frey addresses many of these and future issues concerning rubella virus.

20

Present Status of Rubella Virus

TERYL K. FREY

Rubella virus causes a benign disease known as rubella or German measles that is usually contracted during childhood. The significant complication of rubella virus infection is the constellation of severe birth defects known collectively as congenital rubella syndrome (CRS) that results when infection occurs during the first trimester of pregnancy and the virus readily crosses the placenta and replicates in the fetus. Rubella virus was first isolated in 1962 and since then significant progress has been made in both medical and experimental aspects of the virus. Medically, the significant accomplishment was the development of live attenuated vaccines by 1969 and their use in vaccination programs that subsequently controlled the virus in most developed countries of the world. Progress on the molecular biology of rubella virus came more slowly; however, recent findings have revealed that rubella virus occupies a unique niche with respect to virus taxonomy and evolution. In this chapter, the current status of medical and experimental aspects of rubella virus is summarized and future challenges in this field are explored. The reader is also referred to reviews by Wolinsky (1990) and Frey (1994a), which cover the medical and experimental aspects of rubella virus, respectively.

RUBELLA VIRUS TAXONOMY

Rubella virus belongs to the rubivirus genus of the Togavirus family of animal viruses. Togaviruses have a genome of single-stranded RNA of roughly 4×10^6 daltons (10,000 to 12,000 nucleotides) which is infectious. The basis of infectivity is that the genomic RNA serves as an mRNA; such genomic RNAs have come to be known as "positive-sense" or "plus-strand" RNAs. The Togavirus genomic RNA is encapsidated in an icosahedral nucleocapsid composed of multiple copies of a single virus-specified capsid (C) protein. The nucleocapsid is, in turn, surrounded by a lipid bilayer envelope of cellular origin in which are embedded two virus-specified glycoproteins, E1 and E2, that form spikes projecting from the surface of the virion envelope. It is these spikes that recognize and initiate contact with susceptible cells. These spikes are also the moiety that is recognized by antibodies that neutralize virus infectivity.

Rubella virus is the only member of the rubivirus genus. Alphavirus is the other Togavirus genus, which contains 26 species of arthropod-borne viruses, the type members being Sindbis virus and Semliki Forest virus. Although the genomic organization and replication strategy of rubella virus and the alphaviruses are similar, the biologies of these genera differ dramatically. Alphaviruses replicate in both the vertebrate host and the invertebrate vector (usually a mosquito). In cell culture, alphaviruses replicate cytolytically and to high titer in vertebrate cells, but noncytopathically and persistently in mosquito cells. Alphaviruses replicate efficiently in a wide variety of vertebrate culture cells and thus have been extensively studied at the molecular level and are considered model viruses [the molecular biology of alphaviruses was recently reviewed by Strauss and Strauss (1994)]. In contrast, rubella virus replicates slowly, noncytopathically, and to low titer in most vertebrate cell lines, a feature that has slowed its molecular characterization.

The initial Togavirus family contained a number of other viruses, including the flaviviruses (group B arboviruses such as yellow fever virus and dengue virus), the pestiviruses (hog cholera virus, bovine viral diarrhea virus, border disease virus), the arteriviruses (equine arteritis virus), and ungrouped viruses such as murine lactate dehydrogenase-elevating virus, simian hemorraghic fever virus, and the cell-fusing agent of *Aedes aegypti.* However, molecular characterization of all of these viruses has led to their reclassification into other virus families. Thus, rubella virus occupies a unique taxonomic niche and has no known animal virus relatives. This is a highly unusual situation and hinders research on rubella virus pathogenesis, since no animal model for rubella virus exists and rubella virus replicates only to a limited extent in laboratory animals.

RUBELLA VACCINES AND VACCINATION PROGRAMS

Several live, attenuated rubella virus vaccine strains have been developed and licensed (reviewed in Plotkin 1988). The strategy for development of all of these vaccine strains was similar, namely extensive serial passaging of a wild isolate in culture cells until a virus with an attenuated phenotype was selected. The attenuated phenotype was indicated by a lack or a substantial lessening of symptoms following injection into volunteers but with retention of induction of humoral immunity. Interestingly, despite attenuation, the vaccine viruses can still cross the placenta and infect the fetus. Until 1989, the United States Public Health Service kept a registry of inadvertent vaccinations in the period of the three months before and three months after conception. Of a total of 324 children born to entrants in the registry, no cases of CRS were reported. Thus, the attenuation appears to have been successful in preventing the development of congenital malformations in inadvertently infected fetuses.

Outside of Japan, the universally accepted vaccine strain is the RA 27/3, which through extensive testing has been shown to be superior to the other available vaccine strains in terms of induction of immunity and lack of complications. The RA 27/3 strain was derived from an isolate from a therapeutic abortion by passage 25 times in the human diploid lung cell line, WI-38. In addition to passage in a human cell line, other novel approaches were used in the development of this vaccine, such as passaging the virus several times at 30°C rather than the standard 37°C and using limiting dilutions at some of the passages which has the effect of genetically purifying the strain. Unfortunately, the original isolate and the attenuating passages have been lost and thus the molecular basis of attenuation of this very successful vaccine might never be understood.

Other vaccine strains developed, licensed, and used to some extent in the United States and Europe were the HPV-77 and the Cendehill strains. Five independent live, attenuated vaccine strains (TCRB-19, Matsuura, To-336, Takahashi, and Matsuba) were developed and have been used successfully in Japan.

Rubella vaccination is aimed at preventing CRS. The strategy employed in the United States calls for vaccination in early childhood (12 to 15 months), usually administered in trivalent form with measles and mumps, which has become known as MMR. Childhood vaccination is augmented with vaccination of seronegative adult women. Despite the absence of reports of CRS following inadvertent vaccination during pregnancy, vaccination during pregnancy is contraindicated and seronegative pregnant women are vaccinated postpartum. Such women are at risk and thus vaccination of seronegative health care workers who could come in contact with a seronegative pregnant woman is required in many states. This strategy has been successful in reducing both rubella and CRS to very low levels.

In England and Japan, a different strategy was adopted. Reasoning that immunity stimulated by natural infection was more robust than immunity induced by vaccination, only teenage girls were targeted for vaccination. Thus, rubella persisted in children and males. Unfortunately, this reservoir led to infection of women who had managed to reach adulthood without a rubella titer and CRS was thus not eliminated. In both countries, female teenage vaccination was augmented with universal early childhood immunization by 1988.

CURRENT EPIDEMIOLOGY

Humans are the only host for rubella virus. The virus is spread via the respiratory route. Prior to establishment of a vaccination program, the virus is usually contracted in childhood. In countries with a temperate climate, the peak of rubella activity is in the spring. Although rubella is present every year, epidemics occur every five to nine years. Congenital rubella infection occurs during periods of rubella activity. However, since CRS is not recognized until after birth, peaks of occurrence of CRS lag behind peaks of rubella by nine months to a year.

Vaccination was implemented in the United States in 1969; however, vaccination of adults was not initially aggressively pursued. Thus, although yearly rubella rates fell and the epidemics ceased, CRS still occurred at an average rate of 50 cases per year. Following more aggressive insistence on both childhood and adult vaccination in the late 1970s, CRS rates finally fell dramatically after 1979 and bottomed out at one to two cases per year in the late 1980s. Despite the success of the United States vaccination strategy, a resurgence of rubella occurred between 1989 and 1991 (Lindegren et al 1991). The resurgence was primarily among groups who were never vaccinated, such as Hispanics in the Southwest and Amish religious dissenters in New York and Pennsylvania. However, outbreaks also occurred in foci of seronegative adults on college campuses and prisons. Although the total number of cases never amounted to more than 1500 per year nationally, the proportion of CRS to rubella cases was much higher than during prevaccination years when the virus was endemic, since the average age of infected individuals was the late teens or early twenties; 56 cases of CRS occurred during the resurgence (Mellinger et al 1995).

The lesson learned from the 1989–1991 resurgence was that "herd immunity" as applied to animals is not acceptable in a human population. Even with a high proportion of vaccine coverage, the number of seronegatives will eventually be large enough such that the virus, once reintroduced from endemic areas, can spread. Fortunately, the cases of rubella during the resurgence were primarily among individuals who had never previously contracted the disease or been vaccinated. Thus, there was no evidence of a vaccine failure rate as there was during the concurrent measles epidemic. Since the 1989–1991 rubella resurgence and concurrent measles epidemic, the United States Public Health Service has mandated two doses of MMR during childhood (the second at 5 to 12 years of age) and has attempted to aggressively pursue vaccination through publicity, distribution of vaccine at cost, and a tightening of restrictions where vaccination is required (such as school enrollment).

Rubella is on the wane in the United Kingdom following adoption of early childhood immunization (Miller et al 1993). In Japan, the early childhood MMR vaccination with Japanese vaccine strains of all three components was interrupted in 1991 due to problems with the mumps vaccine strains. Therefore, rubella is still endemic in Japan, with epidemics occurring every five years (S. Katow, personal communication). Some countries, such as Sweden, have eliminated rubella. Rubella vaccination programs have been initiated in some areas outside of the West and Japan, such as Singapore (Epidemiological Notes 1989) and Taiwan (Lin and Chen 1994). However, in many countries, rubella vaccination is not pursued (Miller 1991). Rubella is not included in the World Health Organization's Expanded Programme for Immunization (Bart and Lin 1990). Thus, in most of the world, rubella is still endemic.

CURRENT MEDICAL CHALLENGES OF RUBELLA

VACCINE COMPLICATIONS

Chronic Arthropathy. In terms of complications, the rubella vaccines have been among the most successful virus vaccines developed. However, attention in the United States has recently focused on a proposed association between vaccination of adult women and development of chronic arthritis and neurological complications. Transient arthritis is recognized as a relatively common sequela of natural rubella virus infection, particularly in adult women, and a specific arthritis, rubella arthritis, has been defined. Arthritis following natural rubella occasionally becomes chronic. The first vaccine strain used in the United States vaccination program, HPV-77, was associated with an unacceptably high rate of transient arthritis as well as chronic arthritis and neurological complications, such as "catcher's crotch syndrome," in some children. The HPV-77 vaccine was withdrawn in 1979 and replaced with the RA 27/3 vaccine strain. The expected rate of transient arthritis following RA 27/3 vaccination is 25% as stated by the United States Advisory Committee on Immunization Practices (ACIP) (CDC 1990).

The laboratories of Aubrey Tingle and Janet Chantler of the University of British Columbia presented data indicating that chronic arthritis can also follow RA 27/3 vaccination of adult women (Tingle et al 1986). The chronic arthritis is often accompanied by neurological manifestations, such as parasthesias and carpal-tunnel syndrome, and the complete spectrum of complications has been termed "chronic arthropathy." The condition is often resolved within a year or two, but can also persist indefinitely. The data in one report suggested that the rate of post-vaccination chronic arthropathy could be as high as 5 to 10% (Tingle et al 1986). Dr. Chantler's laboratory reported isolation of rubella virus from peripheral blood lymphocytes of chronic arthropathy patients (Tingle et al 1985), as well as from children with juvenile rheumatoid arthritis (Chantler et al 1985). This is consistent with the finding that rubella virus can be isolated from synovial fluid of patients suffering with rubella arthritis (Ford et al 1988) and suggests that virus persistence plays a major role in the pathogenesis of chronic arthropathy.

The findings of Tingle and Chantler are disputed by the ACIP, particularly the incidence (CDC 1990). The issue was reviewed by the Institute of Medicine (IOM) as part of a review on adverse reactions to childhood vaccines (Howson and Fineberg 1992). The IOM concluded that chronic arthropathy as a complication of rubella vaccination was consistent with sequelae known to follow natural rubella infection. However, the IOM also concluded that the data were scanty, particularly regarding incidence rates, and called for more research. The rubella vaccine has been added to the list of vaccines for which adverse reactions will be compensated by the United States government, given that a list of symptoms is met in a time-frame appropriately following rubella vaccination.

The main need surrounding the controversy about rubella vaccine-related chronic arthropathy is further research. Concerning the incidence, chronic arthropathy as a complication of rubella vaccination was not widely known or publicized in the medical community until the past five years, a situation that would lead to underreporting. However, since 10% of the population routinely suffers from various forms of arthritis, it is difficult to sort out rubella vaccine-associated arthritis from this general background. At least three studies on the incidence of chronic complications are underway or have been completed (see Frey 1994b); unfortunately none so far has been published in refereed journals.

Concerning the pathogenesis of vaccine-associated chronic arthropathy, Chantler's isolation of virus from chronic arthropathy patients is consistent with findings on classic rubella arthritis, but these results have not been reproduced by other labs (Frenkel et al 1996; R.M. Dougherty, personal communication). The polymerase chain reaction has also been employed to detect persistent rubella virus in chronic arthropathy patients; one report was positive on two patients (Mitchell et al 1993), while in a second study, the results were negative on two patients (Frenkel et al 1996). A major problem is finding willing patients who fulfill the case definition of rubella-associated chronic arthropathy (which contradicts the incidence rate of 5% of adult female vaccinees). Ideally, repeated specimens should be taken from such patients, particularly when episodes of symptoms occur.

It is also possible that chronic arthropathy has an immunopathological basis. Rubella virus has been associated with other autoimmune diseases, such as insulin-dependent diabetes mellitus and thyroiditis in CRS patients and progressive rubella panencephalitis. The occurrence of chronic arthropathy almost exclusively in women also indicates an immunopathological mechanism. It also should be considered that some women who reach adulthood seronegative for rubella, despite the great opportunity for exposure either through natural infection or vaccination, may suffer from an immunological abnormality with respect to the virus and may therefore be at risk for development of chronic complications following vaccination with live virus.

Thus, while rubella vaccine-associated chronic arthropathy may be a real clinical entity, its incidence is too rare and its pathogenesis too ill-defined to justify any changes in the adult vaccination strategy. Research, particularly on the pathogenesis, is needed, but unfortunately no animal model for rubella exists. Hopefully, such study will identify risk factors (such as major histocompatibility complex grouping) that would contraindicate vaccination and will lead to therapeutic strategies to attenuate or terminate the complications.

CRS Caused by Reinfection. Cases of CRS in infants born to seropositive mothers have clearly been documented. The immunity induced by vaccination is not as robust as that induced by natural infection. It has therefore been hypothesized that CRS caused by reinfection is more of a threat in a vaccinated population than in a population whose immunity is due to natural infection. However, there is no proof for this hypothesis. CRS caused by reinfection is very rare (only 30 cases have been documented in the literature; S. Katow as reported in Frey 1994b). Reinfection is presumably due to a wild virus. Since childhood vaccination programs substantially reduce the rate of rubella, CRS due to reinfection in a vaccinated population is a self-limiting problem.

RUBELLA AS A MODEL FOR AUTOIMMUNE DISEASE CAUSED BY VIRUSES

A connection between virus infection, particularly persistent virus infection, and chronic and autoimmune diseases such as multiple sclerosis, rheumatoid arthritis, and insulin-dependent diabetes mellitus (IDDM), has been postulated and is supported in some cases by epidemiological and/or immunological evidence (reviewed in Oldstone 1987, 1989). However, the association is not with an individual, but with numerous viruses (rubella virus itself has been associated with all three of these diseases). Furthermore, attempts to demonstrate virus in diseased tissue have generally been negative (Godec et al 1992; Buesa-Gomez et al 1994). In the absence of evidence of direct virus infection, it is thought that viruses may act to trigger aberrant autoimmune responses and disappear by the time symptoms develop. One triggering mechanism could be molecular mimicry between epitopes on virus and cell proteins which stimulate a misdirected cell-mediated response against the cell protein, resulting in autoimmunity.

The best connection between a specific persisting human virus and autoimmune disease is in the CRS population. CRS patients are persistently infected at birth, although detectable virus usually disappears by one year of age. Although congenital infection could lead to tolerance, CRS infants have a robust anti-rubella virus humoral immunity (including neutralizing antibodies) but have defects in the cell-mediated response which allow the virus to persist. In later life, a large percentage of CRS patients develop endocrine dysfunctions such as IDDM (20 to 40% of CRS patients) and thyroiditis. A rare, fatal neurodegenerative disease, progressive rubella panencephalitis (PRP), has also been recognized in the CRS population. The pathogenesis of

PRP appears to be autoimmune in nature. Considering the incidence of autoimmune disease in the CRS population, it would be of great interest to characterize the cell-mediated response to rubella virus in CRS patients. Among other things, it could be determined whether virus epitopes that display mimicry with cell epitopes are recognized.

ERADICATION

Since humans are the only host for rubella virus and an effective vaccine exists, eradication is feasible. However, an organized effort to eradicate rubella, let alone accomplishing eradication, will be a long time in coming. The primary obstacle to organization of an eradication effort is the benign nature of the disease. Countries will necessarily deal with other, more virulent pathogens, such as HIV, influenza, measles, diarrhea viruses and hepatitis viruses, before eradication of rubella is considered. Once undertaken, an eradication campaign against rubella will be difficult to monitor because over half of rubella cases are inapparent.

However, rubella eradication is a worthwhile effort. The load placed on society by CRS is significant, considering that in a developed country, many victims require lifelong institutionalization. The 20,000 CRS patients in the United States from the 1964 epidemic are now only 30 years old and the estimated lifelong cost of maintaining a CRS patient was calculated to be $200,000 in 1985 (White et al 1985). Fortunately, the virus is controlled in most developed countries. However, as evidenced by the 1989–1991 rubella resurgence in the United States, a vigilant vaccination effort is required. Therefore, it is in the best interests of developed countries to assist eradication efforts in less developed countries. Since the cost-benefit ratio of controlling rubella is less striking than with deadlier viruses, development of an inexpensive, DNA-based genetic rubella vaccine may be necessary to provide the needed impetus for less developed countries to undertake rubella vaccination efforts.

THE MOLECULAR BIOLOGY OF RUBELLA VIRUS

Most of the current understanding of the molecular biology of rubella virus has been obtained in the past ten years. The rubella virus genomic RNA is 9759 nucleotides in length. Figure 1 shows the coding strategy. In the positive-polarity, the genome contains two long open reading frames (ORFs). The 5' terminal of these (roughly 2200 amino acids in length) encodes the nonstructural proteins which are only present in infected cells and function primarily as RNA replication enzymes (this ORF is termed the nonstructural protein or NSP-ORF). The 3' proximal ORF (roughly 1000 amino acids in length) encodes the virion (or structural) proteins, C, E2, and E1. The genome contains 5' and 3' untranslated regions of 40 and 58 nucleotides, respectively, and the ORFs are separated by 123 nucleotides. As shown in Figure 1, the rubella virus genomic coding strategy is similar to that of an alphavirus (Sindbis) genome.

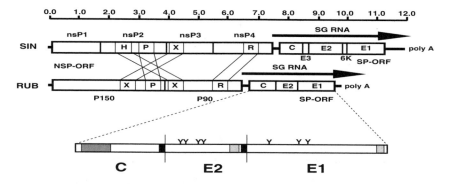

FIGURE 1. TOPOGRAPHY OF THE GENOME RNAS OF RUBELLA VIRUS (RUB) and alphaviruses (Sindbis virus, SIN). The scale at the top is in kilobases. Black lines denote untranslated sequences, open boxes indicate open reading frames (ORFs), and arrows indicate the regions of the genome included in the subgenomic RNA (SG RNA). In both viruses, the 5′ proximal ORF encodes nonstructural proteins (NSP-ORF) and the 3′ proximal ORF encodes structural proteins (SP-ORF). The boundaries of the individual proteins processed from the precursor translated from each ORF are denoted. Within the nonstructural protein ORFs, the location of global amino acid motifs indicative of replicase (R), helicase (H), and cysteine protease activity (P) as well as the small region of homology between the deduced amino acid sequence of rubella virus and Sindbis virus (X motif) are shown. An expanded topography of the rubella virus SP-ORF is shown at the bottom of the diagram. Within the ORF, the positioning of the following domains of the structural proteins are shown: hatched box, the hydrophilic region of C which contains a high concentration of basic amino acids and putatively interacts with the virion RNA; black box, the hydrophobic signal sequences which precede the N-termini of E2 and E1; stippled box, the transmembrane sequences of E2 and E1; Y, potential N-linked glycosylation sites.

Figure 2 diagrams the rubella virus replication cycle in infected cells. Rubella virus replication occurs entirely in the cytoplasm and has no reproducibly detectable effect on cell RNA or protein synthesis. After binding to a receptor on the cell surface (the identity of which has not been determined), the virus enters the cell by receptor-mediated endocytosis (Petruzziello et al 1996). In the lowered pH environment of the primary endosome, the virus membrane fuses with the endosomal membrane, releasing the capsid and virus genome into the cytoplasm. Following release from the capsid, the virion RNA is translated to produce the nonstructural proteins. A protease within the primary translation product of the NSP-ORF cleaves the precursor at a single site to produce the two final nonstructural proteins, P150 and P90. These proteins probably function as a complex to replicate the viral RNA; however, their precise functions are not known. Computer-assisted homology searching revealed amino acid motifs indicative of methyl transferase activity on P150 and motifs indicative of helicase and RNA replicase activity on P90.

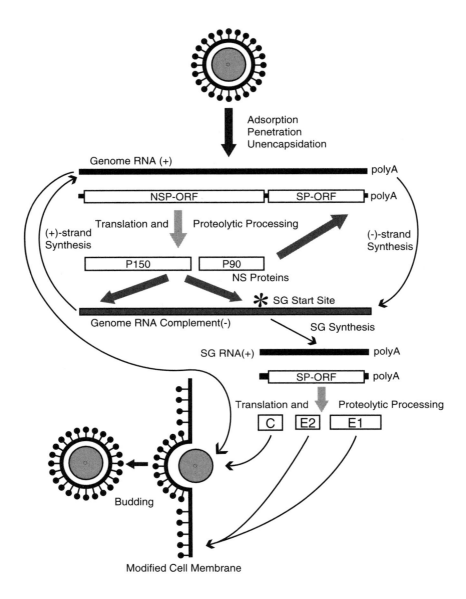

FIGURE 2. RUBELLA VIRUS REPLICATION. The ORFs present are indicated (NSP-ORF = nonstructural protein ORF; SP-ORF = structural protein ORF) under both (+)-polarity RNA species, the genome and subgenome (SG) RNAs (solid black lines). The (-)-polarity RNA species complementary to the genomic RNA (hatched line) is used solely as a template for the two (+)-polarity RNA species.

The viral replicase proteins use the genomic RNA as a template to make a negative-polarity complement, which then serves as template for synthesis for two positive-polarity RNA species, the genomic RNA and a subgenomic RNA consisting of the 3' terminal sequences of the genomic RNA. Cell proteins have been shown to have a function in RNA virus RNA replication and are thought to bind regulatory sequences (or *cis*-acting elements) on the viral RNA and then facilitate the recognition and/or association of the virus replicase components. Few of these cell proteins have been identified. It has recently been shown that calreticulin specifically binds to a prominent stem-and-loop structure near the 3' end of the rubella virus genomic RNA (Singh et al 1994). Human calreticulin is associated with cytoplasmic ribonucleoprotein complexes that contain a small RNA species known as hYRNA and components of the Ro/SSA autoantigen complex. Autoantibodies against Ro/SSA are found in patients suffering from autoimmune diseases. It was recently shown that one member of the Ro/SSA complex, a protein, specifically binds the tertiary structure at the 5' end of the rubella virus genomic RNA (Pogue et al 1996). The possible involvement of Ro/SSA complex proteins in rubella virus replication may explain the pathogenesis of the arthritis associated with rubella infection and vaccination.

The rubella virus structural protein ORF is translated from the subgenomic RNA species. The coding sequences for the structural proteins are in the order NH_2-C-E2-E1-COOH within the ORF. E2 and E1 are preceded by signal sequences of hydrophobic amino acids that direct the translation of both E2 and E1 into the lumen of the endoplasmic reticulum. The C-E2 and E2-E1 cleavages are both mediated by signalase, a cellular lumenal enzyme that removes signal sequences from cellular membrane-associated and secreted proteins. The genomic RNA associates with the C protein, and following coalescence of the capsid, the capsid buds through cellular membranes modified to contain solely E2-E1 complexes to acquire its envelope. Budding occurs primarily in the Golgi apparatus; however, late in infection, budding also occurs at the cell surface.

Although, the strategy for synthesis of the rubella virus and alphavirus structural proteins is similar (Figure 1), there are three notable differences: 1) the alphavirus C protein is an autoprotease that cleaves itself from the precursor while nascent on the ribosome; 2) the alphavirus structural protein ORF contains an additional small protein (the 6K protein) of unknown function between the E2 and E1 sequences; and 3) alphavirus capsids form in the cytoplasm of infected cells and then migrate to the cytoplasmic membrane where budding occurs, whereas rubella virus capsid formation is concomitant with budding. In this latter regard, rubella virus capsid morphogenesis is reminiscent of retroviruses.

RUBELLA VIRUS AND RNA VIRUS EVOLUTION

Computer-assisted homology analysis has revealed homologies between the nonstructural proteins of diverse positive-strand RNA viruses of animals and plants (reviewed by Goldbach 1991). This finding indicates that, despite the diversity of the positive-strand RNA viruses currently in existence, these viruses diverged from ancient common ancestors. One mechanism by which the current diversity of positive-strand RNA viruses is thought to have been generated is interviral recombination. Recombination occurs between RNA viruses by a copy-choice mechanism in which a replicase copying one template in a cell dissociates from the template but remains associated with the product. If this replicase-product complex associates with a heterologous template RNA (an RNA virus co-infecting the same cells or, occasionally, a cell RNA) and replication proceeds, a recombinant product strand will be produced.

Based on these computer-assisted analyses, at least three superfamilies of positive-strand RNA viruses have been defined that each contain both plant and animal viruses. Both rubella virus and the alphaviruses belong to the alphavirus-like superfamily. Interestingly, in phylogenetic analyses of members of this superfamily using sequences surrounding both the helicase and replicase motifs, alphaviruses segregate on one branch with several plant viruses such as tobacco mosaic virus and brome mosaic virus, while rubella virus segregates on a different branch with beet necrotic yellow vein virus (a furovirus) and human hepatitis E virus, a cause of enterically spread, non-A, non-B hepatitis. Other than the presence of homology in the NSP-ORF, hepatitis E virus is unlike rubella virus since it is nonenveloped, encoding only a capsid protein, and has a third ORF that encodes a protein of unknown function.

Considering the similarity of the genomic coding strategy of rubella virus and the alphaviruses, the initial assumption was that these genera diverged from an immediate common ancestor. However, phylogenetic analyses indicate that Togavirus evolution was probably more complicated. Another indication of the complexity of Togavirus evolution is the different order of the global motifs within the NSP-ORF of rubella virus and alphaviruses (Figure 1). Such a rearrangement would require four copy-choice recombinational events to accomplish. A more likely scenario is that either rubella virus was generated by recombination between an alphavirus and hepatitis E virus or that alphaviruses were generated by a recombinational event between rubella virus and a plant virus such as tobacco mosaic virus. Another possibility is that the Togavirus ancestor was the primordial member of the alphavirus-like superfamily and that the alphavirus-rubivirus divergence was the first to occur, with all of the other superfamily members deriving from the subsequent alphavirus and rubivirus lines.

Although this revised version of Togavirus evolution is appealing, it is based entirely on computer-assisted homology analysis, as are many other hypotheses about RNA virus evolution. This version of Togavirus evolution predicts that since the NSP-ORF of alphaviruses and rubiviruses originated from different sources, the encoded proteins might function differently, which could in part explain some of the differences in the interactions of these viruses with infected cells and in their biologies. Thus, it will be of great interest to compare the nonstructural proteins of rubella virus with alphaviruses as well as with hepatitis E virus, which is predicted to be more closely related to rubella virus than are alphaviruses. The NSP-ORF of rubella virus is only in the initial stages of characterization. However, proteolytic processing is distinctly different from alphaviruses in which the NS protease works in *trans* to produce four proteins from the nonstructural precursor (Figure 1) (the NSP-ORF of hepatitis E virus has not been characterized at all). Recently it was shown that the rubella virus NS protease functions only in *cis* (Chen et al 1996). This evolutionary scheme also predicts that the SP-ORFs of alphaviruses and rubella virus are more closely related than are the NSP-ORFs. Comparative study of the three-dimensional structure of the alphavirus and rubella virus structural proteins should reveal any similarities that exist.

ACKNOWLEDGMENTS

Research on rubella virus in my laboratory is supported by grant AI-21389 from the National Institutes of Health.

REFERENCES

Bart KJ, Lin KF (1990). Vaccine-preventable disease and immunization in the developing world. Pediatr Clin North Am 37: 735-756.

Buesa-Gomez J, de la Torre JC, Dyrberg T, Landin-Olsson M, Mauseth RS, Lernmark A, Oldstone MBA (1994). Failure to detect genomic viral sequences in pancreatic tissues from two children with acute-onset diabetes mellitus. J Med Virol 42: 193-197.

CDC (1990). Rubella prevention: Recommendations of the Immunization Practices Advisory Committee (ACIP). MMWR 39: RR-15.

Chantler JK, Tingle AJ, Petty RE (1985). Persistent rubella virus infection associated with chronic arthritis in children. N Engl J Med 313: 1117-1123.

Chen JP, Strauss JH, Strauss EG, Frey TK (1996). Characterization of the rubella virus nonstructural protease domain and its cleavage site. J Virol 70: in press.

Ford DK, Tingle AJ, Chantler JK (1988). Rubella arthritis. In: Espinoza L, ed. Infections in the rheumatic diseases. Orlando: Grune and Stratton, 103-106.

Epidemiological Notes (1989). Introduction of measles-mumps-rubella (MMR) into the national childhood immunisation programme. Epidemiological News Bulletin (Singapore) 15: 66-68.

Frenkel LM, Nielsen K, Garakian A, Wolinsky JS, Cherry JD (1996). A search for persistent rubella infection in persons with chronic symptoms after rubella and rubella immunizations and in patients with juvenile rheumatoid arthritis. Clin Infect Dis 22: 287-294.

Frey TK (1994a). Molecular biology of rubella virus. Adv Virus Res 44: 69-160.

Frey TK (1994b). Report of an international meeting on rubella vaccines and vaccination, 9 August 1993, Glasgow, United Kingdom. J Infect Dis 70: 507-509.

Godec MS, Asher DM, Murray RS, Shin ML, Greenham LW, Gibbs CJ, Gajdusek DC (1992). Absence of measles, mumps, and rubella viral genomic sequences from multiple sclerosis brain tissue by polymerase chain reaction. Ann Neurol 32: 401-404.

Goldbach RW (1990). Genome similarities between positive strand RNA viruses. In: Brinton MA, Heinz FX, eds. New aspects of positive strand RNA viruses. Washington, DC: American Society for Microbiology, 1-12.

Howson CP, Fineberg HV (1992). Adverse events following pertussis and rubella vaccine. Summary of a report of the Institute of Medicine. JAMA 267: 392-396.

Lin DB, Chen CJ (1994). Seroepidemiology of rubella virus infection among female residents on the offshore islets of Taiwan. J Trop Med Hyg 97: 75-80.

Lindegren ML, Fehrs LJ, Hadler SC, Hinman AR (1991). Update: Rubella and congenital rubella syndrome, 1980-1990. Epidemiol Rev 13: 341-348.

Mellinger AK, Cragan JD, Atkinson WL, Williams WW, Kleger B, Kimber RG, Tavris D (1995). High incidence of congenital rubella syndrome after a rubella outbreak. Pediatr Infect Dis J 14: 573-578.

Miller CL (1991). Rubella in the developing world. Epidemiol Infect 107: 63-68.

Miller E, Waight PA, Vurdien JE, Jones G, Tookey PA, Peckham CS (1992). Rubella surveillance to December 1992: second joint report from the PHLS and National Congenital Rubella Surveillance Programme. CDR Review 1993 3: 35-40.

Mitchell LA, Tingle AJ, Shukin R. Sangeorgan JA, McCune J, Braun DK (1993). Chronic rubella vaccine-associated arthropathy. Arch Intern Med 153: 2268-2274.

Pogue GP, Hofmann J, Duncan R, Best JM, Etherington J, Southheimer RD, Nakhasi HL (1996). Autoantigens interact with cis-acting elements of rubella virus RNA. J Virol 70: in press.

Oldstone MBA (1987). Molecular mimicry and autoimmune disease. Cell 50: 819-820.

Oldstone MBA (1989). Viral persistence. Cell 56: 517-520.

Plotkin SA (1988). Rubella vaccine. In: Plotkin SA, Mortimer EA, eds. Vaccines. Philadelphia: W.B. Saunders, 235-262.

Singh NK, Atreya CD, Nakhasi HL (1994). Identification of calreticulin as a rubella virus RNA binding protein. Proc Natl Acad Sci USA 91: 12770-12774.

Strauss JH, Strauss EG (1994). The alphaviruses: Gene expression, replication, and evolution. Microbiol Rev 58: 491-562.

Tingle AJ, Chantler JK, Pot KH, Paty DW, Ford DK (1985). Postpartum rubella immunization: Association with development of prolonged arthritis, neurological sequelae, and chronic rubella viremia. J Infect Dis 152: 606-612.

Tingle AJ, Allen M, Petty RE, Kettyls GD, Chantler JK (1986). Rubella-associated arthritis. I. Comparative studies of joint manifestations associated with natural rubella infection and RA 27/3 rubella immunisation. Ann Rheum Dis 45: 110-114.

Wolinsky JS (1990). Rubella. In: Fields BN, Knipe DM et al, eds. Virology. Ed 2. New York: Raven Press, 815-838.

White CC, Koplan JP, Orenstein WA (1985). Benefits, risks and costs of immunization for measles, mumps, and rubella. Am J Public Health 75: 739-744.

PLANT VIRUSES

...today we demand with a great hue and cry more laboratories,
more microbe hunters, better paid researchers to free us
from the diseases that scourge us.....

PAUL DE KRUIF—*Microbe Hunters*

Vignette

Plant Viruses

Mankind is highly dependent on plants for survival and welfare. Diseases of plants can cause economic losses and shortage of food. In the mid-nineteenth century, plant pathology became established as the scientific study of plant diseases, which were shown to be caused first by fungi and later by bacteria, viruses, and nematodes. In 1892, the agent of tobacco mosaic disease was found to be smaller than all known bacteria, which could be observed by a light microscope and cultured *in vitro*. In the first half of the twentieth century, (plant) viruses were defined on the basis of their ability to cause pathological conditions (their infectivity), their small size relative to other pathogens, and their inability to multiply *in vitro* with ordinary bacteriological methods. By 1940, more than 80 plant viruses had been described and a further 50 or so were thought to be potential candidates. At present, more than 850 different plant viruses are known. The chapter by Ton van Helvoort recounts the early history of plant virus discovery.

The study of plant viruses has two main aspects. The first is fundamental research on the nature of plant viruses and plant virus diseases. Plant viruses were, and still are, used as models for structural and functional research in general virology. The results of these studies have been of major importance for the study of viruses infecting animal cells. Until the mid-1950s, viruses were characterized by their (pathological) effect on the infected cell and host organism. The study of plant viruses has greatly contributed to the characterization and classification of viruses on the basis of their composition, the organization of the viral genome, and the morphological structure of the virus particle. Currently, fundamental research of plant viruses is resulting in detailed information on the genetic function of viruses and genetic material in general.

The second aspect of plant virus research concerns the control of plant virus diseases. The symptoms of such diseases are often mosaics or mottles, but may also include necrosis (lesions or streaks) and deformation or outgrowths. The importance of these diseases was realized from the 1920s. Some do little harm, but others result in serious crop losses. Plant viruses are most troublesome in plantation crops or in those that are raised by vegetative methods of propagation, e.g., fruit trees, cocoa, sugar cane, and potatoes.

For centuries, classical crossing techniques have been used to improve crops and to make them resistant to plant diseases, pesticides and herbicides. Since the 1970s, the new techniques known as biotechnology, involving genetic modification, have been added to these. By incorporating the genetic information of a toxin or enzyme into a plant cell, resistance against pests, pesticides, or herbicides can be attained. Such genetically modified plant cells can be regenerated into a new so-called transgenic plant. At present, attempts are being made to render plants resistant to particular plant viruses by introducing genetic information of that or a related plant virus into the plant cell. Plant viruses are also used in the construction of transgenic plants, serving as vectors to introduce genetic information into cells and as a source of specific genetic regulatory elements. Hence, although plant viruses may be a scourge to mankind, they are also very useful tools and sources in the new field of biotechnology. The development of these viral tools and their potential usefulness as vaccines is discussed in the chapter by Beachy and Fitchen.

21

The Early Study
of Tobacco Mosaic Virus

TON VAN HELVOORT

Tobacco mosaic disease played an important role in the history of virus research. The etiological agent of this infectious disease was the first to be passed through an ultrafilter. The agent, called tobacco mosaic virus (TMV), was the first virus to be crystallized. Also, the virus was instrumental in research that clarified the functions of nucleic acid and protein—the components of TMV (van Helvoort 1991).

The crystallization of TMV in 1935 won the American Wendell M. Stanley the Nobel Prize in chemistry (1946). However, Stanley was soon criticized by English scientists for his analyses of the virus. Frederick Bawden and Bill Pirie reported a different content of nitrogen in the TMV preparations they had isolated. Furthermore, they found phosphorus, an element that was not reported on in the material isolated by Stanley. The rapid and uncritical acceptance of Stanley's work around 1935 is discussed by the historian of science, Lily Kay (1986). She does not, however, pay attention to the smoldering and protracted controversy between Stanley on one side and Bawden and Pirie on the other about the nature of tobacco mosaic virus. I will argue that this controversy was based upon different approaches to the study of tobacco mosaic disease by these scientists (van Helvoort 1996a).

CRYSTALLIZATION OF TMV

In 1935, *Science* published a paper by Stanley in which he announced the crystallization of TMV. He characterized the virus as a globulin or protein (Stanley 1935). Criticism by Bawden et al (1936) published in *Nature* mentioned, among other things, that their own group had found phosphorus in their liquid crystalline material prepared from tobacco mosaic-infected plants and that this element was present in the form of nucleic acid. Furthermore, they discussed the question whether the liquid crystalline material isolated was the virus itself or not. Was it present as such in the diseased plants? They concluded their paper with the comment that "it is still not proved, nor is there any evidence that the particles we have observed exist as such in infected sap." In an extensive report about their work on TMV, Bawden and Pirie (1937) wrote that the liquid crystalline substances they had isolated were probably present in infected plants as complexes. With purification, the virus underwent changes resulting in the loss of the very property that characterized TMV as a filterable virus: "...the virus undergoes a change that is not readily reversed and loses completely the property that first distinguished TMV as a new type of disease agent, namely, that it should pass fine filters." Thus, Bawden and Pirie took a physiological approach to the problem of the nature of tobacco mosaic virus by posing the question of what tobacco mosaic disease is. This approach was to be a distinguishing feature of the work of these English scientists.

Virus research in the first decades of this century was inspired by a bacteriological tradition. Multiplication of a (filterable) virus was seen as the reproduction of an *autonomous* entity. According to this point of view, the symptoms of virus diseases were seen as the result of a kind of "metabolic fatigue" of the infected cell. That Stanley's research seemed to indicate that TMV is not an organism but a proteinous molecule did not change this view on the nature of disease symptoms. The multiplication of the virus was considered as some kind of (auto)catalytical process, but leading to a similar result of metabolic exhaustion.

In his *Plant Viruses and Virus Diseases* (1939), Frederick Bawden criticized this notion on the symptoms of virus diseases, since the severity of symptoms seldom reflected the quantity of virus that was produced in the diseased plant. For a perhaps more appropriate view on the nature of virus production, Bawden referred to the work of Jules Bordet on phage production. Bordet interpreted phage production as a result of a change in the infected cells. Instead of normal metabolic products, the bacteria produced bacteriophage (van Helvoort 1992a). Bawden thought that such a mechanism was a more plausible explanation for virus multiplication.

PHYSICOCHEMICAL METHODS

After incorporating the criticism concerning the chemical analysis of TMV preparations, Stanley's results and conclusions were soon accepted. Undoubtedly, as is argued by Kay, this was related to the prestige of the physicochemical approach in studying this kind of biological phenomenon at the "edge of life." Stanley, a full member of the Rockefeller Institute, had access to the new physicochemical techniques that were developed in the 1930s and 1940s. Among the new devices were the ultracentrifuge, the electrophoresis apparatus, and the electron microscope.

This chemical characterization of TMV definitely established that this particular virus did not answer to the notion that a filterable virus is analogous to a small bacterium or ultramicrobe. Since bacteria multiply by an increase in size followed by division, one would expect that virus multiplication by such a process of binary fission would result in heterogeneous virus particles. However, Theodor Svedberg found that TMV behaved with almost perfect homogeneity in the ultracentrifuge and in electrophoresis. In Svedberg's words, this was hardly expected to be obtained in the "surface layer of organisms in different stages of development" (Eriksson-Quensel and Svedberg 1936).

The application of the new technologies reinforced Stanley (1938) in his idea that TMV could be characterized as a (nucleo)protein with a well-defined size: "...all of the data available at present [1938] indicate that the virus proteins fulfill the accepted chemical definition of a molecule as the smallest weight which cannot be subdivided without a complete change in the properties." In his Nobel Prize lecture, Stanley (1946) argued that electron micrographs showed that most of the virus particles had a length of 280 ± 8.6 nm.

Pirie criticized the importance attached to such physicochemical methods. In his opinion, plant viruses were composed of nucleoproteins, chemical substances that belong to the *normal* constituents of the cell. Clearly, a plant virus could not be characterized on the basis of its intrinsic properties, but only by its physiological characteristics. One of Pirie's criticisms of the physicochemical approach was that such isolation techniques resulted in an aggregation of the particles. The mean length of the particles was controlled by the "past history and present environment of the preparation." Therefore, it did not seem plausible that a particular physicochemical method would result in the isolation of *the* virus particle (Pirie 1946; van Helvoort 1996b).

A BIOCHEMICAL APPROACH

Pirie held that plant viruses were closely linked to their host cell, based on several lines of evidence. First, the virus was part of the metabolism of the host cell. Second, viruses are strongly associated with cell components *in vivo*. This meant that the process of physicochemical purification might result in a product lacking parts of the active structure of the virus. Therefore, as Pirie (1970) later explained, it was a basic assumption that "the complex form of the virus is the most relevant in any study of the details of virus multiplication and it is unwise to assume that anything removable from a virus preparation, without loss of infectivity, is an impurity."

The approach to the study of plant viruses advocated by Bawden and Pirie was premised on the notion that virus disease resulted from a disturbance of host cell metabolism. This approach was also consistent with the finding that a mosaic-infected tobacco plant could produce large quantities of virus material. The presupposition that large quantities of infected plant sap were needed in order to isolate TMV appeared to be unfounded (Pirie 1973). Bawden and Pirie's view also implied that virus infection not only yielded identical virus particles, but that many other products emerged during the multiplication process that might not be infectious. In his article "A Biochemical Approach to Viruses," Pirie (1950) argued that the study of virus disease must include more than the particles with a well-defined size and form: "...in gaining an understanding of the mechanism of infection, all the anomalous nucleoprotein must be studied."

According to Bawden and Pirie, the course of virus infection was also determined by the physiological condition of the host. The quantification of a plant virus was dependent on the leaf surface, the volume of the plant sap, permeability of the leaf, etc. A plant's susceptibility to a certain virus was dependent on the properties of the plant and thus on environmental factors. Therefore, Bawden (1948) concluded that the virulence of a virus is not an intrinsic property of the virus strain but a reflection of the interaction between virus and host.

Ten years later, Bawden (1959) expressed himself forcefully, asserting that the establishment of an infection is almost as dependent on the physiological state of the plant as on the identity of the virus and the host plant. Of course, the genetic constitution of both the virus and the plant had to allow for an infection, but "nurture comes only a little behind nature in determining what happens when the two are brought together."

SITTING ON THE FENCE

Obviously, the findings with plant viruses and plant virus diseases had repercussions for the prevailing ideas on animal viruses. Until the 1950s, there was a controversy between those who viewed animal viruses as organisms and those who thought they were molecules, a notion borrowed from plant virus studies and research on bacteriophage. This dichotomy becomes apparent from the lectures held at the Second International Poliomyelitis Conference in September 1951. The lecture of Christopher Andrewes was titled "Viruses as Organisms," whereas Stanley's contribution carried the title "Viruses as Chemical Agents." Bawden presented biochemical aspects of virus infections (Andrewes et al 1952). Shortly after this symposium, Andrewes (1953) wrote: "Dr Bawden is one of those queer people who are not sure that viruses are organisms. However, his saving grace is that he is sure that they are not molecules. In fact, he and Pirie are sitting very nicely on the fence, having built up there a strongly entrenched position."

Bawden and Pirie in 1953 reinterpreted their notion of virus infections from one representing a disturbance in host metabolism to one of virus infection as a special case of protein synthesis. According to Bawden and Pirie, a virus disease in a bacterium, plant, or animal was best perceived as "metabolic disturbances of the host, which produce, among other things, more material that can induce similar metabolic disturbances in other hosts." To them, the old notions of virus multiplication were no longer valid, and the analogy with the multiplication of bacteria had to be discarded. Likewise, the view of virus disease as an autocatalytic process of virus molecules could no longer be defended: "...we suggest that virus multiplication is comparable neither with the growth of a bacterium in a culture medium nor with the direct conversion of a precursor into an enzyme" (Bawden and Pirie 1952). This idea of protein synthesis as the conversion of a precursor into active protein was strongly promoted by John Northrop of the Rockefeller Institute, and Stanley was strongly influenced by Northrop's ideas (van Helvoort 1992b).

ENTITY OR COMPLEXITY

In the 1950s, a consensus was formed on the functions of the component parts of TMV. Ribonucleic acid isolated from TMV appeared to be infective and, when introduced into a host cell, gave rise to infective and complete TMV particles. The protein functioned as a protective coat and was thought to be encoded by the nucleic acid. The protein of TMV appeared to consist of identical molecules which could be removed and the virus could be reconstituted at will. Reconstitution resulted in "native" virus. The virus was metabolically inert and used the metabolic machinery of the host cell (Fraenkel-Conrat 1981). Together with the research on bacteriophages, knowledge about TMV contributed to a large extent to the formulation of a new virus concept such as that offered in 1957 by André Lwoff, who defined viruses as:

> ...infectious, potentially pathogenic, nucleoproteinic entities possessing only one type of nucleic acid, which are reproduced from their genetic material, are unable to grow and to undergo binary fission, and are devoid of a Lipmann system (Lwoff 1957).

This definition seems to unify the two approaches discussed above. The virus particles that were studied with the physicochemical approach were the result of virus replication. The physiological or biochemical approach focused on the contribution of host metabolism through which viral nucleic acids and proteins are synthesized.

However, this consensus between the physicochemical and biochemical approaches might only be apparent. To Bawden, the notion of virus multiplication as reproduction of identical particles remained opposed to the notion that it was a complex phenomenon. In 1966, he wrote that for "many years now the old idea has been untenable of virus multiplication being the direct replication of the infecting particles, giving a single end-product resembling the initial inoculum" (Bawden 1966). Those of the "Stanley school," however, never doubted that TMV consisted of a sharply demarcated particle:

> That the virus rod is a molecule of definite dimensions and molecular weight was an article of faith motivating many in Stanley's laboratory. (...) The shorter fragments, however, were worrisome. Were they merely broken rods, or were they the true virus molecules? This question was

not answered for several years. (…) Finally, Williams and Steere (1951), then at the University of Michigan, demonstrated beyond reasonable doubt that the short particles resulted from mechanical breakage of the fundamental unit (Lauffer 1984).

In the mid-1960s, however, it was concluded that pieces smaller than the 300-nm long, cigarette-shaped particles played a role in the course of virus infection. Francki (1966) concluded that these shorter particles enhanced the infectivity of the 300-nm particles since "virus preparations are more infectious than would be expected if all short particles were incapable of infection." Hulett and Loring (1965) likewise noticed the enhancement of infectivity when smaller particles were present.

The importance of heterogeneity in virus preparations was also discussed in the context of the so-called "defective interfering" (DI) particles. Alice Huang and David Baltimore (1970) concluded that such DI particles were important "determinants of the course of acute, self-limiting viral infections and of persistent, slowly progressing viral diseases. In addition, many host reactions may alter the production of DI particles and thus influence the outcome of viral infections." The significance of DI particles in viral infection is an illustration of the biochemical approach of Bawden and Pirie. It emphasized not only the dependence on host metabolism, but also the relevance of complexity in virus infection. This was the vision that Bawden and Pirie had advocated for many decades.

References

Andrewes CH, Stanley WM, Bawden FC (1952). In: Poliomyelitis—papers and discussions presented at the Second International Poliomyelitis Conference. Philadelphia: J. B. Lippincott Co, p. 3-5, 6-8, and 9-12 respectively.

Andrewes CH (1953). Discussion. In: Fildes P, Heyningen WE van, eds. The nature of virus multiplication: second symposium of the Society for General Microbiology. Cambridge, England: University Press for the Society, 41.

Bawden FC, Pirie NW, Bernal JD, Fankuchen I (1936). Liquid crystalline substances from virus infected plants. Nature 138, 1051-1052.

Bawden FC, Pirie NW (1937). The isolation and some properties of liquid crystalline substances from solanaceous plants infected with three strains of tobacco mosaic virus. Proc R Soc Lond Ser B 122: 274-320.

Bawden FC (1939). Plant viruses and virus diseases. Leiden, The Netherlands: Chronica Botanica Company, 262-264.

Bawden FC (1948). Some effects of host-plant physiology on resistance to viruses. Proc R Soc Lond Ser B 135: 187-195.

Bawden FC, Pirie NW (1952). Physiology of virus diseases. Ann Rev Plant Dis 3: 171-188.

Bawden FC, Pirie NW (1953). Virus multiplication considered as a form of protein synthesis. In: Fildes P, Heyningen WE van, eds. The nature of virus multiplication: second symposium of the Society for General Microbiology. Cambridge, England: University Press for the Society, 21-41.

Bawden FC (1959). The establishment and development of infection. In: Holton CS et al, eds. Plant pathology—problems and progress 1908-1958. Madison: University of Wisconsin Press, 503-510.

Bawden FC (1966). Some reflexions on thirty years of research on plant viruses. Ann Appl Biol 58: 1-11.

Eriksson-Quensel I-B and Svedberg T (1936). Sedimentation and electrophoresis of the tobacco mosaic virus protein. J Am Chem Soc 58: 1863-1867.

Fraenkel-Conrat H (1981). Portraits of viruses: Tobacco mosaic virus. Intervirology 15: 177-189.

Francki RIB (1966). Some factors affecting particle length distribution in tobacco mosaic virus preparations. Virology 30: 388-396.

Huang AS, Baltimore D (1970). Defective viral particles and viral disease processes. Nature 226: 325-327.

Hulett HR, Loring HS (1965). Effect of particle length distribution on infectivity of tobacco mosaic virus. Virology 25: 418-430.

Kay LE (1986). W. M. Stanley's crystallization of the tobacco mosaic virus, 1930-1940. Isis 77: 450-472.

Lauffer MA (1984). Contributions of early research on tobacco mosaic virus. Trends Biochem Sci 9: 369-371, esp. p. 370.

Lwoff A (1957). The concept of virus. J Gen Microbiol 17: 239-253, esp. p. 246.

Pirie NW (1946). The viruses. Ann Rev Biochem 15: 573-592, esp. p. 584.

Pirie NW (1950). A biochemical approach to viruses. Nature 166: 495-496, esp. p. 496.

Pirie NW (1970). Retrospect on the biochemistry of plant viruses. Biochem Soc Symp 30: 43-56, esp. p. 53.

Pirie NW (1973). Frederick Charles Bawden 1908-1972. Biogr Mem F R S 119: 19-63, esp. p. 40.

Stanley WM (1935). Isolation of a crystalline protein possessing the properties of tobacco-mosaic virus. Science 81: 644-645.

Stanley WM (1938). Virus proteins—A new group of macromolecules. J Phys Chem 42: 55-70.

Stanley WM (1946). The isolation and properties of crystalline tobacco mosaic virus. In: (1964) Nobel lectures chemistry 1942–1962. Amsterdam: Elsevier, 137–159.

van Helvoort T (1991). What is a virus? The case of tobacco mosaic disease. Stud Hist Phil Sci 22: 557-588.

van Helvoort T (1992a). Bacteriological and physiological research styles in the early controversy on the nature of the bacteriophage phenomenon. Med Hist 36: 243-270.

van Helvoort T (1992b). The controversy between John H. Northrop and Max Delbrück on the formation of bacteriophage: Bacterial synthesis or autonomous multiplication? Ann Sci 49: 545-575.

van Helvoort T (1994). History of virus research in the twentieth century: The problem of conceptual continuity. Hist Sci 32: 185-235.

van Helvoort T (1996a). When did virology start? ASM News 62: 142-145.

van Helvoort T (1996b). A comment on the early influenza virus vaccines: The role of the concept of virus. In: Plotkin S, Fantini B, eds. Vaccinia, vaccination and vaccinology: Jenner, Pasteur and their successors. Amsterdam: Elsevier, 193-197.

22

Pathogen-Derived Resistance to Plant Viruses and Plant Viruses as Vaccines

Roger N. Beachy and John H. Fitchen

During the 1980s, the technologies of transgenesis led to the development of plants that have increased resistance to diseases and insect pests. The first successful example of disease resistance was the development of transgenic plants that accumulated the capsid (coat protein) of tobacco mosaic tobamovirus (TMV) and, as a consequence, were resistant to TMV. This success in "pathogen-derived resistance" (PDR) was followed by the use of transgenes that encode a variety of viral sequences, some of which led to the production of proteins and others that produced transcripts but not proteins. Recently, regulatory agencies of the United States approved for commercial release the first food crop that is protected by PDR; many more are expected within the next several years.

In another application, three different viruses have been studied for their possible use as vehicles to deliver animal and human vaccines. TMV, cowpea mosaic comovirus (CPMV), and Johnsongrass mosaic potyvirus (JMV) were genetically modified to carry peptides as linear extensions or loop structures exposed on the surface of virion or assembled capsid molecules; animals injected with the modified viruses produced antibodies against the added peptides. The potential use of plant virus-based vaccines is discussed.

PDR Against Virus Infection

A number of strategies have been developed to control one or more steps in the uptake, release and expression of viral genes, virus replication, transmission to adjacent cells, and assembly of progeny (Figure 1). However, the cellular and molecular mechanisms responsible for PDR are generally poorly understood.

COAT PROTEIN-MEDIATED RESISTANCE

The first example of PDR in transgenic plants was developed in tobacco plants by expressing a gene encoding the capsid or coat protein (CP)

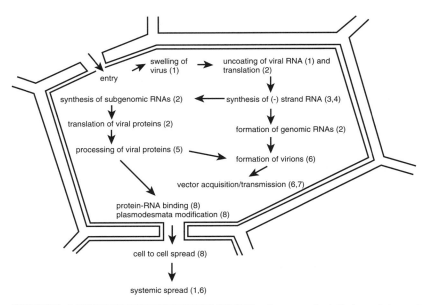

FIGURE 1. A GENERALIZED VIRUS LIFE CYCLE and gene products that may interrupt the cycle. Gene products that may interfere with different steps of virus infection in transgenic plants: 1) coat protein; 2) (-) sense RNA (+ribosome); 3) (+) sense RNA (+ribosome); 4) defective replicase; 5) modified protease; 6) defective coat protein; 7) defective transmission factors; 8) defective movement protein (From Beachy 1993).

of TMV (Powell-Abel et al 1986). Since the first report of CP-mediated resistance, similar approaches have been used to develop resistance against viruses in many taxons in a variety of different plants (reviewed by Fitchen and Beachy 1993). However, most of the studies to elucidate the mechanisms of CP-mediated resistance have been carried out with TMV.

CP-mediated resistance against TMV depends on the amount of transgenic CP and is characterized by reduced numbers of sites of infection in CP-expressing plants, compared with control plants, and by susceptibility to infection by "swelled" virus particles (lacking 10-20 capsids from the virus) and viral RNA (Register and Beachy 1989). Resistance is greatest against TMV and closely related tobamoviruses, with decreased levels of resistance against more distantly related tobamovirus and no resistance against viruses in other taxons (Nejidat and Beachy 1990). The capacity to overcome resistance was found to depend on the sequence of the CP of the challenge virus rather than the sequence of the viral RNA (Clark et al 1995a). These results led us to propose that the transgenic CP interferes with disassembly of the challenge virus by reversing the rate of release of capsid from the virus, thereby causing re-assembly and preventing the binding of ribosomes to viral RNA to initiate virus infection. If the sequence of the CP of the challenge virus is sufficiently similar to that of the transgenic CP, there is a positive interaction and binding of the transgenic CP to that of the challenge virus, either before or after the release of a limited number (probably less than five) capsid molecules from the virus. While resistance might involve a receptor-mediated event, there are currently no data to support this suggestion (Clark et al 1995b). Our recent studies of transgenic plants containing mutant CP molecules with selected amino acid modifications indicate that proteins capable of limited assembly confer resistance so long as the CP can form certain types of intersubunit interactions (Clark et al 1995a).

CP-mediated resistance has been applied to control a variety of plant viruses; however, not all viruses are as well controlled as others by transgenic CP. For example, CP of a potyvirus can provide resistance to the virus from which the gene is derived, and in some cases to different potyviruses (e.g., see Stark and Beachy 1989), but the degree of resistance is not directly related to the amount of transgenic CP that accumulates in the plant (reviewed by Fitchen and Beachy 1993). A similar situation was reported for the tomato spotted wilt tombusvirus, a member of the Bunyaviridae. In this case, resistance is provided by the nucleocapsid protein, which is analogous to the capsid protein in nonenveloped viruses. In such cases, physical features of the CP molecule may affect its stability, its subcellular localization, or its capacity to attain the necessary secondary and tertiary structure required to provide resistance. In the cases of the potyviruses and tombusviruses, the (nucleo) capsid provides resistance against the virus from which the gene was obtained, but not to related viruses in other serogroups. Because of the lack of strong positive correlation between the amount of protein and resistance in these cases, it is important for researchers to demonstrate that the molecule responsible for resistance is, in fact, the CP, rather than the gene transcript (as discussed below).

REPLICASE–MEDIATED RESISTANCE

In 1989, Zaitlin and colleagues first reported that transgenes encoding a portion of the replicase/polymerase *(rep)* of TMV in transgenic tobacco plants could confer extremely high levels of resistance against TMV (Golemboski et al 1990); some plant lines were essentially immune to infection. Similar results were obtained with *rep* gene sequences of other viruses (Anderson et al 1992; Braun and Hemenway 1992; MacFarlane and Davies 1992). However, once again, the degree of resistance could not be predicted based upon the amount of transgenic protein or RNA transcript that accumulated (in some cases, no protein was detected), and the role of protein in providing resistance was placed in doubt. Recently, plant lines that contain transgenes encoding nontranslatable mRNAs were developed; in some instances, resistance apparently required an open reading frame.

Replicase-mediated resistance is characterized by a restricted range, i.e., resistance is effected only against the virus strain from which the gene was obtained. Because the mechanisms of resistance are poorly understood, it is difficult to understand how such specificity is determined or how a broader type of resistance might be conferred using this strategy of PDR.

MOVEMENT PROTEIN-MEDIATED RESISTANCE

Movement proteins (MP) are nonstructural viral proteins that are required for local (cell-to-cell) and/or long-distance movement of the virus or viral nucleic acids. The identity and characterization of an MP was first described for TMV (Deom et al 1987; Meshi et al 1987). It was shown that transgenic plants containing an MP complement the local spread of a mutant TMV lacking a functional MP (Deom et al 1987; Holt and Beachy 1991). Studies of several MPs have demonstrated one or more of the following characteristics of this protein: 1) it alters the size exclusion limits of plant plasmodesmata (the cytoplasmic bridges that interconnect plant cells and through which plant viruses pass); 2) it binds nucleic acids under *in vitro* reaction conditions; and 3) as judged by comparisons of amino acid sequences, it is not related between virus groups. Based on these characteristics and the presumption that plant viruses pass between plant cells mainly through plasmodesmata, Deom et al (1992) suggested the possibility of designing a mutant MP that acts in a dominant manner in transgenic plants to block the function of MPs produced during virus infection. Indeed, transgenic plants that accumulated a temperature-sensitive mutant of the TMV MP (Malyshenko et al 1993) and a dysfunctional TMV MP (dysMP) (Lapidot et al 1993) showed an increased resistance to infection by TMV and other tobamoviruses (Lapidot et al 1993). However, resistance was not exhibited as lack of infection, as in the case of CP-mediated resistance, but as reduced local and systemic spread of infection. Cooper et al (1995) used transgenic plants that contained the dysMP to demonstrate that resistance extended to certain viruses outside of the tobamovirus group. Furthermore, those viruses against which resistance was demonstrated were more virulent in plant lines that contained the unmodified (wild-type) MP. Beck et al (1994) found that transgenic plants containing a mutant of p13, one of the three proteins of white clover mosaic potexvirus that together provide MP function, were resistant to this virus and to a related potexvirus. While the mechanism of interference with virus infections conferred by mutant MPs remains unclear, resistance likely results from interference with one or more functions that are common amongst the MPs of different viruses, although there are little or no similarities in amino acid sequences among the MPs. Mutant MPs appear to be promising in providing significant levels of resistance to multiple viruses.

RNA-MEDIATED RESISTANCE

During studies of CP-mediated resistance to tobacco etch potyvirus (TEV), it was observed that some plant lines with very low levels of CP exhibited very high levels of resistance to infection by the strain of TEV from which the CP gene was derived, but not to other strains or other potyviruses (Lindbo and Dougherty 1992). It was later shown that resistance was the result of gene expression, but did not rely on an open reading frame. Subsequent research (Smith et al 1994; Mueller et al 1995) demonstrated that such RNA-mediated resistance generally occurred in plant lines with multiple copies of the transgene and for which rates of transcription were high but levels of accumulated transcript were low. These workers proposed that expression of the transgene induced a sequence of events that resulted in selective degradation of the overexpressed transcript as well as the virus from which the transgene was derived. This resulted in a very high level of selective resistance. Although the precise mechanisms of resistance are unknown, it is expected that RNA-mediated resistance will be important in the development of crop varieties with PDR.

PLANT VIRUSES AS VACCINES

The rationale for using plant viruses as carriers of immunogenic epitopes is based on several features of plant viruses. First, most plant viruses are relatively simple structures, comprised of RNA or DNA encapsidated in a single type of protein subunit and lacking a membrane component. The lack of a membrane contributes to the structural stability of the virion: some plant viruses can retain structural integrity and infectivity for many years in dried leaf tissues and as semi-purified or highly purified preparations. Second, certain plant viruses can accumulate to very high levels in their hosts from which they can be purified with relative ease; for example, the TMV CP can comprise from 10 to 40% of the total leaf proteins in some infected plants, and can be purified by a three-step process that is completed over several hours. Furthermore, because TMV can infect leaves of edible plants, including lettuce and spinach, as well as tomato fruits, the possibility of developing vaccines that are delivered via oral ingestion as a component of food can be considered. However, while the use of plant viruses as vaccines for humans and other animals has interesting and intriguing possibilities, a tremendous amount of research is necessary before this can become a reality.

To date, members of three different virus groups have been explored as possible vaccines: TMV, a tobamovirus; cowpea mosaic virus (CPMV), a comovirus; and tobacco etch virus (TEV), a potyvirus. Each has unique advantages and members of such groups may be useful in selected circumstances. TMV is a rigid rod-shaped virus that contains a single type of CP and was among the first viruses to be characterized by crystallography. This structural information has made it possible to target specific structural features for modification and insertion of peptides that may be useful vaccines. Potyviruses are flexuous rod-shaped viruses that also contain a single type of CP; however, structural analysis of members of this virus group has not yet been completed. TEV (and other potyviruses) may have an advantage as a potential vaccine over TMV because it contains a sequence of 60 or more amino acids at its amino-terminus that is dispensable for assembly and can accommodate significant sequence variability while retaining structural integrity. Therefore, it may be possible to add epitopes of 60 amino acids or more to the CP without interfering with virion assembly. CPMV is a spherical particle containing two types of CP. Structural analysis of CPMV has been completed and areas of the particle have been identified where amino acids can be added as loops with minimal effect on virus assembly and stability. Here we briefly describe the results of recent studies to develop TMV, TEV, and CPMV as vaccines.

TMV AS A VACCINE CARRIER

The determination of the TMV nucleotide sequence (Goelet et al 1982) and the high-resolution structural analysis of assembled coat protein and virus (Namba et al 1989) made it possible to consider modifying the exposed surface of assembled virus to carry novel sequences, including immunogenic epitopes. TMV is a rod-shaped particle comprised of 2140 capsids arranged in an elongated helix, 16.4 subunits per helix. Intersubunit assembly can occur in the presence or absence of viral RNA and is more stable in the presence of RNA. Both the carboxy- and amino-termini of the CP are exposed on the surface of the virion and are potential sites for the addition of novel amino acid sequences.

The first report of genetic modification of the TMV CP to carry a novel peptide sequence (Haynes et al 1986) described the addition of a peptide derived from poliovirus. In that study, a cloned cDNA encoding the CP of TMV was modified to carry an 8-amino acid peptide derived from the VP1 protein of poliovirus 3. The modified cDNA was expressed in *E. coli,* and protein extracted from the bacteria were placed under the appropriate *in vitro* conditions to cause intersubunit assembly, producing rodlets consisting of CP carrying the poliovirus epitope. Immunization of rats with semi-purified rodlets elicited antibodies capable of neutralizing poliovirus in *in vitro* infection assays (Haynes at al 1986). Protein that was not assembled to rodlets induced low-titer or no neutralizing antibodies. This result led to the hypothesis that assembly caused the epitope to be exposed and elicited the antibody response following injection of rodlets into the animal.

In subsequent studies by three different groups, mutations made use of full-length cloned cDNAs from which infectious transcripts could be derived; mutations made in capsid genes in these "infectious clones" led to production of modified viruses that carried potential epitopes. Hamamoto et al (1993) carried out mutations that added the 12 amino acids of the angiotensin-I-converting enzyme inhibitor to the carboxyl end of the CP of tomato mosaic tobamovirus. Protoplasts infected with transcripts of the modified virus yielded virus particles assembled by the fusion protein; however, the modified virus was unable to cause a normal infection in plants due to its inability to move locally and systemically.

Studies by Clark et al (1995a) showed that efficient local and systemic spread of TMV carrying a modified CP was apparently dependent on the nature of the sequence extending from the carboxy-terminus (the CP of TMV contains 158 amino acids). When 10 to 15 amino acids derived from sunn hemp mosaic tobamovirus or from the polylinker sequence of the cloning plasmid pUC were added to the TMV CP, infection spread efficiently in the inoculated leaves and only slightly more slowly than wild-type virus to upper leaves. However, when other sequences were added to the CP, infection induced a localized necrotic reaction in a variety of different hosts, confining the infection to the necrotic tissues (J.H. Fitchen and R.N. Beachy, unpublished).

Careful examination of the structure of TMV revealed that amino acids 154-158 are not fixed in relation to other sequences of the protein; this region was therefore selected for subsequent modification, with the hypothesis that limited modifications would not prevent subunit assembly. Mutations were made in the infectious clone so that a variety of different amino acid sequences of varying lengths and sequence were added between CP amino acids 154 and 155. Transcripts of the modified infectious clone were inoculated into tobacco or spinach plants. Most of the modifications did not cause necrotic reactions, although certain modifications caused infections that were more or less severe than infection by wild-type virus. Sequences added to the virus ranged from 9 to 25 amino acids in length, and were derived from well-known immunoreactive sequences of influenza HA, HIV gp120, and murine zona pellucida ZP3 (J. Fitchen, unpublished). For a variety of biological reasons, we selected the latter for further studies, since the role of the target protein in fertility was well known.

ZP3 is one of three glycoproteins that comprise the murine zona pellucida and is the primary binding site for sperm during fertilization (Bleil and Wassarman 1980). An epitope comprising amino acids 336-343, a B-cell epitope, is responsible for inducing antibody-mediated contraception (Millar et al 1989). A mutation was made in the infectious clone of TMV to encode these amino acids between codons 154 and 155 in the TMV CP, and plants were infected with the modified virus. Virus isolated from these plants was parenterally administered to mice and induced antibodies that recognized ZP3. Antibodies produced in response to TMV carrying the CP-ZP3 fusion, but not wild-type TMV, were recruited and bound the zona pellucida in female mice, demonstrating that the TMV-borne epitope can induce a B-cell mediated response (Fitchen et al 1995).

POTYVIRUS AS A VACCINE CARRIER

A single report describes studies with sequences fused to CPs of a potyvirus (Jagadish et al 1993). Fusion proteins were made at the amino-terminus of the CP of Johnsongrass mosaic virus with an octapeptide epitope from *Plasmodium falciparum* or with the 10-amino acid leutinizing hormone-releasing factor. In another construct, the Sj-26-glutathione S-transferase from

Shistosoma japonica replaced 62 amino acids at the amino-terminus of the CP. Fusion proteins that accumulated in *E. coli* assembled to form virus-like particles (VLPs). Injection of the Sj-26:CP fusion protein into mice elicited an antibody response to Sj-26. These studies demonstrated the capacity of modified potyvirus CP to assemble to form VLPs, including CP that carry large foreign proteins. However, it is unknown whether virions encapsidated by these mutant CP molecules are capable of causing local and systemic infections in plants.

COWPEA MOSAIC VIRUS AS A VACCINE CARRIER

The structure of CPMV was determined at the atomic level in the late 1980s (reviewed by Chen et al 1990) and revealed that the virus surface includes several exposed loops. Of particular interest was a loop comprising amino acids 20-27 of the S capsid subunit which lies on the virus surface and is not involved in intersubunit interactions. It was proposed that these amino acids could be deleted and replaced by new sequences, including those whose immunogenicity requires a loop structure. Usha et al (1993) inserted amino acids 136-160 from VP1 of foot-and-mouth disease virus (FMDV) between amino acids 18 and 19 of the S peptide, inoculated both protoplasts and whole plants with the modified virus, and recovered chimeric virus particles from the inoculated leaves but not the systemically infected upper leaves. These particles reacted to antibodies raised against FMDV, confirming that the FMDV sequence was retained in the virus capsid.

It was subsequently found that the FMDV sequences inserted in CPMV were rapidly lost upon serial passaging of the virus, leading the investigators to redesign the chimeric protein in order to stabilize the inserted sequences (Porta et al 1994). Mutations were made such that foreign sequences were placed between amino acids 22 and 23 of the S protein. Sequences that were stable included amino acids 141-159 of VP1 of FDMV, amino acids 85-98 of human rhinovirus 14 (HRV-14), and amino acids 731-752 of gp41 from human immunodeficiency virus III (HIV-III). These viruses were reactive with antibodies directed against the specific peptides, and in the case of the HRV-14 epitope, the virus was active as a vaccine in rabbits and induced antibodies directed against HRV-14.

The chimeric virus containing the HIV gp41 sequences was then used to inject C57/BL6 mice. Antibodies recovered from injected mice reacted with purified peptide and, when tested in neutralization assays, were shown to neutralize HIV-1 strain IIIB as well as HIV-1 strains RF and SF2 (McLain and Dimmock 1994). Titers of neutralizing antibodies declined to undetectable levels several weeks after the second injection but were stimulated upon re-injection.

CONCLUSIONS

Recent advances in virology have led to a greater understanding of virus replication and the cellular and molecular basis of their pathologies. In the fields of plant virology and biotechnology, this has led to the applications of strategies of pathogen-derived resistance; the first transgenic plant varieties rendered resistant to virus diseases by coat protein-mediated resistance techniques were marketed in 1995 by the Asgrow Seed Company (Kalamazoo, Michigan, USA) which offered a variety of squash plants with coat protein-mediated resistance against two viruses. Increasing numbers of transgenic virus-resistant crops will reach the marketplace in coming years.

As a direct consequence of the recent advances in molecular virology and structural biology, there is a growing understanding of how virus particles are assembled and stabilized. From this information has come the first examples of virus particles that carry immunogenic epitopes and that can act as nonreplicating vaccines upon injection into animals. While early studies have concentrated on confirming the stability of the modified viruses and their capacity to induce the predicted antibody response, current and subsequent studies will focus on demonstrating that the induced antibodies can provide protective immunity against disease agents. Future challenges include better understanding how the viruses induce B-cell and T-cell responses and how to maintain protective immunity. Although an important future goal is the delivery of plant virus-based vaccines as a component of food, considerable experimentation will be required before it is clear how to accomplish such immunization and how to avoid immune tolerization in the process.

ACKNOWLEDGMENTS

The authors are grateful to Lyric VanNess for assistance in preparation of the manuscript and to the Scripps Family Chair, Keith and Jean Kellogg, and the National Institutes of Health for supporting this research (#RO1 AI27161).

REFERENCES

Anderson JM, Palukaitis P, Zaitlin M (1992). A defective replicase gene induces resistance to cucumber mosaic virus in transgenic tobacco plants. Proc Natl Acad Sci USA 89: 8759-8763.

Beachy, RN (1993). Introduction: Transgenic resistance to plant viruses. Sem Virol 4: 327-328.

Beck DL, Van Dolleweerd CJ, Lough TJ, Balmori E, Voot DM, Andersen MT, O'Brien EW, Forster LS (1994). Disruption of virus movement confers broad-spectrum resistance against systemic infection by plant viruses with a triple gene block. Proc Natl Acad Sci USA 91: 10310-10314.

Bleil JD, Wassarman PM (1980). Mammalian sperm-egg interaction: identification of a glycoprotein in mouse egg zonae pellucidae possessing receptor activity for sperm. Cell 20: 873-882.

Braun CJ, Hemenway CL (1992). Expression of amino-terminal portions or full-length viral replicase genes in transgenic plants confers resistance to potato virus X infection. Plant Cell 4: 735-744.

Chen Z, Stauffacher CV, Johnson JE (1990). Capsid structure and RNA packaging in comoviruses. Sem Virol 1: 453-466.

Clark WG, Fitchen J, Nejidat A, Beachy RN (1995b). Studies of coat protein-mediated resistance to TMV: II. Challenge by a mutant with altered virion surface does not overcome resistance conferred by TMV CP. J Gen Virol 76: 2613-2617.

Clark WG, Fitchen JH, Beachy RN (1995a). Studies of coat-protein mediated resistance to TMV using mutant CP: I. The PM2 assembly defective mutant. Virology 208: 485-491.

Cooper B, Lapidot M, Heick JA, Dodds JA, Beachy RN (1995). Multi-virus resistance in transgenic tobacco plants expressing a dysfunctional movement protein of tobacco mosaic virus. Virology 206: 307-313.

Deom CM, Lapidot M, Beachy RN (1992). Plant virus movement proteins. Cell 69: 221-224.

Deom CM, Oliver MJ, Beachy RN (1987). The 30-kilodalton gene product of tobacco mosaic virus potentiates virus movement. Science 237: 389-394.

Fitchen J, Beachy RN, Hein MB (1995). Plant virus expressing hybrid coat protein and added murine epitope elicits autoantibody response. Vaccine 13: 1051-1057.

Fitchen JH, Beachy RN (1993). Genetically engineered protection against viruses in transgenic plants. Annu Rev Microbiol 47: 739-763.

Goelet P, Lomonossoff GP, Butler PJG, Akam ME, Gait MJ, Karn J (1982). Nucleotide sequence of tobacco mosaic virus RNA. Proc Natl Acad Sci USA 79: 5818-5822.

Golemboski DB, Lomonossoff GP, Zaitlin M (1990). Plants transformed with a tobacco mosaic virus nonstructural gene sequence are resistant to the virus. Proc Natl Acad Sci USA 87: 6311-6315.

Hamamoto H, Sugiyama Y, Nakagawa N, Hashida E, Matsunaga Y, Takemoto S, Watanabe Y, Okada Y (1993). A new tobacco mosaic virus vector and its use for the systemic production of angiotensin-1-converting enzyme inhibitor in transgenic tobacco and tomato. Bio/Technology 11: 930-932.

Haynes JR, Cunningham J, von Seefried A, Lennick M, Garvin RT, Shen S (1986). Development of a genetically-engineered, candidate polio vaccine employing the self-assembling properties of the tobacco mosaic virus coat protein. Bio/Technology 4: 637-641.

Holt CA, Beachy RN (1991). *In vivo* complementation of infectious transcripts from mutant tobacco mosaic virus cDNAs in transgenic plants. Virology 181: 109-117.

Jagadish MN, Hamilton RC, Fernandez CS, Schoofs P, Davern KM, Kalnins H, Ward CW, Nisbet IT (1993). High level production of hybrid potyvirus-like particles carrying repetitive copies of foreign antigens in *Escherichia coli*. Bio/Technology 11: 1166-1170.

Lapidot M, Gafny R, Ding B, Wolf S, Lucas WJ, Beachy RN (1993). A dysfunctional movement protein of tobacco mosaic virus that partially modifies the plasmodesmata and limits virus spread in transgenic plants. Plant J 2: 959-970.

Lindbo JA, Dougherty WG (1992). Untranslatable transcripts of the tobacco etch virus coat protein gene sequence can interfere with tobacco etch virus replication in transgenic plants and protoplasts. Virology 189: 725-733.

MacFarlane SA, Davies JW (1992). Plants transformed with a region of the 201-kilodalton replicase gene from pea early browning virus RNA1 are resistant to virus infection. Proc Natl Acad Sci USA 89: 5829-5833.

Malyshenko SI, Kondakova OA, Nazarova JV, Kaplan IB, Taliansky ME, Atabekov JG (1993). Reduction of tobacco mosaic virus accumulation in transgenic plants producing non-functional viral transport proteins. J Gen Virol 74: 1149-1156.

McLain L, Dimmock NJ (1994). Single- and multi-hit kinetics of IgG-neutralization of human immunodeficiency virus type 1 by monoclonal antibodies. J Gen Virol 75: 1457-1460.

Meshi T, Watanabe Y, Saito T, Sugimoto A, Maeda T, Okada Y (1987). Function of the 30 kD protein of tobacco mosaic virus: involvement in cell-to-cell movement and dispensability for replication. EMBO J 6: 2557-2563.

Millar SE, Chamow SM, Baur AW, Oliver C, Robey F, Dean J (1989). Vaccination with a synthetic zona pellucida peptide produces long-term contraception in female mice. Science 246: 935-938.

Namba K, Pattanayek R, Stubbs G (1989). Visualization of protein-nucleic acid interactions in a virus. Refined structure of intact tobacco mosaic virus at 2.9 Å resolution by X-ray fiber diffraction. J Mol Biol 208: 307-325.

Nejidat A, Beachy RN (1990). Transgenic tobacco plants expressing a coat protein gene of tobacco mosaic virus are resistant to some other tobamoviruses. Mol Plant Microbe Inter 3: 247-251.

Porta C, Spall VE, Loveland J, Johnson JE, Barker PJ, Lomonossoff GP (1994). Development of cowpea mosaic virus as a high-yielding system for the presentation of foreign peptides. Virology 202: 949-965.

Powell-Abel P, Nelson RS, De B, Hoffmann N, Rogers SG, Fraley R, Beachy RN (1986). Delay of disease development in transgenic plants that express the tobacco mosaic virus coat protein gene. Science 232: 738-743.

Register III JC, Beachy RN (1989). Effect of protein aggregation state on coat protein-mediated protection against tobacco mosaic virus using a transient protoplast assay. Virology 173: 656-663.

Smith HA, Swaney SL, Parks TD, Wernsman EA, Dougherty WG (1994). Transgenic plant virus resistance mediated by untranslatable sense RNAs: Expression, regulation, and fate of nonessential RNAs. Plant Cell 6: 1441-1453.

Stark DM, Beachy RN (1989). Protection against potyvirus infection in transgenic plants: evidence for broad spectrum resistance. Bio/Technology 7: 1257-1262.

Usha R, Rohll JB, Spall VE, Shanks M, Maule AJ, Johnson JE, Lomonossoff GP (1993). Expression of an animal virus antigenic site on the surface of a plant virus particle. Virology 197: 366-374.

BACTERIA

Leeuwenhoek is dead, it is too bad, it is a loss that cannot be made good. Who now will carry on the study of the little animals?

PAUL DE KRUIF—*Microbe Hunters*

Tuberculosis

A courtesan, a seamstress, and a convict portrayed in three operas by three different composers reflects the range of people in varying social classes in the nineteenth century who died of tuberculosis. Brought under control by the middle of this century, tuberculosis is now reemerging as a major infectious disease. Its rise is attributed to social factors—increasing poverty, homelessness, substance abuse, and deteriorating health care systems. It is also linked to the AIDS pandemic; because of the effect of the virus on the immune system, people infected with human immune deficiency virus (HIV) are much more susceptible than others to tuberculosis.

Treatment of tuberculosis in recent years is complicated by the emergence of drug-resistant strains, requiring multiple drugs over an extended period of time, thus greatly increasing the cost of treatment and the possibility of interrupted or incomplete treatment, allowing the infectious organisms to mutate and multiply.

The epidemiological profile of tuberculosis is described by William Stead, while the problem of emergence of new strains and difficulties in controlling tuberculosis is addressed by Thomas Shinnick.

23

Epidemiology of the Global Distribution of Tuberculosis

WILLIAM W. STEAD

> *The past is the key to our future.*
> —LOUIS LEAKEY (1992)

When something is as woven into the history and literature of the Western world as tuberculosis (TB), it is often assumed to be universal and to have existed from the beginning of time. Recent revelations and careful study of the origin and history of TB have shown the fallacy of this assumption (Bates and Stead 1993; Stead and Bates 1995; Stead et al 1995). In this chapter, the events are described that antedate effective therapy for TB.

NATURE OF THE EPIDEMIC

All infectious epidemics follow a similar curve if an appropriate time scale is chosen (Grigg 1958) (Figure 1). Highly infectious viral infections with short incubation periods produce an immediate epidemic. However, TB has a delayed ascent because of the long incubation period and the minimal infectiousness of early cases.

Today TB is epidemic worldwide. However, this is not a single epidemic, but instead many concurrent epidemics of different ages, depending upon when *Mycobacterium tuberculosis* first reached a given population. When Grigg described the TB epidemic in several cities in Europe and North America, it was predominantly a highly infectious chronic pulmonary disease of European descendants, among whom the disease was rampant. He showed that the epidemic was already in decline, with children being the principal susceptibles in the population. TB accounted for 25% of all deaths. However, Grigg failed to realize that the various epidemics in most parts of the world were of more recent origin and many are still on the ascending limb of the curve.

In areas where TB has been introduced only recently, the population is much more susceptible, and the disease often runs an acute course to death with little pulmonary involvement to produce infectiousness. Thus, unlike smallpox in a new population, a single case of TB rarely initiates an epidemic.

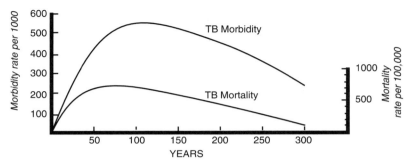

FIGURE 1. A THEORETICAL PLOT OF TUBERCULOSIS morbidity and mortality after the disease is first introduced into a population with no prior exposure. (Adapted from Grigg 1958).

Without the presence of chronic cases to continue seeding the population, there are too few survivors with chronic pulmonary disease to spread the infection. Even when these conditions are met, the disease only becomes epidemic when the element of crowding is added.

ORIGIN OF *MYCOBACTERIUM TUBERCULOSIS*

Mycobacteriacae comprise many species of saprophytic nitrogen-fixing soil bacteria of worldwide distribution, most with no animal hosts. It is presumed that animal pathogenicity began when a chance mutation produced a more specialized form, *M. bovis,* long before humans evolved. This organism was suited to and required a mammalian host. TB then became an endemic disease in many species, spreading from prey to predator.

Archaeologists tell us that domestication of animals began in Europe about 9000 years ago (Clark 1962). This established a closer relationship between humans and domestic animals of several species. During Europe's cold winters, it was common practice for farmers to bring their animals into their "house/barns" at night to conserve heat and for security. Although *M. bovis* produces little disease in humans, it is quite virulent for many animal species, often involving the lungs (Steele and Ranney 1958). With a tuberculous cow and a human family sleeping under the same roof, the stage was set for a mutation conferring human pathogenicity. This mutation was so slight that even present-day DNA technology cannot distinguish *M. tuberculosis* from *M. bovis,* although biologic differences are readily apparent (Stead 1995).

TB AS AN ENDEMIC DISEASE

With a new host and a rural setting, TB became endemic in humans. The oldest human epidemic of TB began several centuries later when crowding into cities began during the Industrial Revolution in the latter half of the eighteenth century (Figure 2). It began to kill susceptibles with a rather acute disease of limited infectiousness (Templeton et al 1995). Gradually, as more individuals were infected, survivors developed chronic pulmonary lesions with less illness, thus furnishing an abundant supply of the organisms to continue the spread of the infection. This social and ecologic change occurred later in Eastern Europe and probably also beyond the Pyrenees Mountains to Spain. Thus the disease became epidemic later in those areas.

By the nineteenth century, TB was called "the captain of all these men of death" and the "White Plague" (Dubos and Dubos 1952). It is not clear whether this was to distinguish it from the Black Death (bubonic plague) or to indicate its high death rate among Europeans. Death from TB was particularly common in children and young adults, making it a strong selection factor favoring survival of individuals with greater innate resistance to this infection (de Vries et al 1979; Stead 1992).

Throughout the world today, TB is a disease that represents a separate relationship with individual populations, largely depending on when the organism was introduced (Bates and Stead 1993; Stead and Bates 1995). The magnitude of the TB problem for a given population is determined by local socioeconomic factors and the proportion of susceptible individuals in the population.

Areas inhabited by descendants of Europeans, such as North America, South Africa, Australia, and New Zealand are far down the descending limb of the epidemic curve (Grzybowski and Allen 1964). In these areas, TB manifests largely as a chronic pulmonary disease in older people.

TB was carried to coastal communities around the world by the exploits of Europeans with chronic pulmonary disease, but did not penetrate into the hinterland (Stead 1992). TB was uncommon in Russia as late as 1880

FIGURE 2. A PLOT OF THE ESTABLISHMENT OF TB as an endemic infection in Europe and its conversion to an epidemic with the onset of the Industrial Revolution. (Adapted from Bates and Stead 1993).

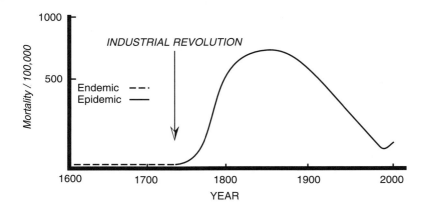

(Metchnikoff et al 1911). It appears to have reached Hawaii in the first half of the nineteenth century and India and China in the middle of the nineteenth century (Wilkinson 1914). Here the disease became epidemic more readily because poverty and crowding preceded its arrival.

Evidence of osseous TB in skeletons of Native Americans dates to the period between 100 BC and 1300 AD (Allison et al 1973; Buikstra 1981). Thus, tuberculosis due to *M. bovis* must have existed in pre-Columbian America (Stead et al 1995), attesting to the great age of this organism. The first major outbreaks of human TB among Native Americans began in the 1880s when they were crowded onto reservations (Bushnell 1921; Ferguson 1955). When several hundred Apache prisoners were confined in the Mount Vernon barracks in 1887, the death rate rose from 5.5% in the first year to 14% in the fourth year, nearly half of the deaths due to TB. These data show that human TB was a new disease to Native North Americans (Diamond 1992). Their greater susceptibility remains to this day.

TB was still practically unknown in sub-Saharan Africa until the twentieth century (Grzybowski 1991; Stead 1992), as averred by several physicians of the time (Cummins 1908, 1920; Lichtenstein 1928; Livingstone 1857; Hirsch 1886). The almost total lack of innate resistance to TB among African natives was clearly shown by the remarkable death rate from disseminated TB in South Africa (Millar 1908) and among Senegalese conscripts in Europe in World War I (Borrel 1920). The most likely reason for this lack of herd immunity to TB is that survival in Africa had never depended on resistance to TB, as it had in Europe (de Vries et al 1979; Hamilton et al 1990). The greater resistance of Africans to malaria, yellow fever, and trypanosomiasis reflects selective pressure from those infections on survival of their ancestors (Hill et al 1991).

African slaves were totally free of TB on arrival in the New World (Stead 1992). Thus, when blacks were first exposed to whites in America, TB tended to be subacute and often mistaken for typhoid fever (Yandell 1831; Osler 1892). However, following emancipation in 1865, blacks were crowded into cities and TB morbidity and mortality began to rise. By 1912 the TB rate among blacks reached 700 per 100,000 (McCarthy 1912).

TB was totally nonexistent in western New Guinea until the 1950s when that area was "discovered" by Europeans (Brown et al 1981). The same is true of areas in Amazonia today (Black 1975; A. DeSouza, personal communication 1995). As TB is introduced there, it resembles typhoid fever and is of minimal infectiousness (Templeton et al 1995).

MECHANISMS OF NATURAL RESISTANCE

Almost by accident in 1987, we discovered that African Americans are significantly less resistant to infection by *M. tuberculosis* than persons of European heritage (Stead et al 1990). Since that time, several groups have found evidence of such a difference at the cellular level (Crowle and Elkins 1992; McPeak et al 1992). Skamene's group has identified and cloned a gene that determines resistance in a strain of inbred mice (Gros et al 1981; Forget et al 1981; Vidal et al 1993). They also found an exact homologue on chromosome 2q in humans (Schurr et al 1990).

American blacks were recently found to have a 4 to 6-fold greater incidence than whites in spread of TB beyond the lung (Stead 1995, unpublished data). This suggests that the difference in susceptibility is not limited to initial resistance to infection, but also involves the response of the immune system.

PROSPECTS FOR THE FUTURE

The present resurgence of TB worldwide is closely associated with the spread of the human immune deficiency virus (HIV) which renders humans at least 100 times more susceptible than healthy individuals to both TB infection and progression of the infection to disease. TB is becoming prevalent especially in areas where HIV infection arrived before the TB epidemic had peaked. This may be the first time the world has seen two concurrent epidemics, with each disease accelerating the course of the other. Moreover, there is the issue of antibiotics and their misuse, including our best tuberculocidal drugs, isoniazid and rifampin. We are now facing a resurgence of TB, only this time the disease may well be incurable. Like many of mankind's woes, the problems with TB stem directly from overpopulation and abuse of the Earth's environment. If we attempt to solve the TB problem without addressing overpopulation and malnutrition, the effort will be wasted. This is well discussed in a recent book, *The Coming Plague: Newly Emerging Diseases in a World Out of Balance* (Garrett 1994).

REFERENCES

Allison MR, Mendoza O, Pezzia A (1973). Documentation of a case of tuberculosis in pre-Columbian America. Am Rev Respir Dis 107: 985.

Bates JH, Stead WW (1993). The history of tuberculosis as a global epidemic. Med Clin North Am 77: 1205-1217.

Black FL (1975). Infectious diseases in primitive societies. Science 187: 515.

Borrel A (1920). Pneumoniae et tuberculose chez les troupes noire. Annals del Inst Pasteur 34: 105.

Brown P, Cathala F, Gajdusek DC (1981). Mycobacterial and fungal sensitivity patterns among remote population groups in Papua, New Guinea, and in the Hebrides, Solomon and Caroline Islands. Am J Trop Med Hyg 30: 1085-1093.

Buikstra JE (1981). Prehistoric tuberculosis in the Americas. Evanston, Ill.: Northwestern University Archeological Program, 89.

Bushnell GE (1920). Epidemiology of tuberculosis. Baltimore: William Wood and Company, 157.

Clark G (1962). World prehistory. Cambridge: Cambridge University Press.

Crowle AJ, Elkins N (1990). Relative permissiveness of macrophages from black and white people for virulent tubercle bacilli. Infect Immun 58: 632.

Cummins SL (1908). Tuberculosis in the Egyptian Army. Br J Tuberc 2: 35.

Cummins SL (1920). Tuberculosis in primitive tribes and its bearing on the tuberculosis of civilized communities. Int J Public Health 1: 137.

de Vries RRP, Meera Kahn P, Bernini LF, van Loghem E, van Rood JJ. (1979). Genetic control of survival in epidemics. J Immunogenet 6: 271-287.

Diamond JM (1992). The arrow of disease. Discover 13: 64-73.

Dubos R, Dubos J (1952). The White Plague. Boston: Little Brown and Company.

Ferguson RG (1955). Studies in Tuberculosis. Toronto: University of Toronto Press, 6.

Forget A, Skamene E, Gros P, Miailhe AC, Turcotte R. (1981). Differences in response among inbred mouse strains to infection with small doses of Mycobacterium bovis BCG. Infect Immun 32: 42.

Garrett L (1994). The coming plague: Newly emerging diseases in a world out of balance. New York: Farrar, Strauss, Giroux, 571-572.

Grigg ER (1958). The arcana of tuberculosis. Am Rev Tb Pul Dis 78: 151-172, 426-453, 583-603.

Gros P, Skamene E, Forget A (1981). Genetic control of natural resistance to *Mycobacterium bovis* (BCG) in mice. J Immunol 127: 2417-2421.

Grzybowski S (1991). Tuberculosis in the third world (editorial). Thorax 46: 689-691.

Grzybowski S, Allen EA. (1964). The challenge of tuberculosis in decline. A study based on the epidemiology of tuberculosis in Ontario, Canada. Am Rev Respir Dis 90: 707-720.

Hamilton WD, Alexrod R, Tanese R (1990). Sexual reproduction as an adaptation to resist parasites. Proc Nat Acad Sci USA 87: 3566.

Hill AVS, Allsopp CEM, Kwiatkowski D, Anstey NM, Twamasi P, Rowe PA, Bennett S, Brewster D, Michael AJ, Greenwood BM (1991). Common West African HLA antigens are associated with protection from severe malaria. Nature 352: 595.

Hirsch A (1886). Handbook of geographical and historical pathology. Vol III. London: The New Sydenham Society.

Leakey, Louis quoted by Leakey R (1992). Origins reconsidered: In search of what makes us human. New York: Doubleday.

Lichenstein H (1928). Travels in Africa, 1803, 1804, 1805, 1806. A reprint of the translation from the original German by A. Plumptre. Cape Town: The Van Rierberck Society.

Livingstone D (1857). Missionary travels and researches in South Africa. London: Ward Lock.

McNeill WH (1976). Plagues and peoples. New York: Anchor Press, Doubleday.

McPeak M, Salkowitz J, Laufman H, Pearl D, Zwilling BS (1992). The expression of HLA-DR by monocytes from black and from white donors: different requirements for protein synthesis. Clin Exp Immunol 87: 163.

Metchnikoff E, Burnet E, Tarassevitch L (1911). Recherches sur l'epidemiologie de la tuberculose dans les steppes des Kalmouks. Ann L'instit Pasteur 25: 785-803.

Millar JG (1908). On the spread and prevention of tuberculosis disease in Pondoland, South Africa. Br Med J 1: 380.

Osler W (1892). The principles and practice of medicine. New York: D Appleton & Co.,198-200.

Schurr E, Skamene E, Morgan K, Chu ML, Gros P (1990). Mapping of Col3al and Col6a3 to proximal murine chromosome 1 identifies conserved linkage of structural protein genes between murine chromosome 1 and human chromosome 2q. Genomics 8: 477-486.

Stead WW (1992). Genetics and resistance to tuberculosis. Could resistance be enhanced by genetic engineering? Ann Intern Med 116: 937-941.

Stead WW, Lofgren JP, Senner JW, Reddick WT (1990). Racial differences in susceptibility to infection with M. tuberculosis. N Engl J Med 322: 422.

Stead WW, Bates JH (1995). Geographic and evolutionary epidemiology of tuberculosis. In: Rom WN, ed. Tuberculosis (In press). Boston: Little Brown & Co.

Stead WW, Eisenach KD, Cave MD, Beggs ML, Templeton GL, Thoen CO, Bates JH. (1995). When did *Mycobacterium tuberculosis* first occur in the New World? An important question with public health implications. Am J Respir Crit Care Med 151: 1267-1268.

Steele JH, Ranney AF (1958). Animal tuberculosis. Am Rev Tuberc 77: 908-922.

Templeton GL, Illing LA, Young L, Bates JH, Stead WW (1995). Comparing the risk for transmission of tuberculosis at bedside and during an autopsy. Ann Intern Med 122: 922-925.

Vidal SM, Malo D, Vogan K, Skamene E, Gros P (1993). Natural resistance to infection with intracellular parasites: Isolation of a candidate for Bcg. Cell 73: 469-485.

Wilkinson E (1914). Notes on the prevalence of tuberculosis in India. Proc R Soc Med 8: 195.

Yandell LP (1831). Remarks on struma africana or the disease called negro poison or negro consumption. Transylvania J Med Assoc Sci 4: 83.

24

Tuberculosis—Present and Future

Thomas M. Shinnick

Global Burden of Tuberculosis

Tuberculosis (TB) has been one of the most persistent diseases of humans since prehistoric times and is still an important source of mortality and morbidity throughout the world today (reviewed in Daniel et al 1994; Snider et al 1994; Sudre et al 1992; Murray et al 1990; Raviglione et al 1993). In April 1993, the World Health Organization (WHO) declared TB a global public health emergency—the only disease so far to earn that distinction. It is estimated that one-third of the world's population (about two billion persons) are, or have been, infected with *Mycobacterium tuberculosis,* the causative agent of TB. Each year, there are eight to ten million new cases of TB and about three million deaths due to TB. Indeed, TB is the leading cause of death in adults due to a single infectious agent and accounts for about 26% of all preventable adult deaths in the world. TB is also an enormous social and economic problem in developing countries because about 95% of new cases occur in developing countries and because about 80% of cases affect persons of child-bearing age and during their most economically productive years, ages 15 to 59.

TB is also reemerging as an important public health problem in many industrialized countries (Raviglione et al 1993; Cantwell et al 1994). That is, during the last 5 to 10 years, many industrialized countries experienced significant increases in the number of new cases diagnosed each year (Raviglione et al 1993). For example, the United Kingdom had a 5% increase in annual TB cases from 1987 to 1991, Ireland had a 9% increase from 1988 to 1991, the Netherlands had a 19% increase from 1987 to 1992, and Italy had a 27% increase from 1988 to 1992 (Raviglione et al 1993). To illustrate the change in the epidemiology of TB and the reasons for the change, the current status in the United States will be described in detail. The epidemiology of

TB is somewhat similar in most industrialized countries, although the exact impact of the various contributing factors varies for each country.

TUBERCULOSIS IN THE UNITED STATES

For most of the past four decades, there was a consistent decline in the number of TB cases reported to the Centers for Disease Control and Prevention (CDC), from 84,000 in 1953 when CDC began collecting nationwide statistics on TB to a low of 22,201 in 1985 (reviewed in Snider et al 1994). From 1985 to 1988, the number of cases remained about the same, and then increased by more than 20% to 26,673 new cases in 1992 (Cantwell et al 1994). The number of TB cases reported to CDC declined to 25,313 cases in 1993 (CDC 1994) and to 24,301 cases in 1994 (CDC 1995). Another way to look at the change in the incidence of TB is to compare the observed number of cases to the number of cases expected if the decline in new cases had continued at the 5.6% yearly rate observed between 1953 and 1984. Such an analysis suggests that there were more than 63,600 excess cases in the United States between 1985 and 1993 (CDC 1994).

In some regions of the United States, the increase in TB was even more dramatic (Cantwell et al 1994; CDC 1994; Brudney and Dobkin 1991). For example, the number of reported cases in New York City more than doubled between 1978 and 1992, before declining by 15% in 1993 (CDC 1994; Brudney and Dobkin 1991). Surprisingly, the incidence of TB in some parts of New York City was greater than that in many developing countries, with case rates approaching 200 cases per 100,000 persons. Several states also reported large increases in the number of new cases between 1985 and 1992, including New York (84.4% increase), New Jersey (80.6% increase), California (54.2% increase), and Texas (32.7% increase) (Cantwell et al 1994). In contrast, during this same time period, the number of new tuberculosis cases in the rest of the states actually declined by 6.9%

Tuberculosis has historically been a disease of the poor and socially disadvantaged, and much of the increase in TB occurred among minority populations and in persons 25 to 44 years of age (Cantwell et al 1994). Among blacks and Hispanics, there was a 70% increase in cases between 1985 and 1992 in persons 25 to 44 years of age. Among non-Hispanic whites, there was a 30% increase in cases in the same age group during this time period. Both groups had a decline in the number of cases in the elderly and an increase in cases in children. The increase in pediatric TB is not too surprising since it parallels the increase in TB in persons of child-bearing age.

Several factors contributed to the increase of TB in the United States and other industrialized countries (reviewed in Snider et al 1994; Cantwell et al 1994; Smith and Moss 1994; Ellner et al 1993). These include human immunodeficiency virus (HIV) infection and AIDS, a deteriorating public health infrastructure, transmission in congregate settings, and increased immigration from countries with a high incidence of TB. For example, TB among the foreign-born increased from 4925 cases in 1986 to 7627 cases in 1994, and TB in foreign-born persons now accounts for about 32% of all new cases in the United States (CDC 1995). Similar increases account for much of the increase in TB in Western European countries, including the Netherlands where TB in foreign-born persons now accounts for about 40% of all cases (Raviglione et al 1993).

TUBERCULOSIS AND HIV INFECTION

A major contributing factor to the increase in TB has been the increase in HIV infection and AIDS (Smith and Moss 1994; Ellner et al 1993; Narain et al 1992; Styblo and Enarson 1991; Hopewell 1992; Bloom and Murray 1992). In the United States, the increase in TB cases paralleled the increase in AIDS cases, and the states with the greatest increase in AIDS cases also reported the greatest increase in TB cases. That is, New York had the largest increase in the number of AIDS cases and the largest increase in the number of TB cases. California had the second largest increases in both, while Florida, Texas, and New Jersey had the third through fifth largest increases in each disease, although the order was slightly different for the two diseases. The strongest evidence linking HIV infection and TB comes from a study of methadone patients in New York City in which the risk of developing TB among persons co-infected with *M. tuberculosis* and HIV was calculated to be about 8% per year as compared to a reported lifetime risk of 5 to 10% among persons infected with *M. tuberculosis* but not HIV (Comstock 1982; Selwyn et al 1989).

On the global scale, the potential deleterious interactions between the HIV and TB epidemics are enormous because the epidemics overlap biologically, demographically, and geographically. Biologically, HIV infection can destroy the cellular immune system, and it is the host's cellular immune response to the tubercle bacillus that is responsible for controlling the infection and preventing the development of active TB. To date, HIV infection is the largest known risk factor for progression from latent *M. tuberculosis* infection to active TB (Styblo and Enarson 1991; Selwyn et al 1989). In addition, *M. tuberculosis* infection may promote the replication of HIV, thereby accelerating destruction of the cellular immune system and the development of AIDS and consequently active TB. Demographically, both epidemics are concentrated among persons in the age group of 15 to 49 years and among the socially disadvantaged. The epidemics also overlap geographically. Sub-Saharan Africa has more than 1 million TB cases and 3.8 million HIV-seropositive persons, and in many African countries, 40% of patients with active TB are also seropositive for HIV (De Cock et al 1992; Narain et al 1992). Overall, it is estimated that more than five million persons are co-infected with HIV and *M. tuberculosis*. This number is likely to grow considerably in the future, because Asia has two-thirds of the world's *M. tuberculosis*-infected persons and the HIV epidemic is just entering this area.

For example, between 1988 and 1992, HIV seroprevalence rates among TB patients increased from 2% to 15% in Bombay and from 5% to 15% in Thailand.

DRUG-RESISTANT TUBERCULOSIS

The most serious aspect of the TB problem in the United States is the emergence of multidrug-resistant TB (National MDR-TB Task Force 1992; Dooley et al 1992), a term that usually refers to TB caused by a strain of *M. tuberculosis* resistant to isoniazid, rifampin, and perhaps to other anti-TB drugs. Multidrug-resistant has also been used to designate a case that is resistant to any two or more anti-TB drugs.

During the past seven years, CDC has assisted local health departments in investigating eight outbreaks of multidrug-resistant TB in hospitals and correctional facilities (Ellner et al 1993; National MDR-TB Task Force 1992; Dooley et al 1992; CDC 1991). The individual outbreaks involved from 7 to 70 cases, and the outbreak strains were resistant to as many as seven of the first- and second-line anti-TB drugs (e.g., isoniazid, rifampin, streptomycin, ethambutol, ethionamide, rifabutin, and kanamycin). So far, the nosocomial outbreaks of multidrug-resistant TB have occurred primarily among HIV-infected persons. In most hospitals, 80 to 100% of the cases were in patients with AIDS, and among these patients, mortality was unusually high (70 to 90%), and the course of the disease was very rapid, with an interval of only 4 to 16 weeks between the time of diagnosis and time of death. The high death rates and rapid time course were probably due to the severity of TB in persons with impaired cellular immune systems as well as to the resistance of these infections to conventional drug regimens. An additional troubling aspect of these outbreaks was the transmission of *M. tuberculosis* from patients to health care workers. For example, tuberculin skin test conversions were detected after outbreak exposure in 22 to 50% of healthcare workers in three of the outbreak hospitals. Twenty of these health care workers have developed multidrug-resistant TB, and nine have died with TB.

Several factors contributed to the nosocomial outbreaks of multidrug-resistant TB (Ellner et al 1993; National MDR-TB Task Force 1992; Dooley et al 1992; CDC 1991). First, there was the problem of adherence with medication regimens. In areas of the United States, less than 80% of patients who start chemotherapy actually complete the entire course of therapy within 6 to 12 months, and in one large New York City hospital during the time of the outbreaks, the completion rate was only about 11%. Such poor adherence rates pose a serious public health problem, because failure to take the medication as prescribed for the entire duration of treatment facilitates the development of drug-resistant TB and allows for continued transmission. One way to examine this concern is by comparing the rates of drug resistance in isolates from TB patients previously treated for the disease with the rates from TB patients not previously treated. In one study of *M. tuberculosis* isolates from

New York City (Freiden et al 1993), the frequency of isoniazid resistance in previously treated patients was 36% compared to 15% in previously untreated patients; the frequency of rifampin resistance was 33% in the previously treated group and 10% in the previously untreated group; and the frequency of doubly resistant strains was 30% and 7%, respectively. Thus, previously treated patients are two to three times more likely to have drug-resistant TB than are previously untreated persons. Similarly, in Korea and Hong Kong, previously treated patients are about three times more likely to have drug-resistant TB than are previously untreated persons.

Another contributing factor to nosocomial TB outbreaks was the prescription of inadequate drug regimens by physicians. Regimens must contain at least two drugs to which the patient's organisms are susceptible; however, in many cases, drug susceptibility testing and reporting were absent or delayed such that the physician had little information on which to base the choice of drugs. A third factor was that patients infected with bacilli resistant to isoniazid and rifampin remained infectious for prolonged times, because such drug-resistant bacilli are less susceptible to the standard anti-TB drug regimens and because the second-line anti-TB drugs are less efficient than rifampin and isoniazid at killing *M. tuberculosis*. A fourth contributing factor was inadequate or delayed infection control procedures, and at the time of the outbreaks, many hospitals did not have any proper acid-fast isolation rooms. In some hospitals, TB patients were housed in the same areas as susceptible contacts. For example, TB patients were allowed to share common rooms with HIV-seropositive patients. A fifth factor was the likelihood that HIV-infected persons will rapidly develop active disease if infected with *M. tuberculosis,* which compressed the time frame of the outbreaks and facilitated their detection and spread.

The problem of drug-resistant TB is not restricted to the hospital setting, but rather it is a prominent feature in the general community. For example, a survey of drug-resistant TB in the United States revealed that 14.2% of TB cases were resistant to one or more drugs, 9.1% were resistant to isoniazid, 3.9% were resistant to rifampin, and 3.5% were resistant to both isoniazid and rifampin (Bloch et al 1994). Also, one study of 466 *M. tuberculosis* isolates from New York City (Freiden et al 1993) found that 33% of the isolates were resistant to one or more drugs, 26% were resistant to isoniazid, 22% were resistant to rifampin, and 19% were resistant to both isoniazid and rifampin. Similarly, a study of 961 *M. tuberculosis* isolates from Atlanta (Miller et al 1996) showed that 5.6% of the isolates were resistant to isoniazid, 3.4% were resistant to rifampin, 1.5% were resistant to both isoniazid and rifampin, and 2.3% were resistant to any two anti-TB drugs.

To collect data on drug susceptibilities of *M. tuberculosis* isolates throughout the United States, a mycobacteriology module was recently added to the Public Health Laboratory Information System (PHLIS). PHLIS is a computer network through which state public health laboratories electronically report data to CDC on the various pathogens isolated in their laboratories (Bean et al 1992). The mycobacteriology module requests information on all *Mycobacterium* species encountered in the state public health laboratories and the drug susceptibilities of the *M. tuberculosis* isolates. A limited amount of patient information is also entered to eliminate duplicate isolates from the same patient and to allow identification of potential clusters of cases. During 1994, data on 16,400 *M. tuberculosis* isolates were reported to CDC, and 12.1% of these were resistant to one or more drugs, 7.6% were resistant to two or more anti-TB drugs, 10.8% were resistant to isoniazid, 6% were resistant to rifampin, and 4.5% were resistant to both isoniazid and rifampin. These data also revealed that 1) drug resistance is increasing in the United States in that the frequency of isolating a strain of *M. tuberculosis* resistant to one or more anti-TB drugs has nearly doubled from about 7% in 1986 (Snider et al 1991) to 12.1% in 1994 and 2) drug resistance is widespread, with 43 states reporting at least one case of multidrug-resistant TB during 1994. These data on susceptibilities of *M. tuberculosis* isolates parallel the data on drug-resistant tuberculosis (Snider et al 1991).

TUBERCULOSIS THERAPY

The prompt initiation of an effective multidrug regimen is crucial to preventing the development and spread of drug-resistant tuberculosis. In the United States, it is recommended that, in general, TB patients should be started on an empiric four-drug regimen (American Thoracic Society 1994; CDC 1993). although a three-drug regimen may be acceptable in communities where isoniazid resistance is documented to be less than 4%, and a five-drug or six-drug regimen may be appropriate in certain situations, such as in institutions that are experiencing outbreaks of multidrug-resistant TB. (Several treatment options are available, and the reader is referred to the above-noted references for a thorough discussion of treatment recommendations.) The recommended initial four-drug regimen includes isoniazid, rifampin, pyrazinamide, and ethambutol or streptomycin (CDC 1993), and the regimen should be modified when the results of drug susceptibility tests become available to ensure that the regimen contains at least three effective antituberculosis drugs. For example, if the tubercle bacilli are documented to be susceptible to isoniazid, rifampin, and pyrazinamide, the initial four-drug regimen may be changed to isoniazid, rifampin, and pyrazinamide. The intensive initial phase of treatment should continue for a total of two months and be followed by a continuation phase, which also employs multidrug regimens. For example, in one recommended treatment option, TB patients without HIV infection receive at least four more months of isoniazid and rifampin for a total of six months of therapy, although therapy is continued for at least three months after the patient's culture has become negative. TB patients co-infected with HIV can be treated similarly but must be monitored closely to ensure response to therapy. Most of the therapy during the continuation phase can be administered on a twice or thrice weekly basis, if directly observed therapy is used. Finally, the overall treatment regimen should be tailored to the characteristics of the individual patient (e.g., drug tolerance, response to therapy) and the infecting bacilli (e.g., drug resistance).

Adherence to the chemotherapeutic regimen is essential to cure the patient and to prevent development of drug-resistant TB. Thus, therapy should be directly observed by a health care provider unless adherence to self-administered medication can be assured. When self-administration is used, health care providers should monitor adherence by periodically interviewing the patient, taking pill counts, and performing urine tests.

Multidrug short-course regimens are also recommended by the International Union Against Tuberculosis and Lung Disease (IUATLD) and by WHO for the treatment of TB in developing countries (IUATLD; WHO TB Programme 1993). For example, IUATLD recommends that smear-positive cases be treated with daily isoniazid, rifampin, pyrazinamide and ethambutol for two months followed by daily isoniazid and thiacetazone for a further six months. To ensure adherence to the chemotherapeutic regimen and reduce the likelihood of the development of drug-resistant TB, therapies should be given under supervision.

It must be emphasized that TB is a treatable disease and that treatment is affordable. More than 95% of the cases in the world could be cured if treated with the recommended short-course regimens, and the cost of the drugs needed to cure these patients is only $30 to $100 in developing countries. On average, the total program cost per year of life saved is $1 to $4 for curing a case of smear-positive TB, which makes it one of the most cost-effective public health strategies available for any disease (Murray 1994). The use of the recommended multidrug regimens and directly observed therapy is important to maintain the cost-effectiveness of TB chemotherapy, because the cost to cure skyrockets and cure rates drop dramatically if drug resistance develops. For example, the cost of treating a single TB case resistant to the four first-line anti-TB drugs can easily exceed $100,000 in the United States and the cure rate is less than 50% (Mahmoudi and Iseman 1993).

FUTURE TRENDS IN TUBERCULOSIS

History teaches us that the numbers of new cases and deaths will increase if we continue to neglect TB. As the incidence of TB in the United States decreased in the 1960s and 1970s, the perception arose that it was a conquered disease, which resulted in drastically decreased funding for TB control programs and research, reduced emphasis on teaching about TB in medical schools, and deterioration of the public health infrastructure necessary to detect, treat, and follow TB cases. These factors, in turn, led to the resurgence of TB during the 1980s and 1990s. A similar neglect of TB on the global scale may lead to an increase in the annual burden of TB to more than 12 million new cases and four million deaths by the year 2005 (Dolin et al

1994). Furthermore, the HIV pandemic is likely to exacerbate the situation since 1) HIV seroprevalence in TB patients is expected to increase in sub-Saharan Africa and to double or triple in Asia in the next decade, and 2) the number of TB cases in HIV-infected persons is predicted to increase from an estimated 315,000 in 1990 to more than 1.4 million by the year 2000 (Dolin et al 1994). Such dramatic increases in the number of new TB cases threaten to overwhelm the health care systems of several developing countries.

Fortunately, controlling the spread of TB is possible and cost-effective using existing methods (reviewed in WHO TB Programme 1994). In general, two key features of successful control programs are the early detection of persons with active TB and the prompt initiation and completion of an effective chemotherapeutic regimen. Timely case detection and treatment is important because a person with untreated, smear-positive TB will on average infect 10 to 15 persons per year with the tubercle bacillus (Murray 1994), and the recommended regimens rapidly render patients noninfectious and prevent transmission and future TB cases.

The World Health Organization (WHO TB Programme 1994) has identified several additional important steps necessary to ensure the control of TB throughout the world:

1) Strong advocacy for aggressive control programs is needed from national and international agencies to convince individual governments of the large economic and medical burdens TB imposes on their countries and the need to act now to establish effective control programs;

2) The development of a public health infrastructure is needed at both the national and local levels to ensure the timely identification of infectious TB cases, provision of a reliable supply of effective anti-TB drugs, and supervision of treatment;

3) The use of short-course, multidrug regimens and directly observed therapy is needed to improve completion and cure rates and to prevent the development of drug-resistant TB;

4) Increased funding from industrialized countries for TB control programs in developing countries is needed, especially for the purchase of effective drugs;

5) Improvements in diagnostic methods are needed to facilitate identification of active TB cases. Relatively small improvements in the specificity and sensitivity of tests to identify smear-negative TB cases may reduce the cost per year of life saved by a factor of 2;

6) Research into the biology of *M. tuberculosis* and the pathogenesis of TB is needed to develop new reagents

and methods for the detection, treatment, and prevention of this disease. A controversial issue to be resolved is the role, if any, of Bacillus Calmette-Guérin (BCG) vaccination in a TB control strategy.

In summary, continued neglect of TB will undoubtedly lead to an increased global burden, with particularly devastating effects in developing countries. Action must be taken now to prevent this increase. Aggressive action along the lines discussed above should prevent the growth in TB and may even significantly reduce the deaths caused by TB by the turn of the century.

REFERENCES

American Thoracic Society (1994). Treatment of tuberculosis and tuberculosis infection in adults and children. Am J Respir Crit Care Med 149: 1359-1374.

Bean NH, Martin SM, Bradford H (1992). PHLIS: an electronic system for reporting public health data from remote sites. Am J Public Health 82: 1273-1276.

Bloch AB, Cauthen GM, Onorato IM, Dansbury KG, Kelly GD, Driver CR, Snider DE (1994). Nationwide survey of dug-resistant tuberculosis in the United States. JAMA 271: 665-671.

Bloom BR, Murray CJL (1992). Tuberculosis: commentary on a reemergent killer. Science 257: 1055-1064.

Brudney K, Dobkin J (1991). Resurgent tuberculosis in New York City: human immunodeficiency virus, homelessness and the decline of tuberculosis control programs. Am Rev Respir Dis 144: 745-749.

Cantwell MF, Snider DE, Cauthen GM, Onorato I (1994). Epidemiology of tuberculosis in the United States, 1985 through 1992. JAMA 272: 535-539.

Centers for Disease Control and Prevention (1993). Initial therapy for tuberculosis in the era of multidrug resistance. MMWR 42: 1-8.

Centers for Disease Control and Prevention (1994). Expanded tuberculosis surveillance and tuberculosis morbidity—United States, 1993. MMWR 43: 361-366.

Centers for Disease Control (1991). Nosocomial transmission of multidrug-resistant tuberculosis among HIV-infected persons—Florida and New York, 1988–1991. MMWR 40: 585-591.

Centers for Disease Control (1995). Tuberculosis morbidity—United States, 1994. MMWR 44(RR-20): 387-395.

Comstock GW (1982). Epidemiology of tuberculosis. Amer Rev Respir Dis 125(Suppl): 8-16.

Daniel TM, Bates JH, Downes KA (1994). History of tuberculosis. In: Bloom BR, ed, Tuberculosis: pathogenesis, protection, and control. Washington DC: American Society for Microbiology Press, 13-24.

De Cock KM, Soro B, Coulibaly IM, Lucas SB (1992). Tuberculosis and HIV infection in sub Saharan Africa. JAMA 268: 1581-1587.

Dolin PJ, Raviglione MC, Kochi A (1994). Global tuberculosis incidence and mortality during 1990–2000. Bull WHO 72: 213-220.

Dooley SW, Jarvis WR, Martone WJ, Snider DE (1992). Multidrug-resistant tuberculosis. Ann Intern Med 117: 257-259.

Ellner JJ, Hinman AR, Dooley SW, Fischl MA, Sepkowitz KA, Goldberger MJ, Shinnick TM, Iseman MD, Jacobs WR (1993). Tuberculosis symposium: emerging problems and promise. J Infect Dis 168: 537-551.

Freiden TR, Sterling T, Pablos-Mendez A, Kilburn JO, Cauthen GM, Dooley SW (1993). The emergence of drug-resistant tuberculosis in New York City. N Engl J Med 328: 521-526.

Hopewell PC (1992). Impact of human immunodeficiency virus infection on the epidemiology, clinical features, management, and control of tuberculosis. Clin Infect Dis 18: 540-546.

International Union Against Tuberculosis and Lung Disease (1988). Antituberculosis regimens of chemotherapy. Recommendations from the Committee on Treatment of the IUATLD. Bull Int Union Tuberc Lung Dis 63: 60-64.

Mahmoudi A, Iseman MD (1993). Pitfalls in the care of patients with tuberculosis. JAMA 270: 65-68.

Miller LM, Blumberg HH, Shinnick TM (1996). A molecular and epidemiologic study of rifampin-resistant strains of *Mycobacterium tuberculosis* in an inner city hospital, 1991–1994. in preparation.

Murray, CJL, Styblo K, Rouillon A (1990). Tuberculosis in developing countries: burden, intervention and cost. Bull Int Union Tuberc Lung Dis 65: 2-20.

Murray, CJL (1994). Issues in operational, social, and economic research on tuberculosis. In: Bloom BR, ed, Tuberculosis: pathogenesis, protection, and control. Washington DC: American Society for Microbiology Press, 583-622.

Narain JP, Raviglione MC, Kochi A (1992). HIV-associated tuberculosis in developing countries: epidemiology and strategies for prevention. Tuberc Lung Dis 73: 311-321.

National MDR-TB Task Force (1992). National action plan to combat multidrug-resistant tuberculosis. MMWR 41: 1-48.

Raviglione MC, Sudre P, Rieder HL, Spinaci S, Kochi A (1993). Secular trends of tuberculosis in Western Europe. Bull WHO 71: 297-306.

Selwyn PA, Hartel D, Lewis VA, Schoenbaum EE, Vermund SH, Klein RS, Walker AT, Friedland GH (1989). A prospective study of the risk of tuberculosis among intravenous drug users with human immunodeficiency virus infection. N Engl J Med 320: 545-550.

Smith PG, Moss AR (1994). Epidemiology of tuberculosis. In: Bloom BR, ed. Tuberculosis: pathogenesis, protection, and control. Washington DC: American Society for Microbiology Press, 47-59.

Snider DE, Cauthen GM, Farer LS, Kelly GD, Kilburn J, Good RC, Dooley SW (1991). Drug-resistant tuberculosis. Am Rev Respir Dis 144: 732.

Snider DE, Raviglione MC, Kochi A (1994). Global burden of tuberculosis. In: Bloom BR, ed. Tuberculosis: pathogenesis, protection, and control. Washington DC: American Society for Microbiology Press, 3-11.

Styblo K, Enarson DA (1991). Epidemiology of tuberculosis in HIV prevalent countries. In: Selected papers. Vol 24. The Hague: Royal Netherlands Tuberculosis Association, 116-136

Sudre P, Ten Dam G, Kochi A (1992). Tuberculosis: a global overview of the situation today. Bull WHO 70: 149-159.

World Health Organization Tuberculosis Programme (1993). Treatment of tuberculosis: guidelines for national programmes. Geneva: World Health Organization.

World Health Organization Tuberculosis Programme (1994). TB: A global emergency. Geneva: World Health Organization.

Vignette

Lyme Disease

Erythema migrans, the disease described in Sweden and Austria early in the twentieth century, was "rediscovered" in the United States in Lyme, Connecticut, in the 1970s and named Lyme disease. The association between arthropods and spirochetes made it possible to identify the causative agent as a *Borrelia* species. It was named *Borrelia burgdorferi* after the medical entomologist who first recognized the spirochetes in the midgut of the deer tick, *Ixodes ricinus*. In the United States, the large increase in the deer population resulted in the spread of Lyme disease among humans. The causative agent of the disease is susceptible to antibiotics, but prolonged treatment is required for a cure. In some cases, the debilitating form of the disease may lead to chronic invalidism of the infected person. The *B. burgdorferi* antigen OspA expressed as a recombinant in *E. coli* is being used in clinical trials, but the results are rather disappointing and it is clear that more research is needed to produce an efficacious vaccine that protects against infection.

25

Etiology of Lyme Disease: Lessons for the Present and Future

ALAN G. BARBOUR

The infection known around the world as Lyme disease was first described during the early years of the twentieth century as erythema migrans. There was much in the European medical literature on the characteristic clinical features of erythema migrans, if less on the epidemiology and treatment of the disorder, before an outbreak of arthritis among the residents of Lyme, Connecticut, was reported in the 1970s. Thus, it is to the chagrin of European investigators and clinicians that the disorder first reported by physicians in Austria and Sweden decades ago now takes its name from a town in North America.

A rash that starts as a small red patch and then expands over the succeeding days to weeks to become 20 or more centimeters across with central clearing or an inner ring could be little else but erythema migrans (Figure 1). The association of the rash with a prior tick bite, especially an *Ixodes ricinus* tick, was noted in the European literature. Later, a curious, if foolhardy, investigator demonstrated the transmissible nature of erythema migrans by transferring a sample of a patient's skin to that of another person. Anecdotal reports of successful treatment of the condition with antibiotics, such as penicillin and tetracycline, was evidence of a bacterial cause of the disorder. Also associated with exposure to *I. ricinus* was a neurologic disorder comprising chronic or subacute meningitis with signs of inflammation of nerve roots. Again, the first descriptions came from Europe.

Erythema migrans remains the name of the hallmark rash of the infection, but the entire set of manifestations of the infection are now included under the name Lyme disease. The first use of "Lyme" was in a report by Allen Steere and colleagues of a cluster of polyarticular arthritis among children in a New England town (Figure 2). As the investigation of the original and additional cases continued, the frequent occurrence of a skin rash preceding the onset of arthritis was noted. The similarity of the rash to the erythema migrans of the European literature was clear. Later, the association of the rash

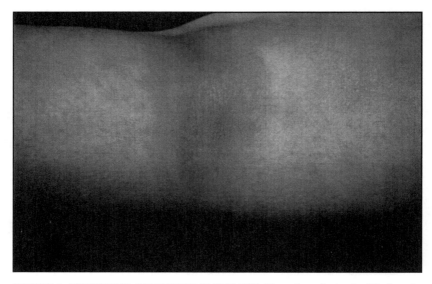

FIGURE 1. ERYTHEMA CHRONICUM MIGRANS skin rash on the back of the leg of a patient with Lyme disease. Photo credit: American Lyme Disease Foundation, Rye, New York.

in Connecticut to a prior tick bite was made in what was to become the most comprehensive studies of the epidemiology and clinical features of the disease. Included in this set of studies from the United States was a description of a potentially life-threatening carditis that not infrequently followed erythema migrans.

By the late 1970s, it was recognized that Lyme disease in North America and erythema migrans and associated disorders in Europe were the same. The seasonality of the acute disease, the geographic clustering of cases, the association with ticks, and the lack of secondary transmission of the disease among humans all suggested a tick-borne zoonosis. The choice was between a bacterium and a virus; a parasite or fungus was considered unlikely. Among bacteria, a possible candidate was a rickettsia or rickettsia-like agent, because of the well-recognized associations between arthropods and these obligate intracellular microorganisms. However, serologic studies failed to implicate any of the known rickettsiae. Among viruses, an arbovirus was a consideration, but again studies of sera from patients and attempts at virus isolation from human and tick specimens were not revealing.

The recognition that the agent of Lyme disease and erythema migrans was another type of tick-borne agent came in 1981. Besides carrying rickettsiae and viruses, ticks were also known to carry bacteria of the phylum spirochetes. The arthropod-borne spirochetes are all in the genus *Borrelia*. Priority for the discovery of spirochetes in the tick vector and for the first cultivation of the organism was given to Willy Burgdorfer and the author, respectively,

FIGURE 2. TOWN HALL OF LYME, Connecticut, a town in the outbreak of oligoarticular arthritis among children in 1975.

but two other individuals independently came to the conclusion that a *Borrelia* sp. was a possible cause. One was Klaus Weber in Munich, who proposed this theory on the basis of the known responsiveness of erythema migrans to penicillin and the recognized association between ticks and members of the genus *Borrelia*. Rickettsial agents, one of the other tick-borne bacteria, would not be predicted to be susceptible to penicillin.

The other investigator on the trail of spirochetes in Lyme disease was Rudolf Ackermann in Cologne, who with his colleagues found that patients with erythema migrans alone or with neurologic complications often had antibodies to *Borrelia duttoni*, a species that causes relapsing fever. Had either Dr. Weber or Ackermann provided more tangible evidence of this proposed etiologic relationship, the name of the agent might be different today. As it happened, the initial discovery of spirochetes in the *Ixodes scapularis* ticks from Long Island, New York, was serendipitous. Willy Burgdorfer, a medical entomologist, was examining the midguts of ticks under the microscope. The ticks had been sent to him by Jorge Benach of New York. Dr. Burgdorfer, an expert in tick-borne diseases, had as a graduate student in Switzerland studied *Borrelia* spp. that caused relapsing fever and in the course of his studies had examined many tick dissections under the microscope. He was one of the few people in the world who was prepared to recognize spirochetes in that preparation. His knowledge of the field of Lyme disease in general and the renewed appreciation of the effectiveness of penicillin for treatment in particular provided the additional information for his conclusion about the possible role of these spirochetes in the etiology of Lyme disease.

At the time, I was working in another department of the Rocky Mountain Laboratories on one of the relapsing fever *Borrelia* spp., namely *B. hermsii*, and routinely used a complex medium for cultivating the spirochetes. With some tick specimens provided by Dr. Burgdorfer, I was able to isolate what came to be called *B. burgdorferi* in this medium. Using patient sera provided by Dr. Benach and Edgar Grunwaldt, a practitioner in a high-risk area for Lyme disease in New York, Dr. Burgdorfer and I showed that patients with Lyme disease had antibodies to the newly isolated spirochete significantly more frequently than did healthy controls.

Publication of the organism's identification and the serologic evidence for its relationship to Lyme disease prompted Allen Steere and Jorge Benach and colleagues to look for *B. burgdorferi* in patients. These two groups independently cultivated the microorganism from blood, skin, and cerebrospinal fluid specimens of patients with Lyme disease. Thereafter, investigators in Europe isolated the spirochete from erythema migrans skin lesions, cerebrospinal fluid, and from the heart of patients on that continent. Initially there was no direct detection of *B. burgdorferi* in the joints of patients with Lyme arthritis, but the demonstration of *B. burgdorferi* in the arthritic joints of mice was evidence that the direct invasion by spirochete was the cause of this sequela of erythema migrans (Figure 3). Eventually, the direct role of the spirochete in Lyme arthritis patients was shown by the polymerase chain reaction (PCR).

Within ten years of the first description of *B. burgdorferi*, a whole-cell killed vaccine for protection of dogs was on the market. Twelve years after the discovery, human field trials of a vaccine against Lyme disease began. The basis for the human vaccine is a single recombinant protein. The antigen is OspA, an abundant outer membrane lipoprotein, which is highly conserved among strains of *B. burgdorferi* isolated from humans in North America. Marcus Simon and Erol Fikrig and their colleagues in Germany and the United States independently showed that mice could be protected from infection as well as arthritis by active immunization with recombinant OspA. The gene for expressing the recombinant protein in the first human field trial of the vaccine is from strain B31, the original isolate of *B. burgdorferi* from 1981.

LESSONS

The progression from agent identification to human trials shows the power of recombinant DNA technology in the 1980s. Had there been an earlier consensus on the justification for the vaccine and sooner appreciation of the potential commercial market for a human vaccine, the speed from agent identification to field testing would have been more rapid. OspA itself was identified with monoclonal antibodies and expressed in recombinant form by *Escherichia coli* within five years of the agent's discovery.

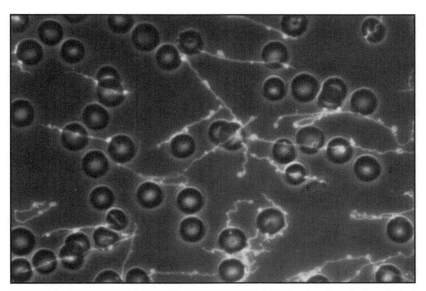

FIGURE 3. INDIRECT IMMUNOFLUORESCENCE ASSAY using *Borrelia burgdorferi*. Washed rat red blood cells were mixed with a suspension of cultivated *B. burgdorferi* in 2% bovine serum albumin in phosphate-buffered saline. A thin smear was made. Sera, followed by a fluorescein-conjugated second antibody, were applied to the fixed, dried slides. The picture was taken under phase contrast with both transmitted illumination and epifluorescence. Fluorescent spirochetes with membrane blebs are interspersed with red blood cells. Magnification, 800X.

The technology of the last few years, in particular PCR and related methods, shortens the time between agent discovery and production of candidate vaccines. Recently, new agents, which are uncultivable, are being identified by PCR. The experience with HIV shows that possession of even the complete sequence of an organism does not assure a successful vaccine. But if animal models show efficacy of the candidate antigen and if there is a perceived market large enough to justify human field trials costing tens of millions of dollars, the progression from laboratory to market can be very fast indeed.

The advantages of application of these technologies is clear. Certainly, few doubt the rationale for and value of a vaccine against HIV. With every year that passes, the death toll from AIDS rises. While there is comparatively greater controversy about the justification for a vaccine against Lyme disease, the human field trials at least will test the feasibility of a vaccine.

A disadvantage of rapid progression from agent identification to vaccine through recombinant DNA and PCR technologies is that a vaccine may be widely used before a more complete understanding of the immunology and pathogenesis of the infection is understood. In the case of Lyme disease, only after large human field trials are underway is it being recognized that OspA, the vaccine basis, probably is not expressed in mammalian hosts during the early phases of infection. The vaccine may need to provide circulating anti-bodies against OspA above a certain minimum concentration. The antibodies would alone or with complement or phagocytes kill OspA-bearing spirochetes as they come out of the ticks. Antibodies to OspA may not provide protection once the infection is underway in a person or other mammal. Had this phe-nomenon been appreciated, the end points for human phase I and II studies might have been different. The goal of a vaccine against OspA is prevention of infection to begin with. The requirements for such a vaccine are obviously different from those for a vaccine designed to prevent disease but not neces-sarily infection. A vaccine against tuberculosis probably would use the latter approach.

ACKNOWLEDGMENTS

My research on Lyme disease reported here was supported by funding from the National Institutes of Health, most recently grant AI37248.

Helicobacter pylori

For millions of people suffering from gastritis, for hundreds of thousands who have stomach ulcers, and for those unfortunates who have gastric cancer, the discovery of *Helicobacter pylori* as the underlying cause of all these syndromes has provided a bright ray of hope.

Although the presence of bacteria-like structures in gastric mucosa of patients suffering from stomach disease was recognized in the early 1900s, humanity owes much to the Australian duo of Marshall and Warren for the proof that *H. pylori* is the cause of all "gastric troubles." Their discovery was rapidly followed by the establishment of an ingenious diagnostic test based on the fact that urease produced by the bacteria in the infected host can convert urea to ammonia. This test obviated the need for endoscopy. Treatment of infection with a combination of antibiotics and bismuth has a remarkable eradication rate of the bacteria in 90 to 95% of the cases. If a link between chronic gastritis and stomach cancer is ever firmly established, the antibacterial drugs will become a cancer prevention therapy.

26

The Discovery of *Helicobacter pylori*

T. ULF WESTBLOM

Some scientific discoveries turn out to have profound effects far beyond initial expectations. The first reports of the new bacterium *H. pylori* started as a series of brief letters to the editor of the *Lancet* but came to dramatically change the way we look at peptic ulcer disease. Before these reports of "unidentified curved bacilli on gastric epithelium in active chronic gastritis" (Warren and Marshall 1983), the dogma of the day had been "no acid, no ulcer." Like all revolutionary concepts, the new idea that peptic ulcer disease may be an infectious disease had a hard time taking hold.

In 1982 Barry Marshall, a young gastroenterology fellow in Perth, Australia, was desperately looking for a suitable research project for his fellowship. He happened to run into Robin Warren, a local pathologist who for many years had been observing what looked like bacteria in the stomach of people with gastritis. Warren was convinced that they somehow played a role in gastric disease, but he had not told anyone in the scientific community about his findings. This new idea suited Marshall perfectly and he talked Warren into helping him make this his research project. Little did he know that this would be the start of years of battling the scientific establishment that was not yet ready for his work. That same year, Marshall submitted an abstract about *H. pylori* to the annual meeting of the Australian Gastroenterology Association. It was flatly rejected, but he was reassured that he was in good company. With such an abundance of good submissions, the organizers had to draw the line somewhere and many quality abstracts had to be rejected. It was only years later that Marshall learned that his abstract was only one of less than a dozen that had been rejected that year.

Fortunately, Marshall was a junior researcher with a passion that did not allow him to become discouraged. He had the foresight to realize that if the gastroenterologists were not ready to recognize a new infectious disease, maybe the infectious diseases community was. Since the bacteria looked as though they could be a *Campylobacter* species, he promptly submitted the same abstract to the International Workshop on Campylobacter Infections in Brussels that same year. This time he was much more successful. The abstract was accepted and Barry Marshall presented the first official report on *H. pylori* to a bewildered group of microbiologists and infectious diseases people. Though most were skeptical, many went home and promptly looked for the same bacteria. Within a year, multiple groups around the world confirmed his reports. The stomach was not the sterile environment everyone had believed for such a long time.

Even more surprising was the fact that this was not a new discovery. The bacteria we know today as *H. pylori* had been observed multiple times in the past, but those observations had also been dismissed. At the turn of the century, Krienitz had reported seeing the same spiral-shaped bacteria in patients with gastric cancer (Krienitz 1906), and 30 years later, *H. pylori* was again observed and described in two independent studies (Doenges 1939; Freedberg and Barron 1940). Freedberg and Barron had looked at 35 patients undergoing partial gastrectomies. They found bacteria, described as "spirochetes," in the stomach of 37% of these patients. Freedberg and Barron also made the astute observation that these bacteria were most often present in the patients that had gastric ulcers or cancer. However, when other investigators looked for these bacteria, they did not always see them. The final blow to this new discovery came in 1954 when E.D. Palmer, a leading authority in gastroenterology, investigated more than 1000 patients with gastric biopsies without seeing any signs of infection (Palmer 1954). He concluded that what people had been looking at was "simple contamination of the mucosal surface by swallowed spirochetes." When Palmer could not see any evidence of infection, most people accepted that it did not exist. The *H. pylori* story went into hibernation for the next three decades.

We now know that the primary reason why Palmer could not verify the existence of the organism was the histologic stain he was using. He stained all his biopsies with hematoxylin and eosin (H&E), a stain which is very poor at visualizing the bacteria. In contrast, both Freedberg and Barron, and Warren and Marshall 30 years later, used silver stains which show the bacteria very well (Westblom 1991).

PATHOGENIC CHARACTERISTICS OF *H. PYLORI*

In his first clinical study, Marshall found that *H. pylori* was highly correlated with mucosal inflammation (Marshall and Warren 1984). The bacteria were seen in 95% of patients with gastritis and 100% of patients with duodenal ulcer, while only 6% of patients with normal histology were infected. This proved an association but not causality. Perhaps *H. pylori* was an opportunist that thrived in diseased mucosa.

To prove the pathogenicity of the organism, Marshall looked back in time for guidance. Like many medical pioneers, he decided to become his own guinea pig. In order to fulfill Koch's postulates, Marshall ingested live *H. pylori*. The inoculation produced abdominal discomfort and vomiting, and his previously normal gastric mucosa developed characteristic gastritis (Marshall et al 1985). Marshall was able to clear his infection, but the second researcher to try the same experiment was not as lucky. While the pathogenic role of *H. pylori* was once again demonstrated, it took the second subject more than five years and multiple courses of antibiotics to get rid of the infection (Morris and Nicholson 1987; Morris et al 1991). After this experiment, no one volunteered to be the third subject.

The association between *H. pylori* and duodenal ulcer is very strong, but early researchers found it hard to conclusively prove the causality of the infection. Instead, the strongest evidence for the pathogenic role of *H. pylori* in peptic ulcer disease came from observations once the organism was eradicated. In a group of patients with verified duodenal ulcer treated with cimetidine, 92% eventually had a recurrence of their ulcer within a year. If patients were given cimetidine and the antibiotic tinidazole, 82% relapsed. Of patients given bismuth subcitrate, 53% relapsed, but only 25% of patients given both bismuth subcitrate and tinidazole had any recurrence of their ulcer (Marshall et al 1988). This proved that antibiotic therapy aimed at eradicating *H. pylori* dramatically changed the natural history of peptic ulcer disease.

Similar findings were reported by other investigators (Patchett et al 1992; Graham et al 1992; Coghlan et al 1987; George et al 1990; Morris et al 1991; Sung et al 1994). In some studies, the one-year relapse rate fell dramatically from 68% to 1% (Labenz and Borsch 1994) and from 85% to 2% (Hentschel et al 1993). Such strong evidence of the role of *H. pylori* in the development of peptic ulcer disease prompted the National Institutes of Health in 1994 to recommend that all *H. pylori*-infected patients with gastric or duodenal ulcers be treated with an appropriate antibiotic regimen (Anonymous 1994) (Table 1).

The precise mechanisms by which *H. pylori* exerts its effect on the gastric and duodenal mucosa are less well understood despite a decade of intensive research. It is believed that the bacteria can compromise the integrity of the gastric mucosa through the action of toxins and mucolytic enzymes. Possible inducing agents include a vacuolating cytotoxin (Phadnis et al 1994; Leunk et al 1988; Cover and Blaser 1992), a heat-labile toxin (Hupertz and

TABLE 1. NIH GUIDELINES for antimicrobial treatment of *H. pylori* infection

Patient status	*H.pylori* −	*H. pylori* +
Asymptomatic (no ulcer)	No	No
Non-ulcer dyspepsia	No	No
Gastric ulcer	No	Yes
Duodenal ulcer	No	Yes

TABLE 2. MUCOSAL DAMAGE by *H. pylori*

Possible mechanisms
• Vacuolating cytotoxin
• Heat-labile toxin
• Phospholipases
• Ammonia
• Acetaldehyde

At least five bacterial products have the potential for causing mucosal damage and could play a role in the development of ulcers.

Czinn 1988), phospholipases (Langton and Cesareo 1992; Weitkamp et al 1993), local ammonia produced by the bacteria's strong urease (Murakami et al 1990; Desai and Vadgama 1993), and the production of acetaldehyde (Roine et al 1992) (Table 2). The effects on the duodenal mucosa are thought to be linked to development of gastric metaplasia. Local injury of the duodenal mucosa, often caused by excess acid, results in areas of gastric-like cells in the duodenum. When *H. pylori* finds its way from the antral region of the stomach to these areas in the duodenum, it gives rise to duodenitis and ultimately duodenal ulcer (Wyatt et al 1987; Goodwin 1988).

The third major pathogenic role for *H. pylori* is in the development of gastric cancer. For many years, it was known that chronic gastritis was associated with the development of gastric cancer. After the discovery of *H. pylori,* several studies investigated its role in causing cancer. Parsonnet et al (1991) at the Centers for Disease Control reported a study of 128,992 persons who had been followed over a 20-year period. Of 109 patients with adenocarcinoma of the stomach, 84% of them had been chronically infected with *H. pylori.* The overall risk of developing gastric cancer was four-fold higher in the presence of *H. pylori* infection, and in some groups, such as women, it was 18 times the expected rate. At the same time, another similar study showed almost identical results among Japanese American men living in Hawaii (Nomura et al 1991). Ninety-four percent of patients with gastric carcinoma had been chronically infected with *H. pylori.* The odds ratio for developing gastric cancer if the individual was infected with *H. pylori* reached as high as 12.0 in some groups, a link of similar strength as that seen between smoking and lung cancer. Later studies have also found a link between *H. pylori* and gastric lymphomas (Parsonnet et al 1994). One type, MALT lymphoma, has in some cases been successfully eradicated through antibiotic treatment of the infection (Wotherspoon et al 1993).

The carcinogenic mechanisms of *H. pylori* infection may be many-fold. The continued inflammatory response from polymorphonuclear leukocytes

TABLE 3. POSSIBLE CARCINOGENIC MECHANISMS

- Oxidative bursts from polymorphonuclear leukocytes
- Reduced gastric levels of ascorbic acid
- Local production of ammonia

Three different mechanisms could play a role in the development of gastric cancer following decades of chronic gastritis.

gives rise to oxidative bursts that are known to cause DNA damage (Correa 1991). *H. pylori* also reduces gastric levels of ascorbic acid (Sobala et al 1991), an antioxidant which has been shown to have protective effects against gastric cancer (Correa 1991). Through the powerful urease enzyme, the bacteria produce large amounts of ammonia which is capable of inducing mutagenesis (Tsujii et al 1992) (Table 3).

EPIDEMIOLOGY OF *H. PYLORI* INFECTION

H. pylori infection has been reported from more than 35 countries, spanning all six major continents. It is estimated that one-third of the world's population is infected (Taylor and Blaser 1991). The only population known to have almost no prevalence of antibodies to *H. pylori* are Australian aborigines (less than 1%) (Dwyer et al 1988). This is also the only population almost free of duodenal ulcer. In all other populations, acquisition of *H. pylori* infection seems to start as early as childhood. Megraud et al (1989) found that 45% of Algerians and 55% of people living in the Ivory Coast developed antibodies in their first decade of life. In Western countries such as France (Megraud et al 1989), England (Jones et al 1986), and the United States (Perez-Perez et al 1988), antibodies to *H. pylori* are rare below the age of 20, but increase with age. In the United States, around 50% of the population aged 60 years or older have a positive antibody test (Perez-Perez et al 1988). This suggests an annual incidence of infection of 0.5 to 2%, but it has been suggested that acquisition in childhood is the most common mode of spread in these countries also. The observed rise in antibodies with age would then be a cohort effect that reflects the higher prevalence of *H. pylori* infection when these older patients were children (Banatvala et al 1993b).

How *H. pylori* infection is transmitted remains unknown. Humans are the only recognized host of *H. pylori*. Several studies give indirect evidence for person-to-person spread. In an analysis of a group of gastroscopists, Mitchell et al (1989) found that 52% of them were infected with *H. pylori*, compared with 21% in an age-matched group of blood donors (P<0.01). They concluded that the endoscopists' close contact with their patients put them at increased risk for person-to-person transmission of *H. pylori*. Two studies have shown a significantly increased infection rate among institutionalized mentally retarded individuals, a group known to be at increased risk for person-to-person spread of other pathogens. In this population, a longer duration of institutionalization was associated with increased prevalence of antibodies to *H. pylori* (Berkowicz and Lee 1987; Lambert et al 1990), consistent with data on other infections spread by person-to-person contact in the same population, such as hepatitis A and hepatitis B.

Further indirect evidence for person-to-person spread of *H. pylori* comes from studies of families of *H. pylori*-infected subjects. Mitchell et al (1987) looked at 14 family contacts of four index cases, aged 7 to 16 years. The carriage rate of IgG antibodies to *H. pylori* was 64% among the contacts compared to 13% in 166 age-matched controls (P<0.001). Drumm et al (1990) found *H. pylori* antibodies in 74% of parents of infected children, compared to only 24% of parents of noninfected children (P<0.001). Of 22 siblings of infected children, 82% had antibodies present, compared to only 14% of controls (P<0.001).

Though these studies suggest person-to-person spread of *H. pylori,* they do not provide information about the route of infection. There is strong epidemiologic evidence against sexual transmission of *H. pylori* (Blaser 1990). Data favoring fecal-oral transmission include age-adjusted prevalence curves for *H. pylori* in Third World countries. They closely parallel infections spread by the fecal-oral route such as hepatitis A (Graham et al 1989). However, in industrialized countries, there does not seem to be the same link (Hazell et al 1994). Still, *H. pylori* has been identified in stool samples (van Zwet et al 1994; Mapstone et al 1993a) and sewage (Westblom et al 1993a) using the polymerase chain reaction (PCR). Despite problems with overgrowth of fecal flora, *H. pylori* has been cultured from fecal specimens in at least two studies (Thomas et al 1992; Kelly et al 1994).

Oral-oral transmission must also be considered. Spontaneous regurgitation of gastric contents may set up colonization of the oral cavity with subsequent transmission through kissing or droplet spread. Though there are no conclusive data to support this theory, there are reports of successful culturing of *H. pylori* from dental plaques (Krajden et al 1989; Banatvala et al 1993a). *H. pylori* has also been identified by PCR in oral samples from about 50% of infected patients (Song 1993; Yang 1993; Westblom et al 1991b; Hammar et al 1992).

DIAGNOSIS OF *H. PYLORI* INFECTION

There are many ways that *H. pylori* infection can be diagnosed. Even though culture is not the most sensitive way of diagnosing *H. pylori,* it remains the most specific and is essential for selecting therapy based on antimicrobial susceptibilities. With appropriate techniques, *H. pylori* can readily be cultured from endoscopic biopsies of the stomach and duodenum (Westblom

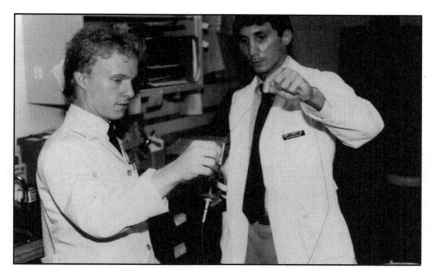

FIGURE 1. COLLECTION OF GASTRIC BIOPSY for diagnosis of *H. pylori* infection. The biopsy is placed in a transport medium before delivery to the microbiology laboratory.

1991) (Figures 1 and 2). Cultivation of *H. pylori* is best done at 37°C and 100% humidity. A number of different media have been used, the most successful of which contain blood (Figure 3) or egg yolk as a basic ingredient (Westblom et al 1991a). *H. pylori* is a slow-growing organism on all media and cultures take from two to five days to become positive (Table 4).

FIGURE 2. THE BIOPSY IS PLACED in a tissue grinder prior to inoculating special selective media.

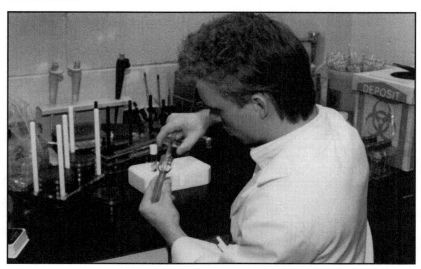

345

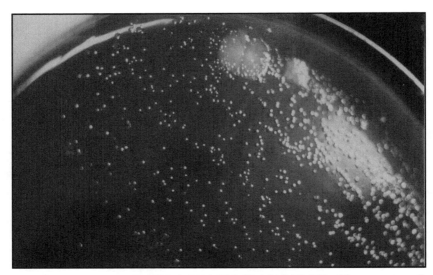

FIGURE 3. *H. PYLORI* **GROWING ON A BLOOD AGAR PLATE.** (Photo courtesy of Barry Marshall, M.D.)

Colonies are about 1 mm in size, pale to transparent in color, and can be either smooth or rough. Identification is made by typical morphology on Gram stain as well as positive reactions for urease, catalase, and oxidase. The specificity of culture is close to 100%, since false-positive results occur only if insufficient cleaning of endoscopy equipment leads to contamination of samples (Fantry et al 1995; Karim et al 1989; Katoh et al 1993). The sensitivity is lower because of the fastidious nature of the organism. Even specialized laboratories rarely obtain higher sensitivities than 80-85%.

As discussed, there are many histologic stains available for diagnosis of *H. pylori* infection, and all require biopsy material obtained by endoscopy. In

TABLE 4. MICROBIOLOGICAL CHARACTERISTICS of *H. pylori*

Basic microbiology

- Gram-negative spiral rod with 4–6 flagella attached to one pole.
- Can assume coccoid morphology if exposed to suboptimal growth conditions.
- Grows best in microaerobic humid atmosphere in the presence of CO_2.
- Requires blood- or egg yolk-containing media for good growth.

their original report, Warren and Marshall (1983) recommended the Warthin-Starry stain, which is very good for visualizing the organism, but is time-consuming and costly and requires an experienced technician. The modified Giemsa stain is a more consistent and less costly alternative. It has been accepted by most institutions as the best stain to use if diagnosis is the prime objective. However, it lacks a counterstain and therefore is less useful for evaluating the status of the gastric mucosa. If diagnosis of pathologic conditions such as metaplastic changes are desired, the newly described Genta stain (Genta et al 1994) is recommended.

The first diagnostic test developed for *H. pylori* was the urease test. Rapid urease tests are based on the ability of bacterial urease to convert urea to ammonia. Increasing levels of ammonia elevate the pH, which can be detected by an indicator such as phenol red. The simplest version of this is Christensen's urea broth, which can be easily manufactured in any microbiology laboratory. A gastric biopsy is dropped into the broth immediately after collection and the vials are visually observed for signs of a color change (Westblom et al 1988). The test is highly specific, but slow compared to later modifications. Its sensitivity is only around 70%. The CLO-test is a commercially available modification of the urea broth developed by Marshall et al (1987). It uses the same phenol red indicator as Christensen's urea broth, but is instead incorporated into a gel. Seventy-five percent of infected biopsies are positive within 20 minutes and more than 90% within 3 hours. Like most other urease tests, it is 100% specific if read within an hour.

Urea breath testing is a variation on the rapid urease tests, but it requires no endoscopy (Logan et al 1991; Marshall et al 1991). When the bacterial urease converts urea to ammonia, CO_2 is formed as a by-product. This CO_2 can be measured as it leaves the body through the exhaled air. In the urea breath test, the patient drinks a small amount of radioactively labeled urea. If *H. pylori* is present in the stomach, it immediately starts to break down the urea and make ammonia. The radioactively labeled CO_2 is collected in a breath sample and subsequently measured in a scintillation counter. The discriminatory power of the urea breath test is very high, with a specificity close to 100% and a sensitivity in the 90-95% range. However, false-negative results may be seen if the patient has consumed any antibiotics, bismuth, or proton pump inhibitors in the preceding two weeks.

H. pylori is a chronic infection if left untreated and most patients develop a strong IgG antibody response. Several commercial tests have been developed to measure *H. pylori* antibodies. They fall into two groups: conventional ELISAs and rapid antibody kits. Available ELISA tests include the Bio-Rad GAP test, Helico-G (Porton Cambridge), Malakit (Biolab), Pylori stat (Whittaker), and the MTP-assay by Roche. Most of these tests have sensitivity and specificity values in the 90% range (Jensen et al 1993; Talley et al

1991). They can therefore be used with a fair degree of accuracy to diagnose *H. pylori* infection prior to treatment. However, because of the long period of time needed for titers to decrease significantly after treatment, these tests have limited usefulness in assessing the results of treatment. Two rapid antibody kits are available: Pyloriset (Westblom et al 1992a) and QuickVue (Westblom et al 1993b). These tests offer the advantage of quick results that can be obtained in the office. However, there is a trade-off in accuracy; both of these tests have positive and negative predictive values that are at best around 80%.

PCR is a highly sensitive technique that can detect very small amounts of specifically targeted DNA. Theoretically, the method can find and identify an organism even if only a few single copies of its DNA are present. The DNA molecule is very stable chemically and can survive in the environment for long periods of time. Thus, PCR has the potential for diagnosing infection even when the target organism is in a nonculturable state or no longer alive. The extreme sensitivity of the assay makes it an excellent diagnostic tool, but it also has disadvantages. A single bacterial cell may be all that is needed for contamination to occur, leading to a false-positive reaction.

PCR remains a research tool, but commercial applications may become available in the near future. When used on gastric biopsies, it has no advantage over conventional diagnostic tests. However, it has been successfully used on gastric juice aspirates which can be obtained by nasogastric tube without the need for endoscopy (Westblom et al 1993c). It has also been used to detect the presence of *H. pylori* in saliva (Mapstone et al 1993b) and stool (van Zwet et al 1994) samples, where regular cultures consistently have had poor results.

TREATMENT OF *H. PYLORI* INFECTION

Therapy for most infectious diseases includes the use of an antimicrobial agent aimed at eradicating the organism. However, several clinical studies of *H. pylori* infection have indicated poor outcome with antibiotics despite good *in vitro* susceptibilities (Glupczynski et al 1987; McNulty et al 1986; Unge and Gnarpe 1988; Westblom et al 1992b). It has become clear that known pharmacokinetics and *in vitro* data on antibiotics cannot be extrapolated and applied directly to the treatment of *H. pylori* infection. The stomach represents a unique environment in the host. The presence of hydrochloric acid in the stomach necessitates the use of antibiotics that can retain activity at a pH much lower than what is seen in other parts of the body. Such antibiotics would also have to be secreted back into the stomach in order to achieve sustained tissue levels.

It is therefore not surprising that single antibiotics, monotherapy, has proven to be ineffective for *H. pylori* eradication. The best eradication rate has been seen with clarithromycin, but still was no higher than 54% (Peterson

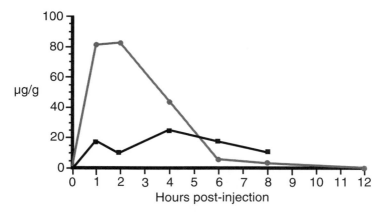

FIGURE 4. H$_2$ BLOCKERS AND ANTIBIOTIC CONCENTRATIONS. Effects of acid-reducing agents on antimicrobial tissue concentrations in gastric mucosa. A weak base such as clindamycin can achieve tissue concentrations ten times higher if an H$_2$-antagonist is given at the same time.

et al 1993). Combination therapy, using two or more drugs, is therefore recommended (Chiba et al 1992). The most extensive experience is with a combination of bismuth, metronidazole and either tetracycline or amoxicillin. This regimen has an eradication rate of 90-95%. Acid-reducing agents, such as H$_2$-blockers, can markedly increase antibiotic concentrations in gastric mucosa (Westblom and Duriex 1991) (Figure 4), and newer regimens therefore include such drugs. A common combination is amoxicillin and omeprazole which can achieve eradication rates as high as 82% (Labenz et al 1993). Recently the combination of omeprazole and clarithromycin has been found to be equally effective (Logan et al 1994). Several ongoing studies are looking at various combinations of these antibiotics both in conventional and in shorter one-week regimens.

THE FUTURE

Ten years of research have revealed that *H. pylori* is just one member of a whole family of bacteria that infect the gastrointestinal tract of humans and animals (Table 5). Development of animals based on these other species will greatly enhance our understanding of how *Helicobacters* cause disease. Although current antibiotic regimens for *H. pylori* infection have eradication results around 80 to 90%, they are far from perfect. Many patients fail therapy, and multiple drug combinations increase the risk of adverse drug reactions. To this can be added the newly recognized problem of drug resistance (Westblom and Unge 1992). Future research is likely to look for new drugs or

TABLE 5. *HELICOBACTER* **SPECIES**

Species	Hosts	Primary site
H. pylori	humans	stomach
H. mustelae	ferret	stomach
H. felis	cat, dog	stomach
H. canis	dog	intestine
H. bizzozeronii	dogs	stomach
H. nemestrinae	pig-tailed macaque	stomach
H. pullorum	chickens, humans	intestine
H. pametensis	birds, swine	intestine
H. acinonyx	cheetah	stomach
H. muridarum	mice, rats	intestine
H. hepaticus	mice	liver, intestine
H. cinaedi	humans, rodents	intestine
H. fennelliae	humans	intestine
Gastrospirillum hominis	humans	stomach

H. pylori is only one in a group of similar organisms that inhabit the gastrointestinal tract. At least one more, *Gastrospirillum hominis (H. heilmannii)*, infects human stomachs and causes gastritis similar to *H. pylori*.

drug combinations and ways to manipulate their local pharmacokinetics in the gastric mucosa. Still, too many people are infected with *H. pylori* to realistically expect that antibiotics will prevent the spread of infection. Vaccines aimed at preventing infection from an early age will therefore become important. This is particularly true in developing countries where *H. pylori* infection is common already in childhood and places people at increased risk for developing gastric carcinoma later in life.

REFERENCES

Anonymous (1994). Helicobacter pylori in peptic ulcer disease. [Review]. NIH Consensus Statement 12: 1-23.

Banatvala N, Lopez CR, Owen R, Abdi Y, Davies G, Hardie J, Feldman R (1993a). Helicobacter pylori in dental plaque [letter]. Lancet 341: 380.

Banatvala N, Mayo K, Megraud F, Jennings R, Deeks JJ, Feldman RA (1993b). The cohort effect and Helicobacter pylori. J Infect Dis 168: 219-221.

Berkowicz J, Lee A (1987). Person-to-person transmission of Campylobacter pylori [letter]. Lancet 2:680-681.

Blaser MJ (1990). Epidemiology and pathophysiology of Campylobacter pylori infections. Rev Infect Dis 12 (suppl 1): S99-106.

Chiba N, Rao BV, Rademaker JW, Hunt RH (1992). Meta-analysis of the efficacy of antibiotic therapy in eradicating Helicobacter pylori. Am J Gastroenterol 87: 1716-1727.

Coghlan JG, Gilligan D, Humphries H, McKenna D, Dooley C, Sweeney E, Keane C, O'Morain C (1987). Campylobacter pylori and recurrence of duodenal ulcers—a 12-month follow-up study. Lancet 2: 1109-1111.

Correa P (1991). Is gastric carcinoma an infectious disease? [editorial]. N Engl J Med 325: 1170-1171.

Cover TL, Blaser MJ (1992). Purification and characterization of the vacuolating toxin from Helicobacter pylori. J Biol Chem 267: 10570-10575.

Desai MA, Vadgama PM (1993). Enhanced H^+ diffusion by NH_4^+/HCO_3^-: implications for Helicobacter-pylori-associated peptic ulceration. Digestion 54: 32-39.

Doenges JL (1939). Spirochetes in the gastric glands of Macacus rhesus and of man without related disease. Arch Pathol 27: 469-477.

Drumm B, Perez-Perez GI, Blaser MJ, Sherman PM (1990). Intrafamilial clustering of Helicobacter pylori infection. N Engl J Med 322: 359-363.

Dwyer B, Sun NX, Kaldor J, Tee W, Lambert J, Luppino M, Flannery G (1988). Antibody response to Campylobacter pylori in an ethnic group lacking peptic ulceration. Scand J Infect Dis 20: 63-68.

Fantry GT, Zheng QX, James SP (1995). Conventional cleaning and disinfection techniques eliminate the risk of endoscopic transmission of Helicobacter pylori. Am J Gastroenterol 90: 227-232.

Freedberg AS, Barron LE (1940). The presence of spirochetes in human gastric mucosa. Am J Dig Dis 7: 443-445.

Genta RM, Robason GO, Graham DY (1994). Simultaneous visualization of Helicobacter pylori and gastric morphology: a new stain. Hum Pathol 25: 221-226.

George LL, Borody TJ, Andrews P, Devine M, Moore-Jones D, Walton M, Brandl S (1990). Cure of duodenal ulcer after eradication of Helicobacter pylori. Med J Aust 153: 145-149.

Glupczynski Y, Labbe M, Burette A, Delmee M, Avesani V, Bruck C (1987). Treatment failure of ofloxacin in Campylobacter pylori infection [letter]. Lancet 1: 1096.

Goodwin CS (1988). Duodenal ulcer, Campylobacter pylori, and the "leaking roof" concept. Lancet 2: 1467-1469.

Graham DY, Adam E, Klein PD, Evans DJ, Jr., Evans DG, Hazell SL, Alpert LC, Michaletz PA, Yoshimura HH (1989). Epidemiology of Campylobacter pylori infection. Gastroenterol Clin Biol 13: 84B-88B.

Graham DY, Lew GM, Klein PD, Evans DG, Evans DJ, Jr., Saeed ZA, Malaty HM (1992). Effect of treatment of Helicobacter pylori infection on the long-term recurrence of gastric or duodenal ulcer. A randomized, controlled study. Ann Intern Med 116: 705-708.

Hammar M, Tyszkiewicz T, Wadström T, O'Toole PW (1992). Rapid detection of Helicobacter pylori in gastric biopsy material by polymerase chain reaction. J Clin Microbiol 30: 54-58.

Hazell SL, Mitchell HM, Hedges M, Shi X, Hu PJ, Li YY, Lee A, Reiss-Levy E (1994). Hepatitis A and evidence against the community dissemination of Helicobacter pylori via feces. J Infect Dis 170: 686-689.

Hentschel E, Brandstätter G, Dragosics B, Hirschl AM, Nemec H, Schütze K, Taufer M, Wurzer H (1993). Effect of ranitidine and amoxicillin plus metronidazole on the eradication of Helicobacter pylori and the recurrence of duodenal ulcer. N Engl J Med 328: 308-312.

Hupertz V, Czinn S (1988). Demonstration of a cytotoxin from Campylobacter pylori. Eur J Clin Microbiol Infect Dis 7: 576-578.

Jensen AK, Andersen LP, Wachmann CH (1993). Evaluation of eight commercial kits for Helicobacter pylori IgG antibody detection. APMIS 101: 795-801.

Jones DM, Eldridge J, Fox AJ, Sethi P, Whorwell PJ (1986). Antibody to the gastric campylobacter-like organism ("Campylobacter pyloridis")—clinical correlations and distribution in the normal population. J Med Microbiol 22: 57-62.

Karim QN, Rao GG, Taylor M, Baron JH (1989). Routine cleaning and the elimination of Campylobacter pylori from endoscopic biopsy forceps. J Hosp Infect 13: 87-90.

Katoh M, Saito D, Noda T, Yoshida S, Oguro Y, Yazaki Y, Sugimura T, Terada M (1993). Helicobacter pylori may be transmitted through gastrofiberscope even after manual Hyamine washing. Jpn J Cancer Res 84: 117-119.

Kelly SM, Pitcher MC, Farmery SM, Gibson GR (1994). Isolation of Helicobacter pylori from feces of patients with dyspepsia in the United Kingdom. Gastroenterology 107: 1671-1674.

Krajden S, Fuksa M, Anderson J, Kempston J, Boccia A, Petrea C, Babida C, Karmali M, Penner JL (1989). Examination of human stomach biopsies, saliva, and dental plaque for Campylobacter pylori. J Clin Microbiol 27: 1397-1398.

Krienitz W (1906). Ueber das Auftreten von Spirochäten verschiedener Form im Mageninhalt bei Carcinoma ventriculi. Dtsch Med Wochenschr 28: 872.

Labenz J, Gyenes E, Rühl GH, Börsch G (1993). Omeprazole plus amoxicillin: efficacy of various treatment regimens to eradicate Helicobacter pylori. Am J Gastroenterol 88: 491-495.

Labenz J, Borsch G (1994). Highly significant change of the clinical course of relapsing and complicated peptic ulcer disease after cure of Helicobacter pylori infection. Am J Gastroenterol 89: 1785-1788.

Lambert JR, Lin SK, Kaldor J, Coulepis AG, Gust I (1990). High prevalence of H. pylori antibodies in institutionalized adults. Gastroenterology 98: A74 [Abstract].

Langton SR, Cesareo SD (1992). Helicobacter pylori associated phospholipase A2 activity: a factor in peptic ulcer production? J Clin Pathol 45: 221-224.

Leunk RD, Johnson PT, David BC, Kraft WG, Morgan DR (1988). Cytotoxic activity in broth-culture filtrates of Campylobacter pylori. J Med Microbiol 26: 93-99.

Logan RP, Polson RJ, Misiewicz JJ, Rao G, Karim NQ, Newell D, Johnson P, Wadsworth J, Walker MM, Baron JH (1991). Simplified single sample [13]Carbon urea breath test for Helicobacter pylori: comparison with histology, culture, and ELISA serology. Gut 32: 1461-1464.

Logan RP, Gummett PA, Schaufelberger HD, Greaves RR, Mendelson GM, Walker MM, Thomas PH, Baron JH, Misiewicz JJ (1994). Eradication of Helicobacter pylori with clarithromycin and omeprazole. Gut 35: 323-326.

Mapstone NP, Lynch DA, Lewis FA, Axon AT, Tompkins DS, Dixon MF, Quirke P (1993a). PCR identification of Helicobacter pylori in faeces from gastritis patients [letter]. Lancet 341: 447.

Mapstone NP, Lynch DA, Lewis FA, Axon AT, Tompkins DS, Dixon MF, Quirke P (1993b). Identification of Helicobacter pylori DNA in the mouths and stomachs of patients with gastritis using PCR. J Clin Pathol 46: 540-543.

Marshall BJ, Armstrong JA, McGechie DB, Glancy RJ (1985). Attempt to fulfill Koch's postulates for pyloric Campylobacter. Med J Aust 142: 436-439.

Marshall BJ, Warren JR, Francis GJ, Langton SR, Goodwin CS, Blincow ED (1987). Rapid urease test in the management of Campylobacter pyloridis-associated gastritis. Am J Gastroenterol 82: 200-210.

Marshall BJ, Goodwin CS, Warren JR, Murray R, Blincow ED, Blackbourn SJ, Phillips M, Waters TE, Sanderson CR (1988). Prospective double-blind trial of duodenal ulcer relapse after eradication of Campylobacter pylori. Lancet 2: 1437-1442.

Marshall BJ, Plankey MW, Hoffman SR, Boyd CL, Dye KR, Frierson HF, Jr., Guerrant RL, McCallum RW (1991). A 20-minute breath test for Helicobacter pylori. Am J Gastroenterol 86: 438-445.

Marshall BJ, Warren JR (1984). Unidentified curved bacilli in the stomach of patients with gastritis and peptic ulceration. Lancet 1: 1311-1315.

McNulty CA, Gearty JC, Crump B, Davis M, Donovan IA, Melikian V, Lister DM, Wise R (1986). Campylobacter pyloridis and associated gastritis: investigator blind, placebo controlled trial of bismuth salicylate and erythromycin ethylsuccinate. Br Med J 293: 645-649.

Megraud F, Brassens-Rabbe MP, Denis F, Belbouri A, Hoa DQ (1989). Sero-epidemiology of Campylobacter pylori infection in various populations. J Clin Microbiol 27: 1870-1873.

Mitchell HM, Bohane TD, Berkowicz J, Hazell SL, Lee A (1987). Antibody to Campylobacter pylori in families of index children with gastrointestinal illness due to C. pylori [letter]. Lancet 2: 681-682.

Mitchell HM, Lee A, Carrick J (1989). Increased incidence of Campylobacter pylori infection in gastroenterologists: further evidence to support person-to-person transmission of C. pylori. Scand J Gastroenterol 24: 396-400.

Morris A, Lane M, Hamilton I, Samarasinghe D, Ali MR, Brown P, Nicholson G (1991). Duodenal ulcer relapse after eradication of Helicobacter pylori. N Z Med J 104: 329-331.

Morris A, Nicholson G (1987). Ingestion of Campylobacter pyloridis causes gastritis and raised fasting gastric pH. Am J Gastroenterol 82: 192-199.

Morris AJ, Ali MR, Nicholson GI, Perez-Perez GI, Blaser MJ (1991). Long-term follow-up of voluntary ingestion of Helicobacter pylori. Ann Intern Med 114: 662-663.

Murakami M, Yoo JK, Teramura S, Yamamoto K, Saita H, Matuo K, Asada T, Kita T (1990). Generation of ammonia and mucosal lesion formation following hydrolysis of urea by urease in the rat stomach. J Clin Gastroenterol 12 (suppl 1): S104-S109.

Nomura A, Stemmermann GN, Chyou PH, Kato I, Perez-Perez GI, Blaser MJ (1991). Helicobacter pylori infection and gastric carcinoma among Japanese Americans in Hawaii. N Engl J Med 325: 1132-1136.

Palmer ED (1954). Investigation of the gastric mucosa spirochetes of the human. Gastroenterology 27: 218-220.

Parsonnet J, Friedman GD, Vandersteen DP, Chang Y, Vogelman JH, Orentreich N, Sibley RK (1991). Helicobacter pylori infection and the risk of gastric carcinoma. N Engl J Med 325: 1127-1131.

Parsonnet J, Hansen S, Rodriguez L, Gelb AB, Warnke RA, Jellum E, Orentreich N, Vogelman JH, Friedman GD (1994). Helicobacter pylori infection and gastric lymphoma. N Engl J Med 330: 1267-1271.

Patchett S, Beattie S, Leen E, Keane C, O'Morain C (1992). Helicobacter pylori and duodenal ulcer recurrence. Am J Gastroenterol 87: 24-27.

Perez-Perez GI, Dworkin BM, Chodos JE, Blaser MJ (1988). Campylobacter pylori antibodies in humans. Ann Intern Med 109: 11-17.

Peterson WL, Graham DY, Marshall B, Blaser MJ, Genta RM, Klein PD, Stratton CW, Drnec J, Prokocimer P, Siepman N (1993). Clarithromycin as monotherapy for eradication of Helicobacter pylori: a randomized, double-blind trial. Am J Gastroenterol 88: 1860-1864.

Phadnis SH, Ilver D, Janzon L, Normark S, Westblom TU (1994). Pathological significance and molecular characterization of the vacuolating toxin gene of Helicobacter pylori. Infect Immun 62: 1557-1565.

Roine RP, Salmela KS, Hook-Nikanne J, Kosunen TU, Salaspuro M (1992). Alcohol dehydrogenase mediated acetaldehyde production by Helicobacter pylori—a possible mechanism behind gastric injury. Life Sci 51: 1333-1337.

Sobala GM, Crabtree JE, Dixon MF, Schorah CJ, Taylor JD, Rathbone BJ, Heatley RV, Axon AT (1991). Acute Helicobacter pylori infection: clinical features, local and systemic immune response, gastric mucosal histology, and gastric juice ascorbic acid concentrations. Gut 32: 1415-1418.

Song M (1993). [Detection of Helicobacter pylori in human saliva by using nested polymerase chain reaction]. [Chinese]. Chung Hua Liu Hsing Ping Hsueh Tsa Chih 14: 237-240.

Sung JJ, Chung SC, Ling TK, Yung MY, Cheng AF, Hosking SW, Li AK (1994). One-year follow-up of duodenal ulcers after 1-wk triple therapy for Helicobacter pylori. Am J Gastroenterol 89: 199-202.

Talley NJ, Newell DG, Ormand JE, Carpenter HA, Wilson WR, Zinsmeister AR, Perez-Perez GI, Blaser MJ (1991). Serodiagnosis of Helicobacter pylori: comparison of enzyme-linked immunosorbent assays. J Clin Microbiol 29: 1635-1639.

Taylor DN, Blaser MJ (1991). The epidemiology of Helicobacter pylori infection. Epidemiol Rev 13: 42-59.

Thomas JE, Gibson GR, Darboe MK, Dale A, Weaver LT (1992). Isolation of Helicobacter pylori from human faeces. Lancet 340: 1194-1195.

Tsujii M, Kawano S, Tsuji S, Nagano K, Ito T, Hayashi N, Fusamoto H, Kamada T, Tamura K (1992). Ammonia: a possible promotor in Helicobacter pylori-related gastric carcinogenesis. Cancer Lett 65: 15-18.

Unge P, Gnarpe H (1988). Pharmacokinetic, bacteriological and clinical aspects on the use of doxycycline in patients with active duodenal ulcer associated with Campylobacter pylori. Scand J Infect Dis (suppl) 53: 70-73.

van Zwet AA, Thijs JC, Kooistra-Smid AM, Schirm J, Snijder JA (1994). Use of PCR with feces for detection of Helicobacter pylori infections in patients. J Clin Microbiol 32: 1346-1348.

Warren JR, Marshall B (1983). Unidentified curved bacilli on gastric epithelium in active chronic gastritis. Lancet 1: 1273-1275.

Weitkamp JH, Perez-Perez GI, Bode G, Malfertheiner P, Blaser MJ (1993). Identification and characterization of Helicobacter pylori phospholipase C activity. Int J Med Microbiol Virol Parasitol Infect Dis 280: 11-27.

Westblom TU, Madan E, Kemp J, Subik MA (1988). Evaluation of a rapid urease test to detect Campylobacter pylori infection. J Clin Microbiol 26: 1393-1394.

Westblom TU (1991). Laboratory diagnosis and handling of Helicobacter pylori. In: Marshall BJ, McCallum RW, Guerrant RL, eds. Helicobacter pylori in peptic ulceration and gastritis. Boston: Blackwell Scientific, 81-91.

Westblom TU, Madan E, Midkiff BR (1991a). Egg yolk emulsion agar, a new medium for the cultivation of Helicobacter pylori. J Clin Microbiol 29: 819-821.

Westblom TU, Phadnis S, Normark S (1991b). Sensitivity of an assay for detection of Helicobacter pylori in human saliva using polymerase chain reaction (PCR). Microbiol Ecology in Health and Disease 4 S: S156[Abstract].

Westblom TU, Madan E, Gudipati S, Midkiff BR, Czinn SJ (1992a). Diagnosis of Helicobacter pylori infection in adult and pediatric patients by using Pyloriset, a rapid latex agglutination test. J Clin Microbiol 30: 96-98.

Westblom TU, Madan E, Subik MA, Duriex DE, Midkiff BR (1992b). Double-blind randomized trial of bismuth subsalicylate and clindamycin for treatment of Helicobacter pylori infection. Scand J Gastroenterol 27: 249-252.

Westblom TU, Fritz SB, Phadnis S, Midkiff BR, Leon-Barua R, Recavarren S, Ramirez-Ramos A, Gilman RH (1993a). PCR analysis of Peruvian sewage water: support for fecal-oral spread of Helicobacter pylori. Acta Gastroenterol Belg 56 (suppl): 84 [Abstract].

Westblom TU, Lagging LM, Midkiff BR, Czinn SJ (1993b). Evaluation of QuickVue, a rapid enzyme immunoassay test for the detection of serum antibodies to Helicobacter pylori. Diag Microbiol Infect Dis 16: 317-320.

Westblom TU, Phadnis S, Yang P, Czinn SJ (1993c). Diagnosis of Helicobacter pylori infection by means of a polymerase chain reaction assay for gastric juice aspirates. Clin Infect Dis 16: 367-371.

Westblom TU, Duriex DE (1991). Enhancement of antibiotic concentrations in gastric mucosa by H2-receptor antagonist. Implications for treatment of Helicobacter pylori infections. Dig Dis Sci 36: 25-28.

Westblom TU, Unge P (1992). Drug resistance of Helicobacter pylori: memorandum from a meeting at the Sixth International Workshop on Campylobacter, Helicobacter, and Related Organisms [letter]. J Infect Dis 165: 974-975.

Wotherspoon AC, Doglioni C, Diss TC, Pan L, Moschini A, de Boni M, Isaacson PG (1993). Regression of primary low-grade B-cell gastric lymphoma of mucosa-associated lymphoid tissue type after eradication of Helicobacter pylori. Lancet 342: 575-577.

Wyatt JI, Rathbone BJ, Dixon MF, Heatley RV (1987). Campylobacter pyloridis and acid induced gastric metaplasia in the pathogenesis of duodenitis. J Clin Pathol 40: 841-848.

Yang HT (1993). [Nested-polymerase chain reaction in detection of Helicobacter pylori in human dental plaque]. [Chinese]. Chung Hua I Hsueh Tsa Chih 73: 750-752, 774.

Vaccines Against Pneumococcal Infections

In this year of the celebration of the 100th anniversary of Louis Pasteur's death, it seems fitting to remember that he was the first to describe *Streptococcus pneumoniae*, also described independently by Sternberg in that same year, 1881. Those bacteria are responsible for various diseases—pneumonia, meningitis, otitis—and despite the use of antibiotics and an existing vaccine, the mortality and morbidity rates remain very high. About 30 of the 85 different serotypes of *Streptococcus pneumoniae* are pathogenic.

The first vaccine (a killed bacteria vaccine) was developed by Wright in 1911 during an outbreak of pneumonia in a miners' community in South Africa; while its efficacy was not well established, further trials by Lister suggested a prophylactic effect.

Later, in the 1930s, a new vaccine concept appeared, using only the outer constituent of the bacteria—the capsular polysaccharide. This vaccine was marketed, but the introduction of antibiotics stopped its use and halted further research on vaccines.

However, during the 1950s, mortality due to pneumonia remained very high because pneumococci had developed resistance to antibiotics. From 1977 to 1983, under the direction of R. Austrian at the Food and Drug Administration, vaccines containing first 14, then 23 different *Streptococcus pneumoniae* purified capsular polysaccharides were released and were shown to be efficient in adults and elderly people.

Unfortunately, that vaccine failed to protect infants who are the main target of *Streptococcus pneumoniae* through otitis (about 20,000 cases per 100,000 infants per year), meningitis, and bacteremia.

A new vaccine concept for infants was introduced by J.B. Robbins in 1980 called "polysaccharide conjugate vaccine," which consists of capsular polysaccharide covalently linked to a carrier protein. The first vaccine following this principle, the *Haemophilus* conjugate vaccine, offers an efficacy in infants of about 90%.

Today, several manufacturers are developing pneumococcal conjugate vaccines containing 7 to 12 different serotypes. The preliminary clinical results are very encouraging, and it is reasonable to expect an efficient vaccine for all age groups by the end of the century.

27

History and Current Status of Pneumococcal Vaccines

MONIQUE MOREAU

Despite the use of antibiotics and an existing vaccine, *Streptococcus pneumoniae,* a diplococcus gram-positive bacterium, continues to be an important cause of infections. Pneumococci are responsible for noninvasive diseases such as pharyngitis, conjunctivitis, otitis media, and for invasive diseases, such as pneumonia, meningitis, bacteremia, septicemia, with mortality due to pneumonia and bacteremia remaining particularly high.

Only encapsulated bacteria are pathogenic, and incidence of diseases differs according to age group (Fedson and Musher 1994; Mufson 1994; Austrian 1985; US Public Health Service 1994); the higher incidence for both noninvasive and invasive diseases is in infants under two years of age (i.e., for otitis 200 per 1000 infants per year), the attack rate is low between 2 years and 40 years of age, but it increases after age 40.

The capsular polysaccharide is the main factor in virulence because of its antiphagocytic properties (Fedson and Musher 1994; Mufson 1994; Austrian 1985) and 85 different serotypes have been identified on the basis of the immunological recognition of capsular polysaccharides. Among the most prevalent serotypes, distribution of pneumococcal serotypes responsible for invasive and noninvasive infections depends on geographical location and age group.

POLYSACCHARIDE VACCINE

The first pneumococcal vaccine designed by Wright in 1911 was a whole-cell killed bacteria vaccine; its efficacy in humans was not well established (Wright et al 1914). Lister continued this vaccinal approach and showed some efficacy in humans.

Typing of pneumococcal bacteria was established first in 1910 by Neufield and Haendel and confirmed by Lister's clinical trial (Lister 1916). Dochez, Avery, and Heidelberger elucidated the carbohydrate nature of the capsule and its antigenic properties. They established that specific serotyping of pneumococcus is related to the capsular antigens (Dochez and Avery 1917; Heidelberger and Avery 1923).

Francis and Tillett (1930) and then Finland demonstrated the immunogenicity of the purified polysaccharide in animals and humans (Finland and Ruegsegger 1935). Those findings led scientists to use pneumococcal capsular polysaccharides as a vaccine for a large vaccination program, which demonstrated the interest in the concept of a purified polysaccharide vaccine. A hexavalent vaccine was released on the market in 1946. The hope brought by antibiotics stopped its use and halted further research on vaccine for more than twenty years, but pneumococcal infections still persisted with a high attack rate due to the emergence of antibiotic-resistant strains.

Under the auspices of R. Austrian (Austrian and Gold 1964), a 14-valent vaccine (1, 2, 3, 4, 6A, 7F, 8, 9N, 12F, 14, 18C, 19F, 23F, 25) with a theoretical coverage of 80% was released in 1977. In 1983, the vaccine was improved by the introduction of a 23-valent vaccine (1, 2, 3, 4, 5, 6B, 7F, 8, 9V, 9N, 10A, 11A, 12F, 14, 15B, 17F, 18C, 19A, 19F, 20, 22F, 33F, 45) with a coverage of 90% of cases. The efficacy of the 23-valent vaccine is about 60%, but it is not uniform according to age group. This vaccine is not immunogenic in infants, but it is a good immunogen in children and adults.

Health authorities have recommended pneumococcal polysaccharide vaccination for the elderly, for high-risk patients with chronic diseases (diabetes, pulmonary diseases, otitis media, and sinusitis) and for immunocompromised patients (Hodgkin's lymphoma, chronic renal failure, organ transplants, and HIV infection). For all of these groups of patients, vaccination has been demonstrated to increase antibodies to a level that may prevent pneumococcal infection (US Public Health Service 1994).

Need for a New Vaccine
and Potential Pneumococcal Vaccine Candidates

Because of their immature immune system, infants do not develop antibodies against T-cell-dependent antigens, such as polysaccharides; T-cell-dependent antigens, either proteins or polysaccharides conjugated to proteins, are required for a pediatric pneumococcal vaccine.

PROTEINS

Pneumolysin is a highly conserved cytoplasmic, cytotoxic thiol-activated toxin contributing to the virulence of the bacteria. Anti-pneumolysin antibodies provide some protection in animals. When genetically modified (Boulnois 1992; Paton 1993), it loses 99.5% of its hemolytic properties but retains its protective character.

Because of incomplete clearance of the bacteria by anti-pneumolysin antibodies, pneumolysin is not considered as a vaccine by itself, but it could improve protection in association with polysaccharide conjugates (Kuo et al 1995; Alexander et al 1994).

Pneumococcal surface protein A (PspA). This surface-exposed protein elicits protective antibodies; because of antigen diversity (McDaniel et al 1991), it is to be expected that more than 4 PspAs will be necessary to cover most of pneumococcal diseases.

The protective efficacy of both proteins has yet to be established in humans.

CONJUGATE VACCINE

In 1929, Avery showed that covalent coupling of polysaccharide to a protein increased the immunogenicity of the polysaccharide. It appeared later that covalent linkage to proteins converts polysaccharides to T-cell-dependent antigens (Avery and Goebel 1929). In 1983, Robbins introduced the idea of polysaccharide conjugate vaccines for infants (Robbins et al 1989; Schneerson et al 1980). If different mechanisms have been suggested for the processing of conjugate antigens (Stein 1994; Siber 1994), none of them has been experimentally confirmed, but the efficacy of the first conjugate vaccine against *Haemophilus influenzae* type B has been extensively demonstrated in infants.

Today, pneumococcal capsular polysaccharide conjugates are the most promising T-cell-dependent candidate vaccines for the prevention of pneumococcal infections in infants. If there is no consensus on the number of polysaccharide conjugates to be included in a pneumococcal conjugate vaccine, it should obviously contain the most frequent serotypes involved in invasive and noninvasive diseases: 6B 14, 19F, and 23F, responsible for 60% of cases; with additional serotypes such as 4, 9V, and 18C, expected coverage should be higher than 75%. For some countries, it would be necessary to add the serotype 1 (developing countries) and serotype 5 (Israel).

CLINICAL STUDIES

Clinical trials are now in progress to evaluate different pneumococcal conjugate vaccines from several manufacturers. All vaccines currently tested are immunogenic and safe in adults. In infants, they are immunogenic with a booster effect after the second or third injection, and no adverse reaction has been observed. Most of the studies have been done with mixtures of 1 to 5 µg of each polysaccharide conjugate per dose. This low dosage precludes issues of prohibitive cost or risk of excessive amounts of injected protein with its potential immunosuppressing effects.

Efficacy studies for invasive and noninvasive diseases are ongoing or will begin soon. It is expected that a new pneumococcal vaccine for infants will be released by the beginning of the next century.

REFERENCES

Alexander JE, Lock RA, Peeters CC, Poolman JT, Andrew PW, Mitchell TJ, Hansman D, Paton JC (1994). Immunization of mice with pneumolysin toxoid confers a significant degree of protection against at least nine serotypes of *Streptococcus pneumoniae*. Infect Immun 59: 222-228.

Austrian R (1964). Pneumococcal bacteremia with especial reference to bacteremic pneumococcal pneumonia. Ann Intern Med 60: 759-776.

Austrian R (1985). Polysaccharide vaccines. Annales Institut Pasteur Microbiology 136B: 295-307.

Avery OT, Goebel WF (1929). Chemo-immunological studies on conjugated carbohydrate-proteins. II. Immunological specificity of synthetic sugar-protein antigen. J Exp Med 50: 533-550.

Boulnois GJ (1992). Pneumococcal proteins and the pathogenesis of diseases caused by *Streptococcus pneumoniae*. J Gen Microbiol 138: 249-259.

Dochez AR, Avery OT (1917). The elaboration of specific soluble substances by pneumococcus during growth. J Exp Med 26: 477-493.

Fedson DS, Musher DM (1994). Pneumococcal vaccine. In: Plotkin SR, Mortimer ER, eds. Vaccines. Philadelphia: WB Saunders Company, 517-564.

Finland M, Ruegsegger JM (1935). Immunization of human subjects with the specific carbohydrates of Type III and the related Type VIII pneumococcus. J Clin Invest 14: 829-832.

Francis TJ Jr, Tillett WS (1930). Cutaneous reactions in pneumonia. The development of antibodies following the intradermal injection of type specific polysaccharide. J Exp Med 52: 573-585.

Heidelberger M, Avery OT (1923). The soluble specific substance of pneumococcus. J Exp Med 38: 73-79.

Kuo J, Douglas M, Ree HK, Lindberg AA (1995). Characterization of a recombinant pneumolysis and its use as a protein carrier for pneumococcal type 18C conjugate vaccines. Infect Immun 63: 2706-2713.

Lister FS (1916). An experimental study of prophylactic inoculation against pneumococcal infection in the rabbit and in man. Pub S Afr Inst Med Res 8: 231-287.

McDaniel S, Sheffield S, Delucchi P, Briles E (1991). PspA, a surface protein of streptococcus pneumoniae, is capable of eliciting protection against pneumococci of more than one capsular type. Infect Immun 59: 222-228.

Mufson MA (1994). Pneumococcal infections. Curr Opin Infect Dis 7: 178-183.

Paton JC (1993). Pathogenesis of pneumococcal diseases. Curr Opin Infect Dis 6: 363-368.

Robbins JB, Schneerson R, Szu SC, Fattom A, Yang Y, Lagergard T, Chu C, and Sørensen US (1989). Prevention of invasive bacterial diseases by immunization with polysaccharide-protein conjugates. Current Top Microbiol Immunol 146: 169-180

Schneerson R, Barrera O, Sutton A, Robbins JB (1980). Preparation, characterization and immunogenicity of *Haemophilus influenzae* type b polysaccharide-protein conjugates. J Exp Med 152: 361-376.

Siber GR (1994). Pneumococcal disease: prospects for a new generation of vaccines. Science 265: 1385-1387.

Stein KE (1994). Glycoconjugate vaccines. What next? Intl J Technology Assessment in Health Care 10-1: 167-176.

US Public Health Service; American Family Physician (1994).

Wright AE, Parry Morgan W, Colebrook L, Dodgson RW (1914). Observations on prophylactic inoculation against pneumococcus infections, and on the results which have been achieved by it. Lancet 1:1-10, 87-95.

PARASITES

Seven years before, Pasteur had foretold: "It is within the power of man to make parasitic maladies disappear from the face of the earth..." And when he said these words, the wisest doctors in the world put their fingers to their heads, thinking: "The poor fellow is cracked!"

PAUL DE KRUIF—*Microbe Hunters*

Malaria

An ancient disease, malaria is a serious, acute, and chronic relapsing infection, causing its victims to suffer with periodic paroxysms of chills and fever, anemia, enlargement of the spleen, and often fatal complications.

Malaria occurs throughout the world and is one of the most prevalent of all infectious diseases. Hyperendemic areas are found in Central and South America, in North and Central Africa, in all countries bordering on the Mediterranean, and in the Middle East and East Asia. In many parts of Africa and South Asia, entire populations are infected more or less constantly.

The search for the mode of transmission and the causative organism was described by de Kruif in *Microbe Hunters* in 1926. Now, 70 years later, we are left with disappointing results for the control and eradication of malaria. Despite grand efforts by the World Health Organization and other agencies, malaria still thrives and is a significant cause of disease and death. New approaches to the prevention of malaria by vaccination are presented by John Tine and Enzo Paoletti.

28

Toward the Development of a Malaria Vaccine

JOHN A. TINE AND ENZO PAOLETTI

Malaria parasites continue to be a major cause of disease and death in many areas of the world. More than two billion people reside in malarious areas, which include all or parts of Africa, the Middle East, Asia, Oceania and Latin America. There are as many as 500 million clinical cases of malaria per year which claim approximately 2.5 million lives, mostly of children under the age of five years. Malaria control measures have historically relied on the use of anti-malarial drugs to treat infections and insecticides to control the anopheline mosquito vectors. However, the appearance and rapid spread of drug-resistant parasites as well as the development of insecticide resistance among mosquitoes has seriously impaired the effectiveness of these tools. As a result, the ability to control malaria in historically endemic areas has been seriously compromised and a resurgence of this disease has occurred in areas that had achieved some level of control. The public health and economic impact of malaria, particularly on the underdeveloped countries of the tropics, is truly staggering. Against this backdrop, the need for an effective malaria vaccine cannot be overstated.

Two fundamental observations support the concept of vaccination as a strategy to control human malaria. First, inoculation with attenuated parasites can confer protection from challenge with infectious parasites in humans and nonhuman primates (Siddiqui 1977; reviewed by Nussenzweig and Nussenzweig 1989). Technological limitations, however, make it impractical to pursue vaccination with attenuated parasites on a large scale. Second, the passive transfer of protection in rodents with immune serum or a parasite antigen-specific monoclonal antibody (Potocnjak et al 1980; Egan et al 1987) demonstrates that specific immune responses elicited by immunization can be protective. These observations are the foundation for the extensive efforts to develop recombinant and synthetic vaccines against human malaria. In this review, we outline the evolution in malaria vaccine design that has occurred from the initial single antigen candidates tested in humans to the multi-antigen, multi-stage vaccine candidates currently undergoing clinical evaluation.

Vaccine Targets in the Context of the Parasite Life Cycle

Malaria in humans is caused by four species of protozoan parasites from the genus *Plasmodium*: *P. falciparum, P. vivax, P. malariae,* and *P. ovale,* with *P. falciparum* and *P. vivax* the most widespread. Most vaccine development efforts have focused on *P. falciparum* because this species is responsible for most malaria fatalities.

Any consideration of the development of malaria vaccines must begin with the complex parasite life cycle, the nature of which defines the targets for vaccine intervention (Figure 1). Infection begins when an infected female *Anopheles* mosquito takes a blood meal, during which sporozoites are inoculated into the bloodstream. The sporozoites rapidly invade hepatocytes in the liver. Over the next five to seven days, the infected liver cells develop asexually into liver schizonts, which rupture and release tens of thousands of merozoites into the blood. Merozoites invade red blood cells (erythrocytes) and there develop asexually to again form erythrocytic schizonts, which contain as many as 20 to 30 daughter merozoites. Rupture of these schizonts releases the merozoites into the blood where they infect other red blood cells. This repeated cycle of blood-stage infection causes the clinical symptoms of malaria. Some merozoites in red blood cells differentiate into male and female gametocytes, rather than asexual parasites, and can subsequently be ingested by a mosquito during a blood meal. After fertilization, further development in the mosquito host results in the formation of sporozoites, which make their way to the mosquito salivary glands. The transmission cycle is completed when the infected mosquito takes another blood meal and injects sporozoites into the bloodstream. This complex life cycle provides four general targets for vaccination: the sporozoite, the liver stage, the blood stage, and the sexual stages (Miller et al 1986).

Sporozoite Vaccine Development

The *P. falciparum* sporozoite was the target of the first human malaria vaccine candidates, with the rationale that preventing hepatocyte invasion by sporozoites would prevent blood-stage infection and thus the clinical symptoms of malaria. The ability to protectively immunize with irradiated sporozoites provided support for such an approach (Nussenzweig and Nussenzweig 1989). Further, the identification and cloning of a major surface antigen of the sporozoite, the circumsporozoite protein (CSP), allowed the evaluation of recombinant approaches to vaccine development based on CSP (Dame et al 1984). This advance fostered optimism that an effective malaria vaccine would soon be at hand.

HUMAN CLINICAL TRIALS OF CSP-BASED VACCINES

The goal of the first malaria sporozoite vaccines was to elicit high titers of antibody directed against the immunodominant repeat region of the

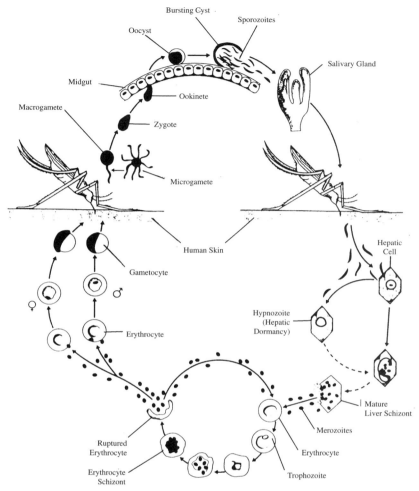

FIGURE 1. THE LIFE CYCLE of the malaria parasite (adapted from Hospital Practice, September 15, 1990, by permission of HP Publishing, New York).

CSP. Studies with rodent malaria model systems indicated that irradiated sporozoite-protected mice developed antibody responses to the CSP repeats (Zavala et al 1983) and that CSP repeat-specific monoclonal antibodies could passively transfer protection to otherwise susceptible animals (Potocnjak et al 1980). Phase I clinical trials in humans were initiated with three *P. falciparum* CSP-repeat based vaccine candidates: an *E. coli*-expressed fusion construct containing 32 four-amino acid CSP repeat units (R32) linked to 32 out-of-frame amino acids from the tetracycline resistance gene (Ballou et al 1987) and two synthetic peptide constructs each consisting of three CSP repeat units conjugated to tetanus toxoid (Herrington et al 1987; Etlinger et al 1988). The results of these initial trials were disappointing. The vaccines elicited relatively low antibody responses that were not boosted by subsequent immunization with vaccine or sporozoites. Protective efficacy was poor, since only two of nine volunteers were protected in the two studies in which efficacy of the vaccines was evaluated by administering a sporozoite challenge to selected individuals (Ballou et al 1987; Herrington et al 1987). Despite the disappointing efficacy data, these studies did demonstrate that it was possible to protect at least some individuals by immunization with CSP repeat constructs. Further, some volunteers exhibited a delay in time to patency after challenge, suggesting that the sporozoite inoculum was significantly reduced by vaccine-induced immune responses (Ballou et al 1987; Herrington et al 1987). Based on this, work has continued with CSP-based vaccine constructs in an attempt to improve their immunogenicity. The strategies pursued include 1) the addition of T-cell epitopes from alternative carrier molecules or from nonrepetitive CSP sequences in an effort to enhance antibody responses to the repeats, 2) the use of experimental adjuvant formulations as an alternative to alum, and 3) the use of live recombinant vectors.

Several studies have evaluated modified formulations of the R32 CSP repeat immunogen. A construct containing R32 fused to 81 amino acids of the influenza A virus NS1 protein was derived and formulated with either alum or Detox, an adjuvant containing monophosphoryl lipid A (MPLA), the cell wall cytoskeleton of *Mycobacterium phlei,* and squalene. Antibody responses to CSP in volunteers inoculated with the construct in Detox were much stronger than those elicited by the vaccine in alum (Rickman et al 1991). However, in a subsequent study, the protective efficacy of this formulation was determined to be poor, with only 2 of 11 volunteers protected from sporozoite challenge (Hoffman et al 1994). Other studies have examined the immunogenicity of the R32/NS1 construct after encapsulation with MPLA in liposomes and adjuvanted with alum (Fries et al 1992b). Antibody responses to CSP in vaccinated individuals were better than those in previous studies using alum-adjuvanted R32 constructs. However, the protective efficacy of this formulation was not significantly enhanced (Magill et al 1995). A construct in which R32 was conjugated to detoxified *Pseudomonas aeruginosa* toxin A (Fries et al 1992a) significantly enhanced antibody responses to CSP in immunized volunteers, but again protective efficacy was poor. Further evaluation of this construct in a field trial in Thailand confirmed the immunogenicity of the construct in both malaria-naive and previously exposed volunteers, but there was no evidence of vaccine efficacy (Brown et al 1994). A phase I trial of a baculovirus-expressed CSP construct containing CSP T-cell epitopes as well as the repeat epitope showed that this construct was poorly immunogenic, and it was not evaluated for protective efficacy (Herrington et al 1992).

Two studies have evaluated the use of yeast-expressed hepatitis B virus surface antigen (HBsAg) particles as a carrier for CSP sequences. A repeat-based construct consisting of 16 repeat units fused to HBsAg and expressed as a particle was evaluated in a phase I study (Vreden et al 1991). This construct elicited better and more sustained CSP repeat-specific antibody responses than the initial CSP-based vaccines, but boosting by subsequent inoculation was observed in only a few individuals. Protective efficacy was not evaluated. A second construct contained both repeat units and sequences derived from the C-terminal nonrepetitive region of CSP fused to HBsAg (Gordon et al 1995). Hybrid particles were evaluated in a phase I study with two adjuvant formulations, alum alone or alum with MPLA. The inclusion of MPLA enhanced antibody responses to CSP, particularly to the repeat region. However, only two of eight individuals inoculated with the alum/MPLA formulation were protected from challenge. A cytotoxic T lymphocyte (CTL) response against CSP was demonstrated in one of the protected individuals (Gordon et al 1995).

Several lessons were learned from these efforts at developing CSP-based vaccines. First, some individuals can be protected by immunization with CSP-based constructs designed to elicit humoral responses. Second, there is no correlation of any measurable immune parameter with protection, although protected individuals generally had higher antibody responses. Finally, the efficacy data suggest that it will be difficult to develop a highly effective vaccine based on a single parasite antigen and thus support the concept that an optimal pre-erythrocytic stage vaccine must include multiple antigens.

EXPERIMENTAL VACCINE STUDIES

Another approach to eliciting high-titer antibody responses against CSP involves the presentation of defined B- and T-cell epitopes as multiple antigen peptides (MAPs). MAP formulations containing such epitopes derived either from CSP or tetanus toxoid elicit strong antibody responses and high levels of protection in mice against challenge with *P. berghei* (Tam et al 1990) or *P. yoelii* (Wang et al 1995) sporozoites.

Although the initial subunit vaccine candidates were geared toward eliciting strong antibody responses, it is now clear from studies with rodent malarias that cellular responses can play an important role in protective immunity. Thus, the focus of pre-erythrocytic stage vaccine development has shifted to include the generation of cellular immune effector mechanisms as well as strong antibody responses.

Bacterial vectors, such as *Salmonella typhimurium,* are being evaluated for their potential as malaria vaccine candidates. A study in the *P. berghei* rodent malaria model has demonstrated the potential of oral immunization with live attenuated *Salmonella typhimurium* recombinants to elicit protective immunity against the pre-erythrocytic stage of the parasite. Mice immunized with a recombinant expressing the *P. berghei* CSP were partially protected from challenge with sporozoites in the absence of detectable CSP-specific antibody responses (Sadoff et al 1988). Subsequent studies confirmed this result and demonstrated that protection was conferred by CD8[+] T cells

(Aggarwal et al 1990). A phase I trial using live attenuated *Salmonella typhi* expressing CSP for oral immunization of humans (Gonzalez et al 1994) showed that *Salmonella*-expressed CSP was poorly immunogenic, probably due in part to low expression levels, although one vaccinee did develop a CSP-specific CTL response.

Several studies have evaluated the protective potential of vaccinia virus recombinants expressing CSP against rodent malaria. Immunization of mice with a Wyeth strain vaccinia recombinant expressing the *P. yoelii* CSP elicited specific antibody responses that were quantitatively comparable to those elicited by irradiated sporozoites, although the epitope specificity of the antibodies was somewhat different (Sedegah et al 1988). This recombinant did not elicit protection from challenge with low doses of *P. yoelii* sporozoites. A WR strain vaccinia recombinant expressing the *P. berghei* CSP also elicited good repeat-specific antibody responses in mice, but only weak CTL responses. Immunization with this recombinant did not protect mice from challenge with *P. berghei* sporozoites (Satchidanandam et al 1991).

In contrast to the results described above, a CSP recombinant based on the highly attenuated NYVAC vaccinia vaccine strain (Tartaglia et al 1992) confers protection against *P. berghei* sporozoite challenge in immunized mice regardless of whether the challenge is by intravenous inoculation or via the bite of infected mosquitoes (Lanar et al 1996). While there was no correlation between repeat-specific antibody responses and protection, depletion of CD8[+] T cells abrogated the protective effect of vaccination.

Another interesting vaccination approach involves immunization of mice with an influenza virus recombinant expressing a CTL epitope from the *P. yoelii* CSP followed by a booster immunization with a vaccinia-CSP recombinant. This regimen elicits partial protection against sporozoite challenge which is mediated by CD8[+] T cells (Li et al 1993). The use of either immunogen alone or priming with the vaccinia recombinant and boosting with the influenza recombinant was not protective (Li et al 1993).

Recently, the technology of direct injection of DNA has been applied to the development of pre-erythrocytic stage malaria vaccines. A plasmid DNA vaccine encoding CSP, when administered to mice by the intramuscular route without adjuvant, conferred good levels of protection against *P. yoelii* sporozoite challenge which was dependent on CD8[+] T cells (Sedegah et al 1994). In addition to the obvious advantages of this approach to vaccination, the use of DNA immunization may hold particular promise for malaria in the development of multi-antigen vaccines.

BLOOD-STAGE VACCINE DEVELOPMENT

The goal of vaccination against the blood stage of the malaria parasite is to prevent the invasion of red blood cells by merozoites and to eliminate the infected cells, thus preventing or alleviating the clinical symptoms of malaria.

Immunization with attenuated blood-stage parasites elicited protection from blood-stage challenge (Siddiqui 1977), demonstrating the feasibility of such an approach and providing a foundation for the identification of blood-stage parasite proteins able to elicit protective immune responses. Candidate blood-stage immunogens have generally been selected on the basis of reactivity with immune human serum, with subsequent evaluation of the protective efficacy of the purified parasite protein in a primate challenge model. Culture of parasites in the laboratory is difficult, and it has not been possible to obtain the large quantities of purified parasite proteins necessary for human immunization studies. Instead, promising candidates must be produced as either recombinant proteins or synthetic peptides. Reproducing the immunogenicity of the native antigens by these means has not been straightforward.

EXPERIMENTAL VACCINE STUDIES

A primary focus of blood-stage vaccine research has been the major surface antigen of the merozoite, known as merozoite surface protein-1 (MSP-1), merozoite surface antigen-1 (MSA-1), gp195, or p190. The MSP-1 precursor protein is expressed on the merozoite surface and is processed to smaller peptides around the time of schizont rupture (reviewed in Holder 1988). Immunization with purified parasite-derived MSP-1 or its processed products elicited complete (Siddiqui et al 1987) or partial (Perrin et al 1984; Hall et al 1984; Patarroyo et al 1987a) protection from *P. falciparum* blood-stage challenge in primates. These results have instigated significant efforts to identify recombinant vaccine candidates based on MSP-1.

The protective efficacy of *E. coli*-expressed peptides derived from both the C- and N-terminal regions of MSP-1 has been evaluated in primates. Immunization of *Aotus* monkeys with a mixture of recombinant MSP-1 fusion peptides, one corresponding to the 42 kDa C-terminal processed MSP-1 product and the second, a smaller C-terminal MSP-1 fusion peptide, elicited modest antibody responses in *Aotus* monkeys and conferred partial protection from blood-stage challenge, with no delay in the prepatent period observed in unprotected monkeys. No correlation was observed between levels of MSP-1-specific antibodies and protection (Holder et al 1988). Studies in laboratory rodents have demonstrated that conformational epitopes on the 42 kDa C-terminal peptide, which contains a cluster of cysteine residues, are expressed by baculovirus-produced 42 kDa peptide, but not when expressed by yeast. Further, recognition of conformational determinants on the 42 kDa fragment appears to be important for the development of parasite inhibitory antibodies (Chang et al 1992). The 42 kDa peptide is proteolytically processed further into a 19 kDa peptide, which is the only MSP-1-derived product remaining on the surface of the merozoite during red blood cell invasion. A yeast-expressed 19 kDa peptide protected *Aotus nancymai* monkeys from blood-stage challenge (Kumar et al 1995). Serum from the protected monkeys had no effect on parasite invasion *in vitro*.

Several *E. coli*-expressed constructs derived from the N-terminus of MSP-1 have been evaluated in primates. A construct consisting of two conserved N-terminal regions of MSP-1 fused to form a single peptide conferred partial protection in *Aotus* monkeys, with no apparent correlation between antibody levels and protection (Herrera et al 1990). A mixture of two different N-terminal peptides evaluated in *Saimiri* monkeys also elicited partial protection, which was poorer than that elicited with native MSP-1, with some correlation between antibody responses to parasites and protection (Etlinger et al 1991). A third study demonstrated no protection of *Aotus* monkeys after immunization with a 200-amino acid peptide derived from the N-terminus (Knapp et al 1988). The addition of a universal T-cell epitope from CSP to a conserved N-terminal peptide of MSP-1 significantly improved the protective efficacy of the peptide in *Aotus* monkeys (Herrera et al 1992). Interestingly, protection correlated with serum levels of γ-interferon elicited by vaccination rather than with antibody responses.

Other protective blood-stage antigens have also been identified in primates. The serine repeat antigen (SERA), also known as SERP I or p126, is expressed as a 126 kDa precursor protein in the parasitophorous vacuoles of developing blood-stage parasites (Delplace et al 1987; Chulay et al 1987; Coppel et al 1988; Knapp et al 1989). Like MSP-1, SERA is proteolytically processed around the time of schizont rupture into smaller peptide fragments. Immunization of both *Saimiri* and *Aotus* monkeys with purified parasite-derived SERA confers partial protection from challenge (Perrin et al 1984; Delplace et al 1988). Recombinant yeast-expressed SERA peptides derived from the N-terminus of the precursor protein elicited partial protection in *Aotus* monkeys (Inselburg et al 1991 1993a, 1993b). Apical membrane antigen-1 (AMA-1) is a protein of 83 kDa that is expressed late in schizogony and distributed on the apex of merozoites developing within the schizont (Peterson et al 1989). AMA-1 is probably exported from the apical complex to the merozoite surface and then processed at the time of merozoite release into a 66 kDa fragment, which may play a role in red cell invasion (Thomas et al 1990). The *P. knowlesi* and *P. fragile* analogs of AMA-1 confer partial protection from challenge in primates (Deans et al 1988; Collins et al 1994). Other blood-stage antigens are also under consideration as candidate vaccine antigens (reviewed in Pasloske and Howard 1994).

The ability of vaccinia virus recombinants expressing blood-stage antigens of *P. falciparum* to elicit antibody responses reactive with the authentic parasite antigens has been described in experimental animals (Tine et al 1993; Theisen et al 1994; Sandhu and Kennedy 1994). Vaccinia recombinants expressing ring-infected erythrocyte surface antigen (RESA), a RESA/S Ag hybrid, MSP-1, merozoite surface antigen-2 (MSA-2), or AMA-1 or a mixture of the RESA, MSP-1, MSA-2, and AMA-1 recombinants did not protect *Saimiri* monkeys from challenge with blood-stage parasites (Pye et al 1991). However, the data in this study are difficult to interpret because analysis of recombinant-directed expression of the parasite antigens was not reported. This is important because the sequences of these genes, derived from the FC27 *P. falciparum* strain, all contain one or more vaccinia early transcription termination signals near the 5' end. Removal of these early transcription termination signals can greatly increase immunogenicity, presumably by increasing expression levels (Earl et al 1990).

Together, these results indicate that recombinant blood-stage proteins are generally less efficacious than the corresponding authentic parasite-derived proteins when evaluated in the monkey challenge models. Similar to studies with pre-erythrocytic vaccine constructs, no measurable parameters of immunity to blood-stage parasites, including *in vitro* functional assays, correlate with protection.

SEXUAL-STAGE VACCINE DEVELOPMENT

Vaccine strategies aimed at interrupting the infectious cycle in humans are complemented by strategies designed to block transmission of sexual forms of the parasite from the human host to the mosquito vector. In areas of low transmission, transmission-blocking vaccines in combination with traditional malaria control measures of drug treatment and mosquito control may be sufficient to significantly reduce the incidence of disease (Kaslow et al 1992). Further, the escape of variant parasites generated by the immune pressure of vaccination against other stages of the life cycle might be prevented by such vaccines.

Numerous studies have demonstrated that immune sera and antibodies directed against particular sexual-stage proteins can block transmission. Three major targets of transmission-blocking antibodies are the Pfs230 and Pfs48/45 proteins, which are expressed on the surface of both male and female gametes, and the zygote/ookinete surface protein Pfs25 (reviewed in Kaslow et al 1992). However, a significant proportion of the human population appears to be unable to generate antibodies to the Pfs230 and Pfs48/45 proteins (Graves et al 1988; Quakyi et al 1989; Carter et al 1989). Humans do not naturally generate antibody responses to Pfs25 because this protein is probably not expressed in the human host, but studies in mice suggest that the ability to generate such responses is not genetically restricted (Good et al 1988; Kaslow et al 1991).

Pfs25 is the most developed transmission-blocking candidate immunogen. Mice immunized with a recombinant vaccinia virus expressing Pfs25 developed antibodies that blocked the transmission of sexual forms of the parasite to mosquitoes (Kaslow et al 1991). Further studies showed that *Aotus* monkeys immunized with a yeast-expressed Pfs25 peptide and muramyl tripeptide (Barr et al 1991; Kaslow et al 1993) or alum (Kaslow et al 1994) as adjuvants developed transmission-blocking antibodies. Recently, an improved, yeast-expressed candidate Pfs25 construct for testing in humans was developed (Kaslow and Shiloach 1994).

MULTI-ANTIGEN VACCINE DEVELOPMENT

The need for a multi-component approach to malaria vaccine development has been recognized for several years to overcome the variability of parasite antigens among isolates and the genetic nonresponsiveness to particular antigens/epitopes observed in individuals from endemic regions. A study with the *P. yoelii* rodent parasite provides strong support for a multi-antigen vaccination approach against the pre-erythrocytic stages, demonstrating that a mixture of transformed cells expressing CSP or a portion of the sporozoite surface protein-2 (SSP2) confers complete protection from sporozoite challenge, a vast improvement over either transformant alone (Khusmith et al 1991). For vaccination against the blood stages, a combination of three synthetic peptides derived from three different blood-stage proteins (SPf66, see below) is also more effective than the individual peptides alone in eliciting protection in monkeys (Patarroyo et al 1987b). Further, segments of SERA and histidine-rich protein II (HRPII) or SERA, HRPII, and MSP-1 expressed in *E. coli* as hybrid MS2-polymerase fusion proteins were more effective than the individual components in *Aotus* monkeys (Knapp et al 1992; Enders et al 1992).

STUDIES WITH THE SPf66 VACCINE CANDIDATE

The SPf66 malaria blood-stage vaccine candidate is the most extensively tested human malaria vaccine, with many thousands of individuals immunized in numerous studies. This vaccine is a synthetic hybrid containing 30 tandem copies of a monomer unit consisting of peptide fragments from three blood-stage antigens—MSP1, Pf55 and Pf33—linked by five amino acid spacer sequences derived from CSP. Initial studies in primates demonstrated that immunization with the combination of three synthetic peptides was more effective than the individual peptides alone at eliciting protective immunity (Patarroyo et al 1987b).

After limited human efficacy testing by controlled laboratory challenge (Patarroyo et al 1988), SPf66 was evaluated in a series of field studies in Latin America, which demonstrated the safety and immunogenicity of this vaccine candidate (reviewed by Tanner et al 1995). A double-blind, placebo-controlled phase III trial in a low-transmission area in Columbia in which individuals over the age of one year were vaccinated with SPf66 reported efficacy of about 33% (Valero et al 1993). Results of double-blind, placebo-controlled phase III trials in areas of high malaria transmission are now being reported. In Tanzania, immunization of children aged one to five years with SPf66 resulted in an estimated efficacy of 31% (Alonso et al 1994). A second study in Gambian infants aged 6 to 11 months found no evidence of significant protection (D'Alessandro et al 1995). Results from a third study performed in Thailand are expected soon. Although the future of this vaccine is controversial (Maurice 1995), the development of second generation constructs is proceeding.

Multi-Antigen, Multi-Stage Vaccine Development

In addition to the antigenic variability among different parasite isolates and the nonresponsiveness of individuals to particular parasite antigens or epitopes, another obstacle to the development of an effective malaria vaccine is the developmental regulation of antigen expression during parasite replication. An extension of the multi-antigen approach is the inclusion of antigens from several different stages of the parasite life cycle. With such a strategy, the inability to mount a fully effective immune response to a particular antigen or to antigens of a given life cycle stage may be compensated by effective responses to other antigens or life cycle stages, resulting in protective immunity. Although it appears that combinations of antigens from single stages of the life cycle are more effective than individual antigens, the problem of overcoming nonresponsiveness to vaccinating antigens derived from a single life cycle stage remains a difficult one, even with multi-component approaches. In field studies with the SPf66 vaccine, 26 to 45% of inoculated individuals did not develop serum antibodies to SPf66 (Valero et al 1993; Noya et al 1994). Thus, a multi-antigen, multi-life cycle stage approach, which elicits immune responses to several antigens from each of the major stages of the parasite life cycle, may provide a more effective vaccination strategy than subunit or single-stage approaches.

Animal studies to evaluate an approach involving inoculation with a cocktail of yeast-expressed peptides derived from CSP, MSP-1, SERA, and Pfs25 (Bathurst et al 1993) showed that each antigen of the cocktail was immunogenic, and sera from immunized animals exhibited biological activity as assessed by *in vitro* assays. Comparison with animals inoculated with individual antigens indicated that there was no suppression of antibody responses to the component antigens when presented as a cocktail (Bathurst et al 1993). As with other approaches using peptides, the identification of more potent adjuvants for use in humans is a priority.

NYVAC-Pf7

Because of its capacity to incorporate and express large quantities of exogenous DNA, vaccinia virus is an appropriate vector for the development of a multi-valent vaccine and its potential as a vaccine vector is well established (Cox et al 1992; Tartaglia et al 1990). The development of highly attenuated vaccinia strains, such as NYVAC (Tartaglia et al 1992), addresses safety concerns that have been raised regarding their use. The ability of NYVAC-based recombinants expressing CSP to confer high levels of protection from *P. berghei* challenge (Lanar et al 1996) supports the development of NYVAC-based recombinants expressing *P. falciparum* antigens as candidate malaria vaccines. Further, no vaccines thus far evaluated in humans have been designed to elicit good CTL responses, despite the extensive data indicating that cellular responses to pre-erythrocytic antigens are protective. The ability of vaccinia recombinants to elicit CTL responses is documented (Cox et al 1992; Tartaglia et al 1990).

NYVAC-Pf7 was derived by inserting the genes encoding the *P. falciparum* antigens CSP, PfSSP2, liver-stage antigen-1 (LSA-1), MSP-1, SERA, AMA-1 and Pfs25 into a single NYVAC genome (Tine et al 1996). This vaccine candidate thus has the potential to elicit immune responses directed against the major life cycle stages of parasite: the sporozoite, liver, blood, and sexual stages. Preclinical studies in rhesus monkeys have demonstrated the safety and immunogenicity of NYVAC-Pf7. Monkeys inoculated with this recombinant develop antibodies against sporozoites, infected liver cells, infected erythrocytes and zygotes, as well as the individual antigens that have been tested (CSP, PfSSP2, MSP-1, SERA, Pfs25). Preliminary results in studies designed to evaluate CTL responses in vaccinated monkeys appear promising. These studies demonstrate that NYVAC-Pf7, the first multi-antigen, multi-stage vaccine candidate for *P. falciparum* malaria, is appropriate for further testing in phase I clinical trials (Tine et al 1996). Such studies are currently underway.

FUTURE PERSPECTIVES

The development of an ideal malaria vaccine is a daunting proposition. Among the desired characteristics of such a vaccine are the ability to elicit immunity against multiple parasite antigens and multiple stages of the parasite life cycle, the ability to elicit both humoral and cellular immune responses, the ability to protect individuals from many different parasite isolates, and the ability to elicit long-lasting immunological memory, so that subsequent exposure to parasites will elicit rapid recall of protective immune responses. Further, given the epidemiology of this disease, it is essential that such a vaccine be economical to produce and distribute.

There is currently no highly effective vaccine for malaria, and it seems unlikely that each of the characteristics of an optimal vaccine will be met with any one technology available today. Further, the clinical complexity of this disease, which is affected by factors such as age, prior exposure and levels of transmission, make it improbable that any one formulation will be effective in all clinical settings. Perhaps a combination of several vaccination strategies, for example, the use of live recombinant vectors or DNA immunization and subunit approaches will allow the development of a vaccine that will have a significant clinical impact in areas where a vaccine is most needed. Surely, the evolution in malaria vaccine design will continue for some time until the goal of an effective vaccine is realized.

References

Aggarwal A, Kumar S, Jaffe R, Hone D, Gross M, Sadoff J (1990). Oral *Salmonella* : malaria circumsporozoite recombinants induce specific CD8[+] cytotoxic T cells. J Exp Med 172: 1083-1090.

Alonso PL, Smith T, Armstrong Schellenberg, JRM, Masanja H, Mwankusye S, Urassa H, Bastos de Azevedo I, Chongela J, Kobero S, Menedez C, Hurt N, Thomas MC, Lyimo E, Weiss, NA, Hayes R, Kitua AY, Lopez MC, Kilama WL, Teuscher T, Tanner M (1994). Randomised trial of efficacy of SPf66 vaccine against *Plasmodium falciparum* malaria in children in southern Tanzania. Lancet 344: 1175-1181.

Ballou, WR, Sherwood JA, Neva FA, Gordon DM, Wirtz RA, Wasserman GF, Diggs CL, Hoffman SL, Hollingdale MR, Hockmeyer WT, Schneider I, Young JF, Reeve P, Chulay JD (1987). Safety and efficacy of a recombinant DNA *Plasmodium falciparum* sporozoite vaccine. Lancet 1: 1277-1281.

Barr PJ, Green KM, Gibson HL, Bathurst IC, Quakyi IA, Kaslow DC (1991). Recombinant Pfs25 protein of *Plasmodium falciparum* elicits malaria transmission-blocking immunity in experimental animals. J Exp Med 174: 1203-1208.

Bathurst IC, Gibson HL, Kansopon J, Hahm BK, Green KM, Chang SP, Hui GSN, Siddiqui WA, Inselburg J, Millet P, Quakyi IA, Kaslow DC, Barr PJ (1993). An experimental vaccine cocktail for *Plasmodium falciparum* malaria. Vaccine 11: 449-456.

Brown AE, Singharaj P, Webster HK, Pipithkul J, Gordon DM, Boslego JW, Krinchai K, Su-archawaratana P, Wongsrichanalai C, Ballou WR, Permpanich B, Kain KC, Hollingdale MR, Wittes J, Que JU, Gross M, Cryz SJ, Sadoff JC (1994). Safety, immunogenicity and limited efficacy study of a recombinant *Plasmodium falciparum* circumsporozoite vaccine in Thai soldiers. Vaccine 12: 102-108.

Carter R, Graves PM, Quakyi IA, Good MF (1989). Restricted or absent immune responses in human populations to *Plasmodium falciparum* gamete antigens that are targets of malaria transmission-blocking antibodies. J Exp Med 169: 135-147.

Chang SP, Gibson HL, Lee-Ng CT, Barr PJ, Hui GSN (1992). A carboxyl-terminal fragment of *Plasmodium falciparum* gp195 expressed by a recombinant baculovirus induces antibodies that completely inhibit parasite growth. J Immunol 149: 548-555.

Chulay JD, Lyon JA, Haynes JD, Meierovics AI, Atkinson CT, Aikawa M (1987). Monoclonal antibody characterization of *Plasmodium falciparum* antigens in immune complexes formed when schizonts rupture in the presence of immune serum. J Immunol 139: 2768-2774.

Collins WE, Pye D, Crewther PE, Vandenberg KL, Galland GG, Sulzer AJ, Kemp DJ, Edwards SJ, Coppel RL, Sullivan JS, Morris CL, Anders RF (1994). Protective immunity induced in squirrel monkeys with recombinant apical membrane antigen-1 of *Plasmodium fragile*. Am J Trop Med Hyg 51: 711-719.

Coppel RL, Crewther PE, Culvenor JG, Perrin LH, Brown GV, Kemp DJ, Anders RF (1988). Variation in p126, a parasitophorous vacuole antigen of *Plasmodium falciparum*. Mol Biol Med 5: 155-166.

Cox WI, Tartaglia J, Paoletti E (1992). Poxvirus recombinants as live vaccines. In: Binns MM, Smith GL, eds. Recombinant poxviruses. Boca Raton: CRC Press, 124-162.

D'Alessandro U, Leach A, Drakely CJ, Bennett S, Olaleye BO, Fegan GW, Jawara M, Langerock P, George MO, Targett GAT, Greenwood BM (1995). Efficacy trial of malaria vaccine SPf66 in Gambian infants. Lancet 346: 462-467.

Dame JB, Williams JL, McCutchan TF, Weber JL, Wirtz, RA, Hockmeyer WT, Maloy WL, Haynes JD, Schneider I, Roberts D, Sanders GS, Reddy EP, Diggs CL, Miller LH (1984). Structure of the gene encoding the immunodominant surface antigen on the sporozoite of the human malaria parasite *Plasmodium falciparum*. Science 225: 593-599.

Deans JA, Knight AM, Jean WC, Waters AP, Cohen S, Mitchell GH (1988). Vaccination trials in rhesus monkeys with a minor invariant, *Plasmodium knowlesi* 66 kD merozoite antigen. Parasite Immunol 10: 535-552.

Delplace P, Bhatia A, Cagnard M, Camus D, Colombet G, Debrabant A, Dubremetz J-F, Dubreuil N, Prensier G, Fortier B, Haq A, Weber J, Vernes A (1988). Protein p126: A parasitophorous vacuole antigen associated with the release of *Plasmodium falciparum*. Biol Cell 64: 215-221.

Delplace P, Fortier B, Tronchin G, Dubremetz J-F, Vernes A (1987). Localization, biosynthesis, processing and isolation of a major 126 kDa antigen of the parasitophorous vacuole of *Plasmodium falciparum*. Mol Biochem Parasitol 23: 193-201.

Earl P, Hugin AW, Moss B (1990). Removal of cryptic poxvirus transcription termination signals from the human immunodeficiency virus type I envelope gene enhances expression and immunogenicity of a recombinant vaccinia virus. J Virol 64: 2448-2451.

Egan JE, Weber JL, Ballou WR, Hollingdale MR, Majarian WR, Gordon DM, Maloy WL, Hoffman SL, Wirtz RA, Schneider I, Woollett GR, Young JF, Hockmeyer WT (1987). Efficacy of murine malaria parasite vaccines: Implications for human vaccine development. Science 236: 453-456.

Enders B, Hundt E, Knapp B (1992). Strategies for the development of an antimalarial vaccine. Vaccine 10: 920-927.

Etlinger HM, Caspers P, Matile H, Schoenfeld H-J, Stueber D, Takacs B (1991). Ability of recombinant or native proteins to protect monkeys against heterologous challenge with *Plasmodium falciparum*. Infect Immun 59: 3498-3503.

Etlinger HM, Felix AM, Gillessen D, Heimer EP, Just M, Pink JRL, Sinigaglia F, Sturchler D, Takacs B, Trzeciak A, Matile H (1988). Assessment in humans of a synthetic peptide-based vaccine against the sporozoite stage of the human malaria parasite *Plasmodium falciparum*. J Immunol 140: 626-633.

Fries LF, Gordon DM, Schneider I, Beier JC, Long GW, Gross M, Que JU, Cryz SJ, Sadoff JC (1992a). Safety, immunogenicity, and efficacy of a *Plasmodium falciparum* vaccine comprising a circumsporozoite protein repeat region peptide conjugated to *Pseudomonas aeruginosa* toxin A. Infect Immun 60: 1834-1839.

Fries LF, Gordon DM, Richards RL, Egan JE, Hollingdale MR, Gross M, Silverman C, Alving CR (1992b). Liposomal malaria vaccine in humans: A safe and potent adjuvant strategy. Proc Natl Acad Sci USA 89: 358-362.

Gonzalez C, Hone D, Noriega FR, Tacket CO, Davis JR, Losonsky G, Nataro JP, Hoffman S, Malik A, Nardin E, Sztein MB, Heppner DG, Fouts TR, Isibasi A, Levine MM (1994). *Salmonella typhi* vaccine strain CVD 908 expressing the circumsporozoite protein of *Plasmodium falciparum:* Strain construction and safety and immunogenicity in humans. J Infect Dis 169: 927-931.

Good M, Miller LH, Kumar S, Quakyi IA, Keister D, Adams JH, Moss B, Berzofsky JA, Carter R (1988). Limited immunological recognition of critical malaria vaccine candidate antigens. Science 242: 574-577.

Gordon DM, McGovern TW, Krzych U, Cohen JC, Schneider I, LaChance R, Heppner DG, Yuan G, Hollingdale M, Slaoui M, Hauser P, Voet P, Sadoff JC, Ballou WR (1995). Safety, immunogenicity, and efficacy of a recombinantly produced *Plasmodium falciparum* circumsporozoite protein-hepatitis B surface antigen subunit vaccine. J Infect Dis 171: 1576-1585.

Graves PM, Carter R, Burkat TR, Quakyi IA, Kuman N (1988). Antibodies to *Plasmodium falciparum* gamete surface antigens in Papua New Guinea sera. Parasite Immunol 10: 209-218.

Hall R, Hyde JE, Goman M, Simmons DL, Hope IA, Mackay M, Scaife J, Merkli B, Richle R, Stocker J (1984). Major surface antigen gene of a human malaria parasite cloned and expressed in bacteria. Nature 311: 379-382.

Herrera MA, Rosero F, Herrera S, Caspers P, Rotmann D, Sinigaglia F, Certa U (1992). Protection against malaria in *Aotus* monkeys immunized with a recombinant blood-stage antigen fused to a universal T-cell epitope: Correlation of serum gamma interferon levels with protection. Infect Immun 60: 154-158.

Herrera S, Herrera M, Perlaza BL, Burki Y, Caspers P, Dobeli H, Rotmann D, Certa U (1990). Immunization of *Aotus* monkeys with *Plasmodium falciparum* blood-stage recombinant proteins. Proc Natl Acad Sci USA 87: 4017-4021.

Herrington DA, Losonsky GA, Smith G, Volvovitz F, Cochran M, Jackson K, Hoffman SL, Gordon DM, Levine MM, Edelman R (1992). Safety and immunogenicity in volunteers of a recombinant *Plasmodium falciparum* circumsporozoite protein malaria vaccine produced in Lepidopteran cells. Vaccine 10: 841-846.

Herrington DA, Clyde DF, Losonsky G, Cortesia M, Murphy JR, Davis J, Baqar S, Felix AM, Heimer EP, Gillessen D, Nardins E, Nussenzweig RS, Nussenzweig V, Hollingdale MR, Levine MM (1987). Safety and immunogenicity in man of a synthetic peptide malaria vaccine against *Plasmodium falciparum* sporozoites. Nature 328: 257-259.

Hoffman SL, Edelman R, Bryan JP, Schneider I, Davis J, Sedegah M, Gordon D, Church P, Gross M, Silverman C, Hollingdale M, Clyde D, Sztein M, Losonsky G, Paparello S, Jones TR (1994). Safety, immunogenicity, and efficacy of a malaria sporozoite vaccine administered with monophosphoryl lipid A, cell wall skeleton of mycobacteria, and squalene as adjuvant. Am J Trop Med Hyg 51: 603-612.

Holder AA (1988). The precursor to major merozoite surface antigens: Structure and role in immunity. Prog Allergy 41: 72-97.

Holder AA, Freeman RR, Nicholls SC (1988). Immunization against *Plasmodium falciparum* with recombinant polypeptides produced in *Escherichia coli*. Parasite Immunol 10: 607-617.

Inselburg J, Bathurst IC, Kansopon J, Barchfeld GL, Barr PJ, Rossan RN (1993a). Protective immunity induced in *Aotus* monkeys by a recombinant SERA protein of *Plasmodium falciparum* : Adjuvant effects on induction of protective immunity. Infect Immun 61: 2041-2047.

Inselburg J, Bathurst IC, Kansopon J, Barr PJ, Rossan R (1993b). Protective immunity induced in *Aotus* monkeys by a recombinant SERA protein of *Plasmodium falciparum*: Further studies using SERA 1 and MF75.2 adjuvant. Infect Immun 61: 2048-2052.

Inselberg J, Bzik DJ, Li W, Green KM, Kansopon J, Hahm BK, Bathurst IC, Barr PJ, Rossan RN (1991). Protective immunity induced in *Aotus* monkeys by recombinant SERA proteins of *Plasmodium falciparum.* Infect Immun 59: 1247-1250.

Kaslow DC, Bathurst IC, Lensen T, Ponnudurai T, Barr PJ, Keister DB (1994). *Saccharomyces cerevisiae* recombinant Pfs25 adsorbed to alum elicits antibodies that block transmission of *Plasmodium falciparum.* Infect Immun 62: 5576-5580.

Kaslow DC, Shiloach J (1994). Production, purification and immunogenicity of a malaria transmission-blocking vaccine candidate: TBV25H expressed in yeast and purified using nickel-NTA agarose. Biotechnology 12: 494-499.

Kaslow DC, Bathurst IC, Keister DB, Campbell GH, Adams S, Morris CL, Sullivan JS, Barr PJ, Collins WE (1993). Safety, immunogenicity, and *in vitro* efficacy of a muramyl tripeptide-based malaria transmission-blocking vaccine in an *Aotus nancymai* monkey model. Vaccine Res 2: 95-103.

Kaslow DC, Bathurst IC, Barr PJ (1992). Malaria transmission blocking vaccines. Trends Biotechnol 10: 388-391.

Kaslow DC, Isaacs SN, Quakyi IA, Gwadz RW, Moss B, Keister DB (1991). Induction of *Plasmodium falciparum* transmission blocking antibodies by recombinant vaccinia virus. Science 252: 1310-1313.

Khusmith S, Charoenvit Y, Kumar S, Sedegah M, Beaudoin RL, Hoffman SL (1991). Protection against malaria by vaccination with sporozoite surface protein 2 plus CS protein. Science 252: 715-718.

Knapp B, Hundt E, Enders B, Kupper HA (1992). Protection of *Aotus* monkeys from malaria infection by immunization with recombinant hybrid proteins. Infect Immun 60: 2397-2401.

Knapp B, Hundt E, Nau U, Kupper HA (1989). Molecular cloning, genomic structure and localization in a blood stage antigen of *Plasmodium falciparum* characterized by a serine stretch. Mol Biochem Parasitol 32: 73-84.

Knapp B, Shaw A, Hundt E, Enders B, Kupper HA (1988). A histidine alanine rich recombinant antigen protects *Aotus* monkeys from *P. falciparum* infection. Behring Inst Mitt 82: 349-359.

Kumar S, Yadava A, Keister DB, Tian JH, Ohl M, Perdue-Greenfield KA, Miller LH, Kaslow DC (1995). Immunogenicity and *in vivo* efficacy of recombinant *Plasmodium falciparum* merozoite surface protein-1 in *Aotus* monkeys. Mol Med 1: 325 - 332.

Lanar DE, Tine JA, de Taisne C, Seguin MC, Cox WI, Winslow JP, Ware LA, Kauffman EB, Gordon D, Ballou WR, Paoletti E, Sadoff JC (1996). Attenuated vaccinia virus-circumsporozoite protein recombinants confer protection against rodent malaria. Infect Immun 64: 1666-1671.

Li S, Rodrigues M, Rodriguez D, Rodriguez JR, Esteban M, Palese P, Nussenzweig RS, Zavala F (1993). Priming with recombinant influenza virus followed by administration of recombinant vaccinia virus induces CD8[+] T-cell-mediated protective immunity against malaria. Proc Natl Acad Sci USA 90: 5214-5218.

Magill AJ, Freis BT, Gordon DM, Wellde BT, Owens R, Krzych U, Schnieder IP, Wirtz RA, Kester K, Ockenhouse CF, Stoute JA, Ballou WR (1995). Protective efficacy of a liposime encapsulated R32NS1 *Plasmodium falciparum* malaria vaccine in a human challenge model. Abstract 137, 44th Annual Meeting of the American Society of Tropical Medicine and Hygiene, San Antonio, TX.

Maurice J (1995). Malaria vaccine raises a dilemma (News and Comment). Nature 267: 320-323.

Miller LH, Howard RJ, Carter R, Good MF, Nussenzweig V, Nussenzweig R (1986). Research toward malaria vaccines. Science 234: 1349-1356.

Noya O, Berti YG, de Noya BA, Borges R, Zerpa N, Urbaez JD, Madonna A, Garrido E, Jimenez MA, Borges RE, Garcia P, Reyes I, Prieto W, Colmenares C, Pabon R, Barraez T, de Caceres LG, Godoy N, Sifontes R (1994). A population-based clinical trial with the SPf66 synthetic *Plasmodium falciparum* malaria vaccine in Venezuela. J Infect Dis 170: 396-402.

Nussenzweig V, Nussenzweig R (1989). Rationale for the development of an engineered sporozoite malaria vaccine. Adv Immunol 45: 283-334.

Pasloske BL, Howard RJ (1994). The promise of asexual malaria vaccine development. Am J Trop Med Hyg 50 (suppl): 3-10.

Patarroyo ME, Amador R, Clavijo P, Moreno A, Guzman F, Romero P, Tascon R, Franco A, Murillo LA, Ponton G, Trujillo G (1988). A synthetic vaccine protects humans against challenge with asexual blood stages of *Plasmodium falciparum* malaria. Nature 332: 158-161.

Patarroyo ME, Romero P, Torres ML, Clavijo P, Andreu D, Lozada D, Sanchez L, del Portillo P, Pinilla C, Moreno A, Alegria A, Houghten R (1987a). Protective synthetic peptides against experimental *Plasmodium falciparum*-induced malaria. In: Chanock RM, Lerner RA, Brown F, Ginsberg H, eds. Vaccines 87. Cold Spring Harbor, N.Y.: Cold Spring Harbor Laboratory, 117-124.

Patarroyo ME, Romero P, Torres ML, Clavijo P, Moreno A, Martinez A, Rodriguez R, Guzman F, Cabezas E (1987b). Induction of protective immunity against experimental infection with malaria using synthetic peptides. Nature 328: 629-632.

Perrin LH, Merkli B, Loche M, Chizzolini C, Smart J, Richle R (1984). Antimalarial immunity in *Saimiri* monkeys. J Exp Med 160: 441-451.

Peterson MG, Marshall VM, Smythe JA, Crewther PE, Lew A, Silva A, Anders RF, Kemp DJ (1989). Integral membrane protein located in the apical complex of *Plasmodium falciparum*. Mol Cell Biol 9: 3151-3154.

Potocnjak P, Yoshida N, Nussenzweig RS, Nussenzweig V (1980). Monovalent fragments (Fab) of monoclonal antibodies to a sporozoite surface antigen (Pb44) protect mice against malarial infection. J Exp Med 151: 1504-1513.

Pye D, Edwards SJ, Anders RF, O'Brien CM, Franchina P, Corcoran LN, Monger C, Peterson MG, Vandenberg KL, Smythe JA, Westley SR, Coppel RL, Webster TL, Kemp DJ, Hampson AW, Langford CJ (1991). Failure of recombinant vaccinia viruses expressing *Plasmodium falciparum* antigens to protect *Saimiri* monkeys against malaria. Infect Immun 59: 2403-2411.

Quakyi IA, Otoo LN, Pombo D, Sugars LY, Menon A, DeGroot AS, Johnson A, Alling D, Miller LH, Good MF (1989). Differential non-responsiveness in humans of candidate *Plasmodium falciparum* vaccine antigens. Am J Trop Med Hyg 41: 125-134.

Rickman LS, Gordon DM, Wistar R, Krzych U, Gross M, Hollingdale MR, Egan JE, Chulay JD, Hoffman SL (1991). Use of adjuvant containing mycobacterial cell-wall skeleton, monophosphoryl lipid A, and squalene in malaria circumsporozoite protein vaccine. Lancet 337: 998-1001.

Sadoff JC, Ballou WR, Baron LS, Majarian WR, Brey RN, Hockmeyer WT, Young JF, Cryz SJ, Ou J, Lowell GH, Chulay JD (1988). Oral *Salmonella typhimurium* vaccine expressing circumsporozoite protein protects against malaria. Science 240: 336-338.

Sandhu JS, Kennedy JF (1994). Expression of the merozoite surface protein gp195 in vaccinia virus. Vaccine 12: 56-64.

Satchidanandam, Zavala F, Moss B (1991). Studies using a recombinant vaccinia virus expressing the circumsporozoite protein of *Plasmodium berghei*. Mol Biochem Parasitol 48: 89-100.

Sedegah M, Hedstrom R, Hobart P, Hoffman SL (1994). Protection against malaria by immunization with plasmid DNA encoding circumsporozoite protein. Proc Natl Acad Sci USA 91: 9866-9870.

Sedegah M, Beaudoin RL, de la Vega P, Leef MF, Ozcel MA, Jones E, Charoenvit Y, Yuan LF, Gross M, Majarian WR, Robey FA, Weiss W, Hoffman SL (1988). Use of a vaccinia construct expressing the circumsporozoite protein in the analysis of protective immunity to *Plasmodium yoelli*. In: Laskey L, ed. Technological advances in vaccine development. New York: Alan R Liss, 295-309.

Siddiqui WA, Tam LQ, Kramer KJ, Hui GSN, Case SE, Yamaga KM, Chang SP, Chan EBT, Kan S-C (1987). Merozoite surface coat precursor protein completely protects *Aotus* monkeys against *Plasmodium falciparum* malaria. Proc Natl Acad Sci USA 84: 3014-3018.

Siddiqui WA (1977). An effective immunization of experimental monkeys against a human malaria parasite, *Plasmodium falciparum*. Science 197: 388-389.

Tam JP, Clavijo P, Lu Y, Nussenzweig V, Nussenzweig R, Zavala F (1990). Incorporation of T and B cell epitopes of the circumsporozoite protein in a chemically defined synthetic vaccine against malaria. J Exp Med 171: 299-306.

Tanner M, Teuscher T, Alonso PL (1995). SPf66—the first malaria vaccine. Parasitol Today 11: 10-13.

Tartaglia J, Perkus ME, Taylor J, Norton EK, Audonnet J-C, Cox WI, Davis SW, Van der Hoeven J, Meignier B, Riviere M, Languet B, Paoletti E (1992). NYVAC: A highly attenuated strain of vaccinia virus. Virology 188: 217-232.

Tartaglia J, Pincus S, Paoletti E (1990). Poxvirus-based vectors as vaccine candidates. CRC Crit Rev Immunol 10: 13-30.

Theisen M, Cox G, Høgh B, Jepsen S, Vuust J (1994). Immunogenicity of the *Plasmodium falciparum* glutamate-rich protein expressed by vaccinia virus. Infect Immun 62: 3270-3275.

Thomas AW, Waters AP, Carr D (1990). Analysis of variation in PF83, an erythrocytic merozoite vaccine candidate antigen of *Plasmodium falciparum*. Mol Biochem Parasitol 42: 285-288.

Tine JA, Lanar DE, Smith DM, Wellde BT, Schultheiss P, Ware LA, Kauffman EB, Wirtz RA, de Taisne C, Hui GSN, Chang SP, Church P, Kaslow DC, Hoffman S, Guito KP, Ballou WR, Sadoff JC, Paoletti E (1996). NYVAC-Pf7: A poxvirus-vectored multiantigen, multistage vaccine candidate for *Plasmodium falciparum* malaria. Submitted for publication.

Tine JA, Conseil V, Delplace P, de Taisne C, Camus D, Paoletti E (1993). Immunogenicity of the *Plasmodium falciparum* serine repeat antigen (p126) expressed by vaccinia virus. Infect Immun 61: 3933-3941.

Valero MV, Amador LR, Galindo C, Figueroa J, Bello MS, Murillo LA, Mora AL, Patarroyo G, Rocha CL, Rojas M, Aponte JJ, Sarmiento LE, Lozada DM, Coronell CG, Ortega NM, Rosas JE, Alonso PL, Patarroyo ME (1993). Vaccination with SPf66, a chemically synthesised vaccine, against *Plasmodium falciparum* malaria in Columbia. Lancet 341: 705-710.

Vreden SGS, Verhave JP, Oettinger T, Sauerwein RW, Meuwissen JHET (1991). Phase I clinical trial of a recombinant malarial vaccine consisting of the circumsporozoite repeat region of *Plasmodium falciparum* coupled to hepatitis B surface antigen. Am J Trop Med Hyg 45: 533-538.

Wang R, Charoenvit Y, Corradin G, Porrozzi R, Hunter RL, Glenn G, Alving CR, Church P, Hoffman SL (1995). Induction of protective polycloncal antibodies by immunization with a *Plasmodium yoelii* circumsporozoite protein multiple antigen peptide vaccine. J Immunol 154: 2784-2793.

Zavala F, Cochrane AH, Nardin EH, Nussenzweig RS, Nussenzweig V (1983). Circumsporozoite proteins of malaria parasites contain a single immunodominant region with two or more identical epitopes. J Exp Med 157: 1947–1957.

* * *

Credit for Figure 1, page 369, THE LIFE CYCLE OF THE MALARIA PARASITE. Hospital Practice 1990; 25:45 ©1990, The McGraw-Hill Companies, with permission of the artist, Nancy Lou Makris Riccio.

OTHER ISSUES

Microbe hunting is a story of amazing stupidities, fine intuitions, insane paradoxes. If that is the history of the hunting of microbes, it is the same with the story of the science, still in its babyhood, of why we are immune to microbes.

PAUL DE KRUIF—*Microbe Hunters*

Vignette

Mucosal Immunity

Mucosal immunity is an important concept in the field of immunology, with momentous implications for the future. The most significant modifiers of human mucosal immunity are events that occur in the neonatal maturation period and, later in life, the interplay between the immune system and the neuroendocrine systems. Since the oral cavity and the gastrointestinal tract are the first portals of entry for many infectious agents, it is logical to assume that protection should be initiated by production of IgA antibodies secreted by cells in the mucous membranes.

Considerable research is being directed toward the use of oral immunization for the prevention of mucosal infections, including otitis media, in infancy and childhood. To date, the most successful vaccines are those against polio for humans and recombinant rabies virus vaccine for immunization of wildlife, both involving oral mucosal immunization. The effectiveness of these vaccines has been proven by epidemiological observation.

Because any global approach to immunization requires that developing countries have access to inexpensive vaccines which can be administered by available health personnel, numerous laboratories in the world are now attempting to produce oral vaccines. Perhaps the most intriguing and promising aspect of these efforts is the possibility of producing these vaccines in plants digestible by human and animal gastrointestinal tracts.

29

The Mucosal Immune System: Recognition and Current Concepts

JOHN J. CEBRA, NICO A. BOS,
ETHEL R. CEBRA, AND KHUSHROO E. SHROFF

Alexandre Besredka, who studied host-resistance to bacterial pathogens and mechanisms of immunopathogenesis at the Pasteur Institute, is considered to be among the first to appreciate local immunization and regionally enriched antibodies, such as copra-antibodies (Besredka 1925). There followed the suggestion that antibodies in respiratory secretions might have a protective role against pneumococci (Francis 1940) and parainfluenza virus (Smith et al 1966). In the latter study, a typical convincing example was provided by the correlation of specific antibodies in upper respiratory secretions, but not of neutralizing serum antibodies, with resistance to intranasal reinfection with the virus (Smith et al 1966).

The antibody isotype most relevant to the humoral mucosal response in most mammalian species was identified among myeloma paraproteins as IgA, ("A" meaning acidic or anodally migrating) by Joseph Heremans and coworkers (Heremans et al 1959). However, the serum version of this isotype was found to lack many of the effector functions of IgM and IgG antibodies (Eddie et al 1971), and its role in protective immunity remained unclear. Within a few years, two groups led by Tomasi and Hanson, respectively, made observations that strongly suggested a major role for IgA in protective immunity against pathogens impinging on the mucosal epithelium—that IgA was the dominant Ig isotype in exocrine secretions, although it was only a minor component of the serum immunoglobulins (Hanson 1961; Chodirker and Tomasi 1963) of many mammalian species. Subsequently, the quaternary structure of secretory IgA (sIgA) was found to consist of four pairs of H- and L-chains (tetravalent) plus a single J-(joining) chain and a single extra polypeptide, the secretory component (SC) (O'Daly and Cebra 1971).

CELLULAR AND MOLECULAR MECHANISMS UNDERLYING SELECTIVE PARTITION OF IgA INTO EXOCRINE SECRETIONS RELATIVE TO SERUM

There are three main bases for the selective partitioning of sIgA into the exocrine secretions. First, IgA plasmablasts comprise the vast majority of secretory immunoblasts in the interstitia of exocrine glands and the lamina propria of the respiratory and gastrointestinal tract (Tomasi et al 1965; Crabbe et al 1965; Crandall et al 1967). They release their product, IgA, proximal to where it appears in the secretions, which arise nearby. Second, IgA plasmablasts in the circulation transit the "high" endothelial-like post-capillary venules and selectively accumulate in exocrine secretory tissue (McWilliams et al 1975). This selective lodging—and possible subsequent antigen-driven expansion (Husband 1982)—is not well understood, but recently the "homing" receptor $\alpha_4\beta_7$ integrin on lymphocytes and the vascular addressin MAdCAM-1 (Ig superfamily) have been suggested as relevant recognition molecules (Nakache et al 1989; Holzmann et al 1989). Finally, the detection of IgA in small vesicles within undifferentiated crypt cells of the small intestine (O'Daly et al 1971), followed by an elegant series of microscopic (Brandtzaeg 1974), electron microscopic (Kraehenbuhl et al 1975) and subcellular/molecular analyses, defined an active transport mechanism, with SC as the poly-Ig-receptor (pIg-R) (Mostov et al 1980). The pIg-R is selectively placed in the basal/lateral plasma membrane of epithelial cells and captures dimeric IgA (IgA$_2$). A significant portion of the IgA$_2$ is transported across the epithelial cell and released by exocytosis, leaving a partially attenuated pIg-R: SC associated with the sIgA. Very recently, an intriguing, protective activity has been suggested for IgA$_2$ present below mucosal epithelium and within mucosal epithelial cells—that this IgA$_2$ can complex with antigens before or after pIg-R-mediated uptake by epithelial cells. The result could be "flushing out" of complexed antigen into the intestinal or glandular lumen, or interference with intracellular replication of viruses (and bacteria?) (Mazanec et al 1992).

GUT LYMPHOID COMPARTMENTS

We will use gut-associated lympoid tissue (GALT) to illustrate the compartmentalization of the mucosal lymphoid system since it has been most extensively studied. Although it is likely not typical of all mucosal lymphoid tissues, GALT generally contains more secretory plasma cells and produces more Ig (mostly IgA) than all other lymphoid tissues of the mammalian body. Further, the content of CD8[+] T leukocytes in the intraepithelial spaces likely exceeds that of the spleen in many mammals. The most salient compartments in GALT are: 1) the lamina propria with conspicuous IgA plasmablasts and T cells in a pronounced $T_H2 > T_H1$ ratio (Taguchi et al 1990); 2) the intraepithelial space between enterocytes, which contains a variety of cell types including γ/δ TCR+ and α/β TCR+, CD8+ T cells and NK cells; and 3) the Peyer's patches (PP), which consist of a single layer of B-cell follicles with interfollicular T cells. Unlike other lymphoid tissues, the B-cell follicles in PP are chronically "secondary," displaying continuous germinal center reactions (GCRs). Efferent lymphatics drain into mesenteric lymph nodes but, unlike proper lymph nodes, there are no afferent lymphatics leading to PPs. Instead, microfold cells overlie the follicles and these pinocytose droplets from the gut lumen and also take up pathogens (Owen 1994; Wolf et al 1987).

THE SITE(S) OF COMMITMENT OF B CELLS TO IgA EXPRESSION

Using adoptive transfer of cells from different lymphoid tissues, we showed that PP were enriched sources of precursors for IgA plasma cells that could repopulate both the spleen and gut lamina propria (Craig and Cebra 1971; Cebra et al 1977). A splenic fragment culture system that allowed T cell-dependent clonal outgrowth of single B cells allowed us to demonstrate that PP ordinarily maintain high frequencies of IgA-memory cells that, when re-stimulated, gave clones that only expressed IgA antibodies (Gearhart and Cebra 1979). These IgA-memory B cells were not IgM+/IgD+ but rather IgA+ B cells, prevalent in PP but also present in other lymphoid tissue (Jones and Cebra 1974; Lebman et al 1987).

THE PROCESS OF IgA COMMITMENT—EARLY STAGES

The chronic germinal center reactions (GCRs) in PP lead to the local proliferation of large numbers of IgA+ B cells (Lebman et al 1987). No other lymphoid tissues generate as many IgA+ B cells. This regional pre-eminence of IgA+ cells in PP is illustrated by immunohistochemical analyses of PP GCRs (see Cebra and Shroff 1994). The GCRs of PP contain a core of peanut agglutinin high-binding GC cells (Rose et al 1980), including many IgA+ cells, which are interspersed with IgM+ cells, whereas IgD+/IgM+ cells are present only in the periphery (corona) of the follicle (i.e., primary B cells) (Cebra and Shroff 1994). Although the lack of viability of GC B cells has made it difficult to assess their potential, *in situ* hybridization readily revealed mRNAα— either productive message for the α-chain of IgA or germline transcripts—in many GC B cells from PP (see below) (Weinstein et al 1991).

Ig C_H-isotype switching seems to involve successive recombination (Honjo and Kaataoka 1978) of $V_H/D/J_H$ gene segments with C_H-genes. However, a variety of other genetic mechanisms could account for expression of IgA by PP cells, such as processing of polycistronic transcripts and transplicing of mRNAs (Shimizu et al 1989; Perlmutter and Gilbert 1984). We counted C_H-genes in PP GC cells and found that most "recombinations" proceeded by deletion of intervening C_H-genes, rather than by processing of polycistronic transcripts or transplicing of mRNAs (Weinstein et al 1991).

HISTOLOGIC SITE *vs* PHYSIOLOGIC STATE:
WHY DO PPs GENERATE IgA-COMMITTED B CELLS?

We hypothesized that the accumulation of IgA-committed B cells in PPs was based on the chronic presence of GCRs there. Each new gut mucosal stimulation by an antigen would be superimposed on ongoing GCRs, perhaps benefiting from activated antigen-presenting cells, bystander cells and activated T_H cells and their released cytokines. Perhaps such a stimulatory state of PP promotes successive isotype-switching and, eventually, IgA expression. Alternatively, the microenvironment of organized lymphoid tissue in the GALT, including PPs and mesenteric lymph nodes, may inherently provide a milieu (cytokines?) that promotes preferred isotype-switching to IgA irrespective of an activated physiologic state. Our experiments utilizing germ-free mice support the latter alternative. Such mice contain PPs with primary B-cell follicles as quiescent as those in draining peripheral lymph nodes (PLN) of conventionally reared mice. Reovirus type 1, a nonpathogenic, non-environmental virus that stimulates an acute humoral and cellular mucosal response when given orally (London et al 1987), was used to stimulate acute GCRs in PP or PLN by oral or footpad intradermal administration, respectively (Weinstein and Cebra 1991). GCRs developed in the draining lymphoid tissue with the same kinetics in both cases, waxing and waning over about 32 days. However, the non-IgM isotypes that arose in the two different sites after local stimulation with reovirus were quite different: within four to five days, IgA was the dominant non-IgM antibody in GCs and secreted by PP fragment cultures, while locally stimulated PLNs exhibited mostly IgG1 at these early stages of GCRs. Interestingly, on secondary challenge with reovirus by the original route, PPs and PLNs exhibited the same original, distinctive antibody isotype profiles. However, the GCRs and output of IgA antibodies were attenuated in PPs compared with the primary responses, while PLNs displayed exaggerated secondary responses.

A MOLECULAR GENETIC CORRELATE
OF SWITCH-RECOMBINATION TO IgA

Stavnezer et al (1988) have demonstrated that transcription of a germline $C\alpha$ transcript, initiated at a site 5' of the switch-recombination region for the $C\alpha$-gene ($S\alpha$), is predictive of B cells proceeding to switch-recombination *in vitro* to give productive $V_H/D/J_H/C\alpha$ transcripts. We found that locally stimulated PPs of formerly germ-free mice contain GC B cells that express germline $C\alpha$ transcripts within three days after oral reovirus infection, while locally stimulated PLNs do not (Weinstein and Cebra 1991). Thus, the microenvironment of PPs seems to favor preferential switching to IgA expression.

CELLULAR AND MOLECULAR REQUIREMENTS
FOR IgA EXPRESSION *IN VITRO*

To understand the basis of the processes leading to IgA expression in the gut mucosa, we sought to define its requirements *in vitro*. Splenic fragment cultures had allowed clonal expression of IgA by both clones generated from primary (IgM+/IgD+) B cells and IgA-memory (IgA+) B cells. However, this culture could not be easily manipulated and resolved into its essential components. We eventually found that both primary B cells and IgA memory B cells generated clones in a single B-cell clonal microculture that expressed IgA, provided that cloned T_H2 cells (D10) and splenic or PP dendritic cells were also added (Schrader et al 1990; George and Cebra 1991). The expression of IgA was dependent on viable dendritic cells and could be induced by the T_H2 cells in either an antigen-dependent, haplotype-restricted fashion or in an antigen-independent, allostimulated (I-A) fashion. This clonal assay has proven effective in following the frequency of specific IgA-memory cells for long periods after original gut mucosal antigen stimulation, since it requires so few B cells (see below). Cytokines important in eventual and terminal IgA expression, both *in vitro* and *in vivo*, include IL-4, IL-5, and IL-6 (Murray et al 1987; Harriman et al 1988; Ramsey et al 1994). The T_H2 line (D10) provides all of these, but dendrite cells still are required for optimal IgA expression (Schrader et al 1990; George and Cebra 1991). Recently, Snapper and coworkers generated *in vitro* a proportion of surface IgA+ B cells from primary IgM+/IgD+ B cells comparable to that observed *in vivo* in PPs (McIntyre et al 1995). They used non-clonal, bulk mini-cultures of primary B cells stimulated with lipopolysaccharide, IL-4, IL-5, and poly-anti-IgD/dextran conjugate in the presence of TGFβ. The cytokine TGFβ has been suggested as a "switch-factor" for IgA expression (Coffman et al 1989). However, the available data suggest that TGFβ may act selectively rather than inductively to lead to expression of IgA by developing clones that have already undergone the crucial induction to make germline Cα transcripts (McIntyre et al 1995; Kim and Kagnoff 1990; Cebra et al 1991; Lebman et al 1994). Finally it must be admitted that no *in vitro* system has yet been devised that exhibits the magnitude of preferred isotype-switching to IgA expression seen *in vivo* in stimulated PPs.

APPROACHES TO A PROTECTIVE MUCOSAL RESPONSE *IN VIVO*

No comprehensive information is yet available on how to deliberately induce protective mucosal immunity. Live attenuated oral viral vaccines, such as the Sabin polio vaccine, appear effective (Ogra and Karzon 1969). Actual infection with certain respiratory viruses, such as parainfluenza, also seems to result in resistance to reinfection (Smith et al 1966). Various animal studies with reovirus or rotavirus also suggest that gut mucosal protection in the context of primary infection, containment, limitation, and resolution can result from a productive local infection (London et al 1987; Cuff et al 1990; Offit and Dudzik 1990; Merchant et al 1991). The global concept of "one mucosal immune system" (Bienenstock 1974) has frequently been invoked to justify oral vaccination against respiratory viruses (see Meitin et al 1994). The most effective, nonreplicating oral vaccine has been found to be cholera toxin (Pierce and Gowans 1975), which is a potent stimulus of specific IgA memory cells in PPs (Fuhrman and Cebra 1981). The presence and frequency of these IgA memory cells clearly correlates with the *in vivo* local gut secondary IgA response to cholera toxin (Fuhrman and Cebra 1981). However, we still do not understand how this antigen effects an exaggerated IgA response. Indeed, the tissue-binding peptide of the toxin, the B-subunit, is one of the most potent mucosal stimuli of systemic tolerance known. The threat of mucosal induction of systemic tolerance has cast a shadow over the application of mucosal vaccines (Mattingly and Waksman 1978; Santos et al 1994).

Currently, two approaches to the delivery of any nonreplicating antigen by the gut mucosal route show some promise. One of these uses a technique for microencapsulation of antigens, including live, attenuated viruses (Offit et al 1994), in an aqueous solvent system—sodium alginate plus antigen added to spermine hydrochloride during sonication, for instance. This method has resulted in effective oral stimulation of both mucosal and systemic immune responses. Another promising approach, which may overcome the difficulties of using conjugates or fusion-proteins of the cholera toxin B subunit, is the possible use of gorse lectins reactive with fucosides as a conjugate carrier to deliver antigens through microfold cells overlying PPs (Giannasca et al 1994).

IS THERE A CELLULAR MUCOSAL IMMUNE SYSTEM?

Earlier experiments on the recirculation of lymphocytes—mainly T cells—in sheep suggested that subsets of such cells showed preferred pathways through peripheral *vs* gut-associated lymphoid tissues (Reynolds et al 1982). Over the years, homing receptors and vascular addressins were suggested to explain such a phenomenon (Nakache et al 1989; Holzmann et al 1989; Reynolds et al 1982). We showed that mice orally infected with reovirus developed a virus-specific, haplotype-restricted population of CD8[+] cytotoxic T-cell precursors (pCTLs) in PP which remained at highest frequencies in gut-associated tissues for up to six months after infection (London et al 1987; Cuff et al 1993). Other than in PPs, these pCTLs remained at high frequency only in the intraepithelial space (Cuff et al 1993). Such pCTLs were α/β TCR[+] and, when passively transferred to susceptible neonatal mice, protected these from a fatal meningoencephalitis caused by oral infection with type 3 reovirus (Cuff et al 1990; Cuff et al 1991). Thus far, clear implication of this promising contributor to a mucosal cellular immune defense against enteric viruses or intracellular bacterial pathogens has not been successful. However, the role of intraepithelial space pCTLs in limiting and containing such infections is suggested.

IS THERE A NONSPECIFIC DRIVE FOR THE ACTIVATION OF CD8⁺ T CELLS AND NK CELLS IN THE INTRAEPITHELIAL SPACE?

Comparison of germ-free, antigen-free and conventionally reared mice revealed a profound increase in the cellularity of the intraepithelial space in the latter (5- to 10-fold) and a shift from mostly γ/δ TCR⁺ T cells to mostly α/β TCR⁺ T cells (Hooper et al 1995). Along with these changes is the acquisition of "constitutive" cytotoxicity by the α/β TCR⁺ T cells (see Kawaguchi et al 1993) on conversion of germ-free or antigen-free mice to a "conventional" environment. It is not known what drives this process—the same for conventionalized, formerly germ-free mice (Kawaguchi et al 1993) and developing neonatal mice (Guy-Grand et al 1991). However, a major component of the murine microflora—the obligate, gram-positive anaerobe known as segmented-filamentous bacteria (SFB)—has been implicated in driving the development of the gut mucosal IgA cells in the lamina propria compartment (Klaasen et al 1993). We have also implicated this same SFB in driving the increase in cellularity and activation of the CD8⁺ T cells in the intraepithelial space (Lee 1995).

Finally, in addition to CD8⁺ T cells, the intraepithelial space is rich in NK cells—large, granular cells. The role of environmental antigens in activating these intraepithelial NK cells can best be seen in severe combined immunodeficient (SCID) mice (George et al 1990). These NK cells are far more activated in the intraepithelial space than in spleen of SCID mice and very much more activated than either splenic or intraepithelial NK cells from immunocompetent mice as judged by quantitative cytotoxicity and target cell range (Cuff et al 1996). A further superimposition of an oral reovirus infection on SCID mice results in a dramatic increase in nonspecific, NK cytotoxicity (Cuff et al 1996), although the infected mice eventually die of a chronic hepatitis (George et al 1990).

To conclude this overview of mucosal immunity, we return to three aspects of the humoral mucosal immune response that may have practical implications for the development of effective oral vaccines.

COMMENSAL ENTERIC BACTERIA ENGENDER A SELF-LIMITING IMMUNE RESPONSE WHILE PERMANENTLY COLONIZING THE GUT

Whether the mammalian host makes a gut mucosal immune response against its own indigenous bacterial flora has been argued for many years (Carter and Pollard 1971; Dubos et al 1965; Pollard and Sharon 1970). To revisit this question, we mono-colonized formerly germ-free mice with an occasional gut commensal, *Morganella morganii* (Potter 1971), which bears a peculiar phosphocholine (PC) determinant as part of its lipopolysaccharide (Young and Richard 1994). Upon oral administration of this gram-negative organism, it rapidly expands in the gut to 10^{10}-10^{11}/g of feces. The hosts remain otherwise healthy, but the PPs of their guts show a transient hypertrophy and the development of GCRs in B-cell follicles (Shroff et al 1995). These wax and wane with about the same kinetics as observed for a transient, totally resolved enteric reovirus infection (see above), although the enteric bacteria persist in the gut lumen in high numbers. The GCRs result in the rapid appearance and only gradually decline of IgA$^+$ B cells, although centers of B-cell proliferation and typical GCRs fade away by about 28 days after colonization. PPs and small intestinal fragment cultures, as well as ELISPOT assays of gut lamina propria cells, show the rapid appearance of specific IgA antibodies against *M. morganii* within 14 days after colonization. This response gradually declines, but persists for over 300 days, long after the GCRs have been resolved in PPs (Shroff et al 1995). Correlates of the appearance of antibodies in serum and the gut lumen include the rapid and continuous endogenous "coating" of luminal bacteria with IgA (see van der Waaij et al 1994) and the complete clearance of translocated bacteria from spleen and mesenteric lymph nodes. Finally, single B-cell clonal analyses for specific germinal center B cells and specific IgA memory B cells show that the former appear transiently, synchronous with the GCRs, while the latter persist, scattered in the follicular and dome areas of PPs for more than 300 days after colonization. Thus, we propose that the chronic, gradually declining mucosal response limits the continuous, acute GCRs in PPs and is maintained by long-lived IgA memory cells (Shroff et al 1995). We do not know the mechanisms by which such memory cells may be occasionally re-activated or how to deliberately "re-awaken" an acute mucosal response to the persistent gut colonizer. However, the next sections may offer some clues.

NEONATAL/MATERNAL INTERACTIONS THAT MODULATE THE HUMORAL MUCOSAL RESPONSIVENESS OF NEONATES

It has long been appreciated that the natural development of mucosal responsiveness by neonates lags behind systemic responsiveness (see Shahin and Cebra 1981). We chose oral reovirus type 1 infection to probe neonatal responsiveness. This non-environmental, nonpathogenic gut mucosal stimulus resulted in a prompt and effective, specific IgA response in 10-day-old neonates, comparable to that displayed by young adults (Kramer and Cebra 1995a). We also noticed that this effective specific response appeared to drive development of the "natural" IgA response in the gut and the rise there of "natural" IgA plasmablasts. To determine whether this premature (by one to two weeks) development of natural IgA in the gut may normally be forestalled by passively acquired maternal IgA antibodies, we analyzed

F_1 neonates from reciprocal crosses of SCID x immunocompetent parents (Kramer and Cebra 1995b). The resulting heterozygotes would be immunocompetent but differ only depending on whether their birth dams were SCID or immunocompetent. We found that F_1 pups born of SCID mothers showed the precocious development of natural IgA secretion by PP and small intestine fragment cultures and in ELISPOT assay for IgA-secreting cells in gut lamina propria, compared with F_1 pups born of immunocompetent mothers. These differences correlated with stomach content of suckled, maternal IgA and early "coating" of gut bacteria in neonates—pups of immunocompetent mothers had easily detectable stomach IgA and heavily coated gut bacteria, whereas pups from SCID mothers did not. Finally, the ability of specific, suckled maternal IgA to inhibit a primary mucosal response against reovirus by pups could be shown using the F_1 pups from the reciprocal crosses (above), orally immunized nurse dams, and foster nursing. Not only did maternal acquired IgA antibodies prevent active immunization of the pups against reovirus, but it also prevented the bystander stimulation of the development of natural gut IgA antibodies, likely against the gut flora (Kramer and Cebra 1995b). A corollary of these findings may be that transient, enteric virus infections may re-awaken gut mucosal responses to chronic or present gut bacterial colonizers.

DO B1 B CELLS CONTRIBUTE TO THE GUT MUCOSAL IGA RESPONSE AND HOW MIGHT THEY BE ACTIVATED/STIMULATED TO SWITCH TO IGA AND EXPRESS THEIR PRODUCT IN MUCOSAL TISSUE?

B1 B cells comprise a distinct subset of B cells in humans, mice, and perhaps other mammals (Herzenberg et al 1992). They have been distinguished by a set of surface markers ($CD5^+$, $Mac1^+$, IgM^{hi}/IgD^{lo}) that are not always definitive and a somewhat activated physiologic state ($IL-5R^+$, responsive to lipopolysaccharide with production of IL-10) (Hitoshi et al 1990; O'Garra et al 1992). In mice, they normally reside in the peritoneal cavity and spleen and seem to be capable of self-renewal, rather than dependent on continuous replenishment from bone marrow stem cells. Finally, clear external antigenic stimuli have not been shown to expand these B cells, although their IgM products do seem autoreactive and widely cross-reactive with bacterial antigens (Hayakawa et al 1984). In immunodeficient mice, the progeny of IgM^+ B cells appear to be capable of populating intestinal lamina propria with IgA plasmablasts (Kroese et al 1989). It has occurred to us that cells of the B1 lineage may play a role in the development of natural IgA in the neonatal gut. We have expanded B1 cells in SCID mice and have found that mesenteric lymph nodes serve as a site for the continuous generation of dividing IgA plasmablasts (Bos et al 1996). These cells apparently can populate the gut lamina propria and serve as productive fusion partners to generate IgA-secreting hybridomas. We have found that all eight IgA hybridomas generated from such sources can "coat" or stain a small but significant proportion of gut bacteria obtained from SCID mice. Analysis of the sequences of the V_H-genes expressed by these hybridomas has indicated that some display germline $V\alpha$ genes, while others show minimal but present point mutations that deviate from the best available germline sequences, and N-additions are present but also limited compared with V_H-genes from typical antigen-expanded B cells. Of somewhat greater import was our observation that most of the $V\alpha$ sequences expressed by these hybridomas corresponded or coincided with $V\mu$ sequences already established for B-cell lines considered rather prototypic for B1 B cells that make IgM. Our data are highly suggestive of the ability of B1 IgM$^+$ B cells to generate IgA$^+$ cells with the same "specificity." Such specificities could contribute to local gut mucosal responsiveness to bacterial colonizers. Since B1 cells seem exquisitely responsive to oral "foreign" lipopolysaccharide (Murakami et al 1994), it seems possible that a polyclonal stimulus delivered via the oral/gut mucosal route could activate a semi-specific protective IgA response.

REFERENCES

Besredka A (1925). Immunisation locale. Paris: Masson & Cie.

Bienenstock J (1974). The physiology of the local immune response and the gastrointestinal tract. In: Brent L, Holborow J, eds. Progress in immunology II. Vol 4. Amsterdam: North-Holland, 197-207.

Bos NA, Bun JCAM, Popma SH, Cebra ER, Deenen GJ, van der Cammen MJF, Kroese FGM, Cebra JJ (1996). Monoclonal immunoglobulin A derived from peritoneal B cells is encoded by both germ line and somatically mutated V_H genes and is reactive with commensal bacteria. Infect Immun 64: 616-623.

Brandtzaeg P (1974). Mucosal and glandular distribution of immunoglobulin components. Differential localization of free and bound SC in secretory epithelial cells. J Immunol 112: 1553-1559.

Carter PB, Pollard M (1971). Host responses to normal microbial flora in germfree mice. J Reticuloendo Soc 9: 580-587.

Cebra JJ, Gearhart PJ, Kamat R, Robertson SM, Tseng J (1977). Origin and differentiation of lymphocytes involved in the secretory IgA response. In: Cold Spring Harbor Symposium on Quantitative Biology: Origins of lymphocyte diversity. Vol XLI. Cold Spring Harbor, N.Y.: Cold Spring Harbor Laboratory, 201-215.

Cebra JJ, George A, Kerlin RL (1991). A cautionary note concerning the possibility that transforming growth factor-beta is a switch factor that acts on primary B cells to initiate IgA expression. Immunol Res 10: 404-406.

Cebra JJ, Shroff KE (1994). Peyer's patches as inductive sites for IgA commitment. In: Ogra PL, Lamm ME, Strober W, McGhee JR, Bienenstock J, eds. Handbook of mucosal immunology: Cellular basis of mucosal immunity. Vol 1. San Diego: Academic Press, Inc, 151-158.

Chodirker WB, Tomasi TB, Jr (1963). Gamma-globulins: Quantitative relationships in human serum and nonvascular fluids. Science 142: 1080-1081.

Coffman RL, Lebman DA, Shrader B (1989). Transforming growth factor β specifically enhances IgA production by lipopolysaccharide-stimulated murine B lymphocytes. J Exp Med 170: 1039-1044.

Crabbe PA, Carbonara AO, Heremans JF (1965). The normal human intestinal mucosa as a major source of plasma cells containing gamma A-immunoglobulin. Lab Invest 14: 235-248.

Craig SW, Cebra JJ (1971). Peyer's patches: An enriched source of precursors for IgA-producing immunocytes in the rabbit. J Exp Med 134: 188-200.

Crandall RB, Cebra JJ, Crandall CA (1967). The relative proportions of IgG-, IgA- and IgM-containing cells in rabbit tissues during experimental trichinosis. Immunology 12: 147-158.

Cuff CF, Cebra CK, Lavi E, Molowitz EH, Rubin DH, Cebra JJ (1991). Protection of neonatal mice from fatal reovirus infection by immune serum and gut derived lymphocytes. In: Mestecky J et al, eds. Immunology of milk and the neonate. New York: Plenum Press, 307-315.

Cuff CF, Cebra CK, Rubin DH, Cebra JJ (1993). Developmental relationship between cytotoxic α/β T cell receptor-positive intraepithelial lymphocytes and Peyer's patch lymphocytes. Eur J Immunol 23: 1333-1339.

Cuff CF, Lavi E, Cebra CK, Cebra JJ, Rubin DH (1990). Passive immunity to fatal reovirus serotype 3-induced meningoencephalitis mediated by both secretory and transplacental factors in neonatal mice. J Virol 64: 1256-1263.

Cuff CF, Molowitz EH, Rubin DH, Lee F, Cebra JJ (1996). Characterization of cytotoxic intraepithelial leukocytes in severe combined immunodeficient mice and modulation by enteric reovirus infection. J Immunol, submitted.

Dubos R, Schaedler RW, Costello R, Hoet P (1965). Indigenous, normal, and autochthonous flora of the gastrointestinal tract. J Exp Med 122: 67-75.

Eddie DS, Schulkind ML, Robbins JB (1971). The isolation and biologic activities of purified secretory IgA and IgG anti-*Salmonella typhimurium* "O" antibodies from rabbit intestinal fluid and colostrum. J Immunol 106: 181-190.

Francis T, Jr (1940). Inactivation of epidemic influenza virus by nasal secretions of human individuals. Science 91: 198.

Fuhrman JA, Cebra JJ (1981). Special features of the priming process for a secretory IgA response. B cell priming with cholera toxin. J Exp Med 153: 534-544.

Gearhart PJ, Cebra JJ (1979). Differentiated B lymphocytes. Potential to express particular antibody variable and constant regions depends on site of lymphoid tissue and antigen load. J Exp Med 149: 216-227.

George A, Cebra J (1991). Responses of single germinal-center B cells in T-cell-dependent microculture. Proc Natl Acad Sci USA 88: 11-15.

George A, Kost S, Witzleben CL, Cebra JJ, Rubin DH (1990). Reovirus-induced liver disease in severe combined immunodeficient (SCID) mice. A model for the study of viral infection, pathogenesis, and clearance. J Exp Med 171: 929-934.

Giannasca PJ, Giannasca KT, Falk P, Gordon JI, Neutra MR (1994). Regional differences in glycoconjugates of intestinal M cells in mice: potential targets for mucosal vaccines. J Am Physiol Soc G1108-G1121.

Guy-Grand D, Cerf-Bensussan N, Malissen B, Malassis-Seris M, Briottet C, Vassalli P (1991). Two gut intraepithelial CD8+ lymphocyte populations with different T cell receptors: A role for the gut epithelium in T cell differentiation. J Exp Med 173: 471-481.

Hanson LA (1961). Comparative immunological studies of the immune globulins of human milk and of blood serum. Int Arch Allergy Appl Immunol 18: 241-253.

Harriman GR, Kunimoto DY, Elliott JF, Paetkau V, Strober W (1988). The role of IL-5 in IgA B cell differentiation. J Immunol 140: 3033-3039.

Hayakawa K, Hardy RR, Honda M, Herzenberg LA, Steinberg AD, Herzenberg LA (1984). Ly-1 B cells: Functionally distinct lymphocytes that secrete IgM autoantibodies. Proc Natl Acad Sci USA 81: 2494-2498.

Heremans JF, Heremans MT, Schultze HE (1959). Isolation and description of a few properties of β_{2A}-globulin of human serum. Clin Chim Acta 4: 96.

Herzenberg LA, Haughton G, Rajewsky K (1992). CD5 B cells in development and disease. Annals of the New York Academy of Sciences, No. 651. New York: New York Academy of Sciences.

Hitoshi Y, Yamaguchi N, Mita S, Sonoda E, Takaki S, Tominaga A, Takatsu K (1990). Distribution of IL-5 receptor-positive B cells. Expression of IL-5 receptor in Ly-1 (CD5)[+] B cells. J Immunol 144: 4218-4225.

Holzmann B, McIntyre BW, Weissman IL (1989). Identification of a murine Peyer's patch-specific lymphocyte homing receptor as an integrin molecule with an α chain homologous to human VLA-4α. Cell 56: 37-46.

Honjo T, Kataoka T (1978). Organization of immunoglobulin heavy chain genes and allelic deletion model. Proc Natl Acad Sci USA 75: 2140-2144.

Hooper DC, Molowitz EH, Bos NA, Ploplis VA, Cebra JJ (1995). Spleen cells from antigen-minimized mice are superior to spleen cells from germ-free and conventional mice in the stimulation of primary *in vitro* proliferative responses to nominal antigens. Eur J Immunol 25: 212-217.

Husband AJ (1982). Kinetics of extravasation and redistribution of IgA-specific antibody-containing cells in the intestine. J Immunol 128: 1355-1359.

Jones PP, Cebra JJ (1974). Restriction of gene expression in B-lymphocytes and their progeny. III. Endogenous IgA and IgM on the membranes of different plasma cell precursors. J Exp Med 140: 966-976.

Kawaguchi M, Nanno M, Umesaki Y, Matsumoto S, Okada Y, Cai Z, Shimamura T, Matsuoka Y, Ohwaki M, Ishikawa H (1993). Cytolytic activity of intestinal intraepithelial lymphocytes in germ-free mice is strain dependent and determined by T cells expressing $\gamma\delta$ T-cell antigen receptors. Proc Natl Acad Sci USA 90: 8591-8594.

Kim P-H, Kagnoff MF (1990). Transforming growth factor $\beta 1$ increases IgA isotype switching at the clonal level. J Immunol 145: 3773-3778.

Klaasen HLBM, Van der Heijden PJ, Stok W, Poelma FGJ, Koopman JP, Van den Brink ME, Bakker MH, Eling WMC, Beynen AC (1993). Apathogenic, intestinal, segmented, filamentous bacteria stimulate the mucosal immune system of mice. Infect Immun 61: 303-306.

Kraehenbuhl JP, Racine L, Galardy RE (1975). Localization of secretory IgA secretory component, and α chain in the mammary gland of lactating rabbits by immunoelectron microscopy. Ann NY Acad Sci 254: 190-202.

Kramer DR, Cebra JJ (1995a). Role of maternal antibody in the induction of virus specific and bystander IgA responses in Peyer's patches of suckling mice. Int Immunol 7: 911-918.

Kramer DR, Cebra JJ (1995b). Early appearance of "natural" mucosal IgA responses and germinal centers in suckling mice developing in the absence of maternal antibodies. J Immunol 154: 2051-2062.

Kroese FGM, Butcher EC, Stall AM, Lalor PA, Adams S, Herzenberg LA (1989). Many of the IgA producing plasma cells in murine gut are derived from self-replenishing precursors in the peritoneal cavity. Int Immunol 1: 75-84.

Lebman DA, Griffin PM, Cebra JJ (1987). Relationship between expression of IgA by Peyer's patch cells and functional IgA memory cells. J Exp Med 166: 1405-1418.

Lebman DA, Park MJ, Hansen-Bundy S, Pandya A (1994). Mechanism for transforming growth factor β regulation of α mRNA in lipopolysaccharide-stimulated B cells. Int Immunol 6: 113-119.

Lee F (1995). Oral listeriosis: murine models for the study of pathogenesis—including central nervous system disease—and for the development of oral vaccines. Ph.D. Dissertation, Philadelphia: University of Pennsylvania, 1-200.

London SD, Rubin DH, Cebra JJ (1987). Gut mucosal immunization with reovirus serotype 1/L stimulates virus-specific cytotoxic T cell precursors as well as IgA memory cells in Peyer's patches. J Exp Med 165: 830-847.

Mattingly JA, Waksman BH (1978). Immunologic suppression after oral administration of antigen. I. Specific suppressor cells formed in rat Peyer's patches after oral administration of sheep erythrocytes and their systemic migration. J Immunol 121: 1878-1883.

Mazanec MB, Kaetzel CS, Lamm ME, Fletcher D, Nedrud JG (1992). Intracellular neutralization of virus by immunoglobulin A antibodies. Proc Natl Acad Sci USA 89: 6901-6905.

McIntyre TM, Kehry MR, Snapper CM (1995). Novel in vitro model for high-rate IgA class switching. J Immunol 154: 3156-3161.

McWilliams M, Phillips-Quagliata JM, Lamm ME (1975). Characteristics of mesenteric lymph node cells homing to gut-associated lymphoid tissue in syngeneic mice. J Immunol 115: 54-58.

Meitin CA, Bender BS, Small PA Jr (1994). Enteric immunization of mice against influenza with recombinant vaccinia. Proc Natl Acad Sci USA 91: 11187-11191.

Merchant AA, Groene WS, Cheng EH, Shaw RD (1991). Murine intestinal antibody response to heterologous rotavirus infection. J Clin Microbiol 29: 1693-1701.

Mostov KE, Kraehenbuhl JP, Blobel G (1980). Receptor-mediated transcellular transport of immunoglobulin: Synthesis of secretory component as multiple and larger transmembrane forms. Proc Natl Acad Sci USA 77: 7257-7261.

Murakami M, Tsubata T, Shinkura R, Nisitani S, Okamoto M, Yoshioka H, Usui T, Miyawaki S, Honjo T (1994). Oral administration of lipopolysaccharides activates B-1 cells in the peritoneal cavity and lamina propria of the gut and induces autoimmune symptoms in an autoantibody transgenic mouse. J Exp Med 180: 111-121.

Murray PD, McKenzie DT, Swain SL, Kagnoff MF (1987). Interleukin 5 and interleukin 4 produced by Peyer's patch T cells selectively enhance immunoglobulin A expression. J Immunol 139: 2669-2674.

Nakache M, Berg EL, Streeter PR, Butcher EC (1989). The mucosal vascular addressin is a tissue-specific endothelial cell adhesion molecule for circulating lymphocytes. Nature 337: 179-181.

O'Daly JA, Cebra JJ (1971). Chemical and physicochemical studies of the component polypeptide chains of rabbit secretory immunoglobulin A. Biochemistry 10: 3843-3850.

O'Daly JA, Craig SW, Cebra JJ (1971). Localization of b markers, α-chain, and SC of SIgA in epithelial cells lining Lieberkühn crypts. J Immunol 106: 286-288.

O'Garra A, Chang R, Go N, Hastings R, Haughton G, Howard M (1992). Ly-1 B (B-1) cells are the main source of B cell-derived interleukin 10. Eur J Immunol 22: 711-717.

Offit PA, Dudzik KI (1990). Rotavirus-specific cytotoxic T lymphocytes passively protect against gastroenteritis in suckling mice. J Virol 64: 6325-6328.

Offit PA, Khoury CA, Moser CA, Clark HF, Kim JE, Speaker TJ (1994). Enhancement of rotavirus immunogenicity by microencapsulation. Virology 203: 134-143.

Ogra PL, Karzon DT (1969). Distribution of poliovirus antibody in serum, nasopharynx and alimentary tract following segmental immunization of lower alimentary tract with poliovaccine. J Immunol 102: 1423-1430.

Owen RL (1994). M cells—Entryways of opportunity for enteropathogens. J Exp Med 180: 7-9.

Perlmutter AP, Gilbert W (1984). Antibodies of the secretory response can be expressed without switch recombination in normal mouse B cells. Proc Natl Acad Sci USA 81: 7189-7193.

Pierce NF, Gowans JL (1975). Cellular kinetics of the intestinal immune response to cholera toxoid in rats. J Exp Med 142: 1550-1563.

Pollard M, Sharon N (1970). Responses of the Peyer's patches in germ-free mice to antigenic stimulation. Infect Immun 2: 96-100.

Potter M (1971). Antigen binding myeloma proteins in mice. Ann NY Acad Sci 190: 306-321.

Ramsey AJ, Husband AJ, Ramshaw IA, Bao S, Matthaei KI, Koehler G, Kopf M (1994). The role of interleukin-6 in mucosal IgA antibody responses in vivo. Science 264: 561-563.

Reynolds J, Heron I, Dudler L, Trnka Z (1982). T-cell recirculation in the sheep: migratory properties of cells from lymph nodes. Immunology 47: 415-421.

Rose ML, Birbeck MSC, Wallis VJ, Forrester JA, Davies AJS (1980). Peanut lectin binding properties of germinal centres of mouse lymphoid tissue. Nature 284: 364-366.

Santos LMB, Al-Sabbagh A, Londono A, Weiner HL (1994). Oral tolerance to myelin basic protein induces regulatory TGF-β-secreting T cells in Peyer's patches of SJL mice. Cell Immunol 157: 439-447.

Schrader CE, George A, Kerlin RL, Cebra JJ (1990). Dendritic cells support production of IgA and other non-IgM isotypes in clonal microculture. Int Immunol 2: 563-570.

Shahin RD, Cebra JJ (1981). Rise in inulin sensitive B cells in ontogeny can be prematurely stimulated by thymus-dependent and thymus-independent antigens. Infect Immun 32: 211-215.

Shimizu A, Nussenzweig MC, Mizuta T-R, Leder P, Honjo T (1989). Immunoglobulin double-isotype expression by trans-mRNA in a human immunoglobulin transgenic mouse. Proc Natl Acad Sci USA 89: 8020-8023.

Shroff KE, Meslin K, Cebra JJ (1995). Commensal enteric bacteria engender a self-limiting humoral mucosal immune response while permanently colonizing the gut. Infect Immun 63: 3904-3913.

Smith CB, Purcell RH, Bellanti JA, Chanock RM (1966). Protective effect of antibody to parainfluenza type 1 virus. N Engl J Med 275: 1145-1152.

Stavnezer J, Radcliffe G, Lin Y-C, Nietupski J, Berggren L, Sitia R, Severinson E (1988). Immunoglobulin heavy chain switching may be directed by prior induction of transcripts from constant region genes. Proc Natl Acad Sci USA 85: 7704-7708.

Taguchi T, McGhee JR, Coffman RL, Beagley KW, Eldridge JH, Takatsu K, Kiyono H (1990). Analysis of Th1 and Th2 cells in murine gut-associated tissues. Frequencies of CD4+ and CD8+ T cells that secrete IFN-γ and IL-5. J Immunol 145: 68-77.

Tomasi TB, Jr, Tan EM, Solomon A, Prendergast RA (1965). Characteristics of an immune system common to certain external secretions. J Exp Med 121: 101-124.

van der Waaij LA, Mesander G, Limburg PC, van der Waaij D (1994). Direct flow cytometry of anaerobic bacteria in human feces. Cytometry 16: 270-279.

Weinstein PD, Cebra JJ (1991). The preference for switching to IgA expression by Peyer's patch germinal center B cells is likely due to the intrinsic influence of their microenvironment. J Immunol 147: 4126-4135.

Weinstein PD, Schweitzer PA, Cebra-Thomas JA, Cebra JJ (1991). Molecular genetic features reflecting the preference for isotype switching to IgA expression by Peyer's patch germinal center B cells. Int Immunol 3: 1253-1263.

Wolf JL, Dambrauskas R, Sharpe AH, Trier JS (1987). Adherence to and penetration of the intestinal epithelium by reovirus type 1 in neonatal mice. Gastroenterology 92: 82-91.

Young MN, Richard JC (1994). Structural characterization of the phosphocholine-bearing antigen from *M. morganii* and its recognition by monoclonal antibodies. Abstracts of the XVIIth International Carbohydrate Symposium, p. 66.

Prion Disease

One of the major discoveries in microbiology in the second half of the twentieth century is that molecules apparently without nucleic acid operate through an infectious protein to cause chronic degenerative diseases of the central nervous system of humans and animals. Prions are formed as the result of a mutation in a normal cellular gene, and prion diseases result from aberrations in protein conformation. The diseases they cause range from scrapie in sheep, bovine spongiform encephalopathy ("mad cow disease"), to Kuru, fatal familial insomnia, Creutzfeld-Jacob and Gerstmann-Sträussler-Scheinker diseases in humans. Some of these diseases are acquired by infection and some are genetically transmitted. Through the implantation of prion genes from hamsters and humans into transgenic mice, Stanley Prusiner, the discoverer of prions, was able to decrease the incubation period of prion-infected mice from more than 300 to less than 100 days, making them an accessible tool for further study of prions.

30

The Prion Saga

Stanley B. Prusiner

Prions, once dismissed as an impossibility, have now gained wide recognition as extraordinary agents that cause a number of infectious, genetic and sporadic disorders. The prion concept, when presented in 1982, evoked a good deal of skepticism when I proposed that the infectious agents causing certain degenerative disorders of the central nervous system (CNS) in animals and, more rarely, in humans might consist of protein and nothing else (Prusiner 1982). At that time, the notion of prions was also thought to be quite heretical. Dogma held that the conveyers of transmissible diseases required genetic material, composed of nucleic acid (DNA or RNA), in order to establish an infection in a host. Even viruses, among the simplest microbes, rely on such material to direct synthesis of the proteins needed for survival and replication.

Later, many scientists were similarly dubious when my colleagues and I suggested that these "proteinaceous infectious particles"—or "prions" as I called the disease-causing agents—could also underlie inherited diseases. Such dual behavior was then unknown to medical science. There was even further resistance to the conclusion that prions (pronounced "preeons") multiply in an incredible way; they convert normal protein molecules into dangerous ones simply by inducing the benign molecules to change shape.

Today, however, a wealth of experimental and clinical data has made a convincing case that we are correct on all three counts. Prions are indeed responsible for transmissible and inherited disorders of protein conformation (Hsiao et al 1989a; Pan et al 1993). They can also cause sporadic disease, in which neither transmission between individuals nor inheritance is evident. It is reasonable to speculate that additional prion diseases will be found in which different proteins are the culprits (Prusiner 1984b). Some of the neurodegenerative diseases that are quite prevalent in humans may eventually be found to be caused by prions. They might even participate in illnesses that attack muscles (Westaway et al 1994b).

TABLE 1. ANIMAL PRION DISEASES

Disease	Host
Scrapie	sheep and goats
Bovine spongiform encephalopathy (BSE)	cattle
Transmissible mink encephalopathy (TME)	mink
Chronic wasting disease (CWD)	mule deer and elk
Feline spongiform encephalopathy (FSE)	domestic and great cats
Exotic ungulate encephalopathy (EUE)	nyala and greater kudu

THE PRION DISEASES

All of the known prion diseases are fatal, degenerative disorders of the CNS. These diseases are sometimes referred to as the transmissible spongiform encephalopathies (Gajdusek 1977). They are so named because they frequently cause the brain to become riddled with vacuoles. These ills, which can brew for years (or even for decades in humans), are widespread in animals.

The most common form is scrapie, found in sheep and goats (Table 1). The disorder has been known for more than 200 years, and by 1938, French workers had shown convincingly that it can be transmitted from one sheep to another by inoculation (Cuillé and Chelle 1939). The demonstration of transmissibility implied that there is an infectious agent capable of reproducing itself in the host animal. Afflicted animals lose coordination and eventually become so incapacitated that they cannot stand. They also become irritable and, in some cases, develop an intense itch that leads them to scrape off their wool or hair (hence the name "scrapie"). The other prion diseases of animals include transmissible mink encephalopathy, chronic wasting disease of mule deer and elk, feline spongiform encephalopathy and bovine spongiform encephalopathy. The last, often called "mad cow disease," is the most worrisome.

Gerald Wells identified the condition in 1986, after it began striking cows in Great Britain, causing them to become uncoordinated and unusually apprehensive (Wells and Wilesmith 1995). The source of the emerging epidemic was soon traced by John Wilesmith to a food supplement that included meat and bone meal from dead sheep. The methods for processing sheep carcasses had been changed in the late 1970s. Where once they would have eliminated the scrapie agent in the supplement, now they apparently did not. The British government banned the use of animal-derived feed supplements in 1988, and the epidemic has probably peaked (Figure 1). Nevertheless, many people continue to worry that they will eventually fall ill as a result of having consumed tainted meat.

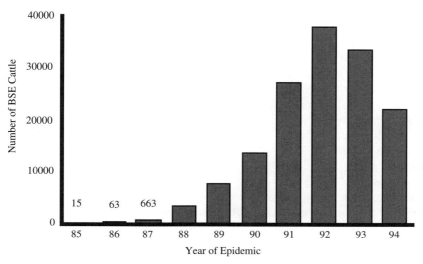

FIGURE 1. ANNUAL INCIDENCE of bovine spongiform encephalopathy in Great Britain. All cases were confirmed clinically and neuropathologically. Statistics compiled by John Wilesmith of the Central Veterinary Laboratory at Weybridge, England.

The human prion diseases are more obscure (Table 2). Kuru has been seen only among the Fore highlanders of Papua New Guinea. It has often been called the "laughing death." Vincent Zigas and D. Carleton Gajdusek described it in 1957, noting that many highlanders became afflicted with a strange, fatal disease marked by loss of coordination (ataxia) and often later by dementia (Gajdusek and Zigas 1957). The affected individuals probably acquired kuru through ritualistic cannibalism: the Fore tribe reportedly honored the dead by eating their brains (Gajdusek 1977). The practice has since stopped, and kuru has virtually disappeared.

TABLE 2. HUMAN PRION DISEASES

Disease	Etiology
Kuru	infection
Creutzfeldt-Jakob disease	
Iatrogenic	infection
Sporadic	unknown
Familial	PrP mutation
Gerstmann-Sträussler-Scheinker disease	PrP mutation
Fatal familial insomnia	PrP mutation
Progressive subcortical gliosis	unknown

William Hadlow suggested that scrapie and kuru might be related (Hadlow 1959) and that kuru might be transmissible to apes after intracerebral inoculation. In 1966, Gajdusek, Clarence Gibbs and Michael Alpers reported the transmission of kuru to apes (Gajdusek et al 1966).

Creutzfeldt-Jakob disease, in contrast, occurs worldwide and usually becomes evident as dementia. Most of the time it appears sporadically, striking one person in a million, typically around age 60. Two years after the experimental transmission of kuru, Gajdusek and Gibbs showed that Creutzfeldt-Jakob disease also could be transmitted to apes (Gibbs et al 1968). Although Creutzfeldt-Jakob disease is a relatively rare human dementia, it came to public notice when it was identified as the cause of death of the choreographer George Balanchine. About 10 to 15% of cases are inherited, and a small number are, sadly, iatrogenic—spread inadvertently by the attempt to treat some other medical problem. Iatrogenic Creutzfeldt-Jakob disease has apparently been transmitted by corneal transplantation, implantation of dura mater or electrodes in the brain, use of contaminated surgical instruments, and injection of growth hormone derived from human pituitaries (before recombinant growth hormone became available) (Brown et al 1994; Fradkin et al 1991).

The two remaining human disorders are Gerstmann-Sträussler-Scheinker disease, which is manifest as ataxia and other signs of damage to the cerebellum, and fatal familial insomnia, in which dementia follows difficulty sleeping. Both these conditions are usually inherited and typically appear in mid-life. Fatal familial insomnia was discovered only recently by Elio Lugaresi, Rossella Medori, and Pierluigi Gambetti (Goldfarb et al 1992; Medori et al 1992).

SLOW VIRUSES

More than two decades after Bjorn Sigurdsson introduced the concept of slow viruses, scrapie continued to be classified as a slow virus illness (Sigurdsson 1954). The related human conditions kuru and Creutzfeldt-Jakob disease were also thought to be caused by a slow-acting virus. Yet, no one had managed to isolate the culprit virus.

In 1967, Tikvah Alper and colleagues suggested that the scrapie agent might lack nucleic acid, which usually can be degraded by ultraviolet or ionizing radiation (Alper et al 1966; Alper et al 1967). When the nucleic acid in extracts of scrapie-infected brains was presumably destroyed by those treatments, the extracts retained their ability to transmit scrapie. In general, ionizing radiation destroys cells and viruses through damage to nucleic acids; the probability of damage is roughly proportional to the size of the target molecules. These investigators had found that extremely high doses of radiation were needed to inactivate the scrapie agent. They concluded that the agent has no nucleic acid and is considerably smaller than a virus. Almost 20 years later, James Cleaver, Ellis Kempner, and I repeated the ultraviolet and ionizing radiation studies using purified preparations and obtained similar results (Bellinger-Kawahara et al 1987a; Bellinger-Kawahara et al 1988). If the organism did lack DNA and RNA, the finding would mean that it was not a virus or any other known type of infectious agent, all of which contain genetic material. What then was it? Investigators had many ideas—including, jokingly, linoleum and kryptonite—but no hard answers. By 1975, more than a dozen hypotheses had been proposed for the nature of the scrapie agent; indeed, there were more hypotheses than there were experimental groups working on the disease (Pattison 1988).

IN SEARCH OF A SCRAPIE-SPECIFIC NUCLEIC ACID

Having established the role of a protein in the infective process as described below, we began to search for nucleic acid. The effort encountered consistent frustration. In many experiments, we attempted to inactivate prions by means of treatments that chemically attack nucleic acids. We exposed fractions enriched for prions to several nucleases, including enzymes that destroy both DNA and RNA, without detecting any significant diminution of prion infectivity.

One possible objection to such experiments is that the enzyme may not be able to gain access to the nucleic acids; many viruses are resistant to nucleases because the protein coat protects the DNA or RNA. In collaboration with John Hearst, we tried treating prions with molecules called psoralens, which can pass through the protein coat of most viruses; on exposure to ultraviolet radiation, the psoralens bind to the nucleic acid and inactivate it (McKinley et al 1983b). Again we observed no loss of prion infectivity. Zinc ions, which catalyze the breakdown of RNA, were likewise ineffective (Bellinger-Kawahara et al 1987b).

Our work with diethylpyrocarbonate (DEP) added further evidence. DEP can inactivate nucleic acids as well as proteins, but only proteins can be restored to functionality by hydroxylamine. Hence the recovery of infectivity we observed after hydroxylamine treatment argues that DEP is not acting on a nucleic acid (McKinley et al 1981). Hydroxylamine not only reverses the modification of proteins by DEP, but also modifies nucleic acids resulting in the inactivation of viruses.

In collaboration with Theodore O. Diener, we compared the effects of various reagents on prions and on viroids. Because both entities seem to be much smaller than viruses, it was once thought that prions might be similar to viroids in structure, that is, they might consist of "naked" RNA. Actually the two kinds of infectious agent are antithetical. Procedures that modify proteins inactivate prions but have no effect on viroids; treatments that attack nucleic acids destroy viroids but not prions (Diener et al 1982).

The question of the prion's size also bears on the question of a scrapie-specific nucleic acid. Individual prions seem to be very small. Therefore the amount of nucleic acid that a prion could encapsulate is probably quite limited. The target size studies done by Alper and colleagues suggested that the infectious scrapie particle might have a molecular mass between 60,000 and 150,000 daltons (Alper et al 1966). The remarkable heterogeneity of prions has made it difficult to determine the size of the smallest infectious particle by more direct methods. After attempting to break up aggregates of prions with detergents and other chemicals, we investigated the size of the particles by sucrose-gradient centrifugation, by timing their passage through a chromatographic column and by passing them through a membrane filter with pores of a known size (Bellinger-Kawahara et al 1988; Gabizon and Prusiner 1990; Prusiner 1982). All the studies have given results consistent with a molecular mass between 50,000 and 100,000 daltons, but each method has potential pitfalls. Because of the many sources of uncertainty, the most that can be said at present is that the smallest infectious form of the prion may be 100 times smaller than the smallest viruses.

If the prion is 50,000 daltons, its diameter would be about 5 nanometers. If it is constructed like a conventional virus, it might take the form of an approximately spherical shell of protein surrounding a core of nucleic acid. The shell could not be less than about a nanometer thick, which would leave room in the core for no more than about 12 nucleotides. Limits on the size of any prion nucleic acid can also be derived from other measurements. The prion's resistance to inactivation by ultraviolet radiation is consistent with a nucleic acid made up of 12 to 50 nucleotides; our experiments with psoralens would not have detected a nucleic acid with fewer than 40 nucleotides (Bellinger-Kawahara et al 1987a; Bellinger-Kawahara et al 1987b).

The failure to detect a nucleic acid in prions could not be taken as proof that it does not exist. It might have still been concealed in some way by a surrounding structure or could be present in quantities too small to be detected. Nevertheless, it seems reasonable to suggest that if the prion has any nucleic acid at all, it is likely to be less than 50 nucleotides long. In the standard genetic code, three nucleotides are needed to specify each amino acid, and so the putative prion "genome" could not encode a protein with more than about a dozen amino acids. This conclusion is strengthened by the fact that assiduous searching for a scrapie-specific nucleic acid, especially by Detlev Riesner and colleagues, produced no evidence that such a polynucleotide is attached to prions (Kellings et al 1992).

APPROACHING THE ENIGMATIC SCRAPIE PATHOGEN

Turning back the clock to 1974 when I focused my attention on trying to solve the mystery of the scrapie pathogen, the first step had to be a mechanical one—purifying the infectious material in scrapie-infected brains so that its composition could be analyzed. The task was daunting; many investigators had tried and failed in the past. By 1982, we had produced extracts of Syrian hamster brains consisting almost exclusively of infectious prions (Prusiner et al 1982). We had also subjected the extracts to a range of tests designed to reveal the composition of the disease-causing component.

In trying to isolate an infectious agent whose structure and chemistry are unknown, it is necessary to take an empirical approach. Typically, a tissue sample from an infected animal is homogenized and then separated into fractions that differ in some physical or chemical property. The concentration of agent in each fraction is then assayed, and the purest fraction is singled out for further attention. In the case of scrapie, the only way to measure the concentration of the agent has been by detecting its ability to induce the disease in animals.

For many years, all such measurements had to be done by the method of end-point titration. Animals were inoculated with progressively more dilute specimens of material; the most dilute specimen capable of inducing the disease gave a measure of the concentration of the agent in the original material. In the early work with sheep and goats, an entire herd of animals and several years of observation were needed to evaluate a single sample.

In 1960, Richard Chandler succeeded in transmitting scrapie to mice (Chandler 1961). End-point titrations in mice typically involved 10 dilutions, with each dilution being 10 times weaker than the one before. Six mice would be inoculated at each dilution; those receiving a large dose would become ill in four to five months, but in the animals given the most dilute solution capable of causing disease, almost a year would pass before symptoms appeared. Hence 60 mice would have to be kept for a year before the end point could be determined. Although end-point titration in mice was an improvement over work in sheep and goats, this method of measuring the scrapie agent was still slower and more cumbersome than the methods used by Pasteur in his studies of viruses almost a century earlier.

In 1978, we found an alternative to end-point titration. Three years earlier, Richard Marsh and Richard H. Kimberlin had described a form of scrapie in hamsters whose onset is about twice as fast as it is in mice (Marsh and Kimberlin 1975). Studying the hamster disease, we found strong correlations between the concentration of scrapie agent and the rapidity of disease onset and between concentration and time of death (Prusiner et al 1980). Thus, instead of determining how much a sample could be diluted and still cause disease, we measured how fast a sample with a known dilution brought on disease symptoms and caused death.

The assay based on incubation times has been found to give an accurate measure of concentration for samples with a high titer of the agent. The gains in speed and economy have had a profound effect; we estimate that they have accelerated our work by a factor of nearly 100. Instead of observing 60 animals for a year, we can assay a sample with just four animals in ~70 days.

PURIFICATION OF SCRAPIE INFECTIVITY

My work on scrapie began as a collaboration with William Hadlow and Carl Eklund. Our initial efforts to purify and isolate the scrapie agent utilized homogenates prepared from the mouse spleen, which we analyzed by ultra-centrifugation and end-point titration bioassays in mice. The bioassays were carried out to determine how much of the infectious agent had sedimented and how much had remained in the supernatant fluid. The procedure was done for a wide range of speeds and times in a series of experiments that took almost two years. When we had finished, we repeated the study to make certain our results were reproducible (Prusiner et al 1977; Prusiner et al 1978a).

In early studies, the greatest purification attained was about a 30-fold enrichment of scrapie agent. One of the factors limiting the degree of purification was also one of our major findings: the infectious particles were shown to be extremely heterogeneous in size and density (Prusiner et al 1978b). Judging from the rate of sedimentation in the centrifuge, some of the particles were almost as large as mitochondria or bacteria, whereas others seemed to be substantially smaller than the smallest viruses. The broad range of sizes implied that the scrapie agent can exist in many molecular forms and raised the possibility that they form aggregates. The use of detergents to obtain more homogeneous preparations had little effect on the observed heterogeneity, but they did aid in purification.

While the story of our endeavor can be told briefly, it represents a decade of painstaking and sometimes frustrating labor. An important turning point in our studies came when we switched from mice to hamsters. In addition to providing a faster assay, the hamster brain has a titer of scrapie agent 100 times greater than that of the mouse spleen. Our purification method again began with detergent extraction and centrifugation, but we added three more steps: exposure to nucleases, exposure to proteases and analysis by gel electrophoresis (Prusiner et al 1980). We had found that the infectivity of the scrapie agent was unaltered by digestion with nucleases, which are enzymes that catalyze the breakdown of nucleic acids. Hence, treating the preparation with a nuclease eliminated most cellular nucleic acids while leaving the scrapie agent intact. Proteases, which cut the chain of amino acids in a protein, were used in a similar way to remove extraneous proteins. The final step in our revised procedure, gel electrophoresis, separates molecules according to the rate at which they migrate through a gel under the influence of an applied electric field. The rate at which a molecule moves is determined primarily by its electric charge, although size and shape also have an effect.

The purification method culminating in electrophoresis gave an overall purification by a factor of about 100, which was enough to establish several vital facts. The most important result was a convincing demonstration that the biological activity of the scrapie agent depends on a protein (Prusiner et al 1981).

Subsequently, we returned to centrifugation as the primary method of purification, but the technique is somewhat different from the one employed in the early studies. The partially purified sample was layered on top of a sucrose gradient and centrifuged. The components of the sample migrated through the gradient until they reached a level where their own density matched that of the sucrose solution.

Our initial version of this sucrose-gradient separation gave ~1,000-fold purification of the scrapie agent (Prusiner et al 1982). We were thereby able to demonstrate that the bulk of the protein consists of a single molecule, which we designated PrP, for prion protein. A larger-scale version of the purification procedure was then developed, based on a centrifuge with a zonal rotor in which substantial quantities of sample could be separated in a sucrose gradient. This purification scheme yielded an enrichment of ~5,000 times (Prusiner et al 1983).

With the greater purification, it was possible to show that the most infectious fractions of the gradient contain essentially one protein that migrated during electrophoresis with an molecular mass of 27-30 kDa (Bolton et al 1982; McKinley et al 1983a). We called this protein PrP 27-30. Subsequently, we learned that PrP 27-30 is derived from a larger protein, designated PrP^{Sc}, by limited proteolysis during purification.

CHARACTERISTICS OF PURIFIED PRION RODS

Electron micrographs of the purified fractions revealed numerous rod-shaped particles much too large to be the individual prions (Figure 2). The rods measured 10-20 nm in diameter and 100-200 nm long, which suggests that a single rod may consist of as many as 1,000 PrP molecules stacked in a paracrystalline array. The rods had been observed in earlier experiments, but we had been unable to determine whether they were aggregates of prions. The possibility remained that the rods might not be the disease agent itself but rather a product of some pathological change brought on by the disease. With our highly purified fractions, we were able to establish that the rods were composed of PrP 27-30 and hence could be considered prion aggregates (Prusiner et al 1983).

The isolation of fractions that were highly enriched for prion infectivity and in substantial quantity proved to be the key to answering many puzzling questions about the prion. Such fractions enabled us to raise antibodies to prions in experimental animals (Bendheim et al 1984), an achievement that had eluded us for several years. The crucial factor turned out to be the quantity of material injected into the animal, which we were able to increase by a factor of 10 (to roughly 100 µg). The availability of α-PrP antibodies greatly changed the pace of and approaches to prion research.

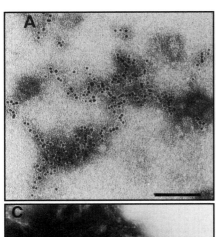
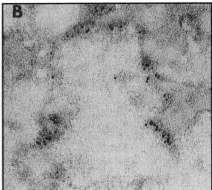
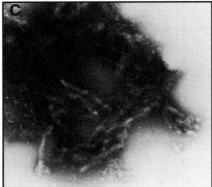

FIGURE 2. ELECTRON MICROGRAPHS of negatively stained and immunogold-labeled prion proteins. (A) PrP^C and (B) PrP^{Sc}. (C) Prion rods composed of PrP 27-30 were negatively stained with uranyl acetate. Bar = 100 nm. Reproduced from Pan et al (1993).

PrP^{Sc} IS A COMPONENT OF INFECTIOUS PRIONS

Our conclusion that a protein is a component of the prion and is necessary for infectivity emerged from our work on purification, which showed that substances known to disrupt proteins diminish prion infectivity, whereas substances that have no effect on proteins leave the infectivity unchanged (Prusiner 1982; Prusiner et al 1981).

The clearest evidence came from experiments with proteases because of their high degree of specificity for proteins. In early studies with crude fractions, protease digestion gave equivocal results, but when enriched fractions became available, we were able to demonstrate convincingly that proteases reduce prion infectivity (Prusiner et al 1981). The use of limited proteolysis to purify prion infectivity and PrP^{Sc} created some confusion. However, subsequent experiments clarified the situation by showing that extended protease digestion destroyed both PrP 27-30 and prion infectivity (McKinley et al 1983a).

Several protein denaturants such as SDS, phenol, urea and certain chaotropic salts reduced prion infectivity irreversibly. Another reagent, diethyl pyrocarbonate (DEP), as noted above, modifies proteins chemically in a way that can be reversed by exposure to hydroxylamine. As in the case of proteases, DEP gave inconsistent results in early experiments with impure mixtures but clearly diminished the titer of infectious agent in purified fractions (McKinley et al 1981). The infectivity could be restored by treatment with hydroxylamine.

Although the presumed target of all these treatments was PrP^{Sc}, the possibility remained that PrP was not a structural component of the prion but a pathological product of the scrapie infection. We therefore undertook a lengthy series of experiments which established that the prion is composed at least in part of PrP molecules.

The first line of evidence is simply that PrP has been found in every sample with a high titer of prions, whether it was prepared by centrifugation or by electrophoresis and even if it was purified from scrapie-infected tissue before the appearance of pathological changes. The concentration of PrP is directly proportional to the titer of prions (McKinley et al 1983a). Furthermore, in purified fractions, PrP 27-30 was the only macromolecule that could be detected (Gabizon et al 1988).

PrP Amyloid

When we noted an extraordinary resemblance of prion rods to purified amyloid from systemic sources (Cohen et al 1982), we stained purified samples with Congo red dye and viewed the specimens by polarizing microscopy (Prusiner et al 1983). Unexpectedly, we found green-gold birefringence which is characteristic of amyloid polymers that are composed of proteins with a high β-sheet content (Glenner et al 1972). These findings suggested that the amyloid plaques seen in some cases of scrapie and Creutzfeldt-Jakob disease are deposits of prions in an aggregated state. With α-PrP antibodies, we were able to establish that such amyloid plaques in these diseases contain PrP (DeArmond et al 1985; Kitamoto et al 1986; Roberts et al 1988; Tagliavini et al 1991).

For more than 60 years, amyloid plaques in the central nervous system were considered accumulations of waste material formed as a result of some disease process. Our findings argued for a quite different interpretation, namely that the plaques were aggregations of prions in a paracrystalline state. The PrP amyloid plaques are analogous to the inclusion bodies that are characteristic of many viral infections; inclusion bodies are crystalline arrays of virus particles.

After we showed that rods were indistinguishable from many purified amyloids (Prusiner et al 1983), some investigators argued that the rods were the same as scrapie-associated fibrils (SAF) (Diringer et al 1983). However, earlier and subsequent papers reported that SAF could be distinguished from amyloids ultrastructurally (Merz et al 1981; Merz et al 1983; Merz et al 1984). SAFs were said to be composed of two or four subfilaments that are helically wound around each other with distinct periodicities. Some investigators thought that SAFs represented the first filamentous animal virus to be discovered (Merz et al 1984).

Despite much evidence to the contrary, some investigators still argue that PrP amyloid deposition participates in the formation of PrP^{Sc} (Gajdusek 1987; Gajdusek 1994; Jarrett and Lansbury 1993; Kocisko et al 1994; Kocisko et al 1995). Studies with transgenic (Tg) mice produced by microinjecting the Syrian hamster (SHa) PrP gene into fertilized oocytes provided an important test of the hypothesis that amyloid deposition features in the formation of nascent PrP^{Sc}. When Tg(SHaPrP) mice were inoculated with SHa prions, amyloid plaques containing SHaPrP were found (Table 3) (Prusiner et al 1990), whereas no amyloid plaques were found in four different lines of these Tg(SHaPrP) mice inoculated with mouse (Mo) prions; thus, amyloid formation is not required for prion propagation. Similar results were obtained for a genetic form of experimental prion disease in Tg mice expressing the Mo PrP gene in which a mutation causing Gerstmann-Sträussler-Scheinker syndrome (GSS) in humans was inserted. These mice, designated Tg(MoPrP-P101L), express the MoPrP that carries the human point mutation (proline→leucine) of GSS at position 102. When the Tg(MoPrP-P101L) mice were crossed onto the null (Prnp$^{0/0}$) background, which was created by disruption of the MoPrP gene through gene targeting using homologous recombination (Büeler et al 1992), numerous PrP amyloid plaques were found in their brains. If the undisrupted wild-type MoPrP gene was present, then relatively few plaques were found (Telling et al 1996). How wild-type MoPrP interacts with mutant MoPrP to prevent amyloid plaque formation is unknown.

Ultrastructural studies demonstrated that purified PrP^{Sc} molecules exist as amorphous aggregates (Figure 2). When these aggregates were partially digested with proteinase K in the presence of detergent, PrP 27-30 was formed and polymerized into rod-shaped particles with the properties of amyloid (McKinley et al 1991a; Pan et al 1993). These findings demonstrate that formation of macroscopic ordered arrays of PrP as amyloid is not required for PrP^{Sc}; once formed, PrP^{Sc} can be converted by limited proteolysis to PrP 27-30 which then assembles into polymers in the presence of detergents. These rod-shaped polymers composed of PrP 27-30 exhibit all the features of amyloid.

While PrP amyloid is diagnostic when present, it is not an obligatory feature of the prion diseases (Table 3). Furthermore, PrP amyloid is neither necessary for PrP^{Sc} formation nor does it participate in the formation of nascent PrP^{Sc}.

TABLE 3. SPECIES-SPECIFIC PRION INOCULA determine the distribution of spongiform change and deposition of PrP amyloid plaques in transgenic mice

Animal	Syrian hamster prions					Mouse prions			
	n^c	Spongiform change[a] Grey	Spongiform change[a] White	PrP plaques[b] Frequency	PrP plaques[b] Diameter[d]	n^c	Spongiform change[a] Grey	Spongiform change[a] White	PrP plaques[b] Frequency
Non-Tg	6	N.D.[e]		N.D.		10	+	+	-
Tg 69	5	+[f]	-	numerous	6.5 ± 3.1 (389)	2	+	+	-
Tg 71	7	+	-	numerous	8.1 ± 3.6 (345)	2	+	+	-
Tg 81	3	+	-	numerous	8.3 ± 3.0 (439)	3	+	+	Few
Tg 7	3	+[g]	-	numerous	14.0 ± 8.3 (19)	4	+	+	-
SHa	3	+	-	numerous	5.7 ± 2.7 (247)		N.D.		N.D.

[a] Spongiform change evaluated in hippocampus, thalamus, cerebral cortex and brainstem for grey matter and the deep cerebellum for white matter.
[b] Plaques in the subcallosal region were stained with SHaPrP mAb 13A5, anti-PrP rabbit antisera R073 and trichrome stain.
[c] n = number of brains examined.
[d] Mean diameter of PrP plaques given in microns ± standard error with the number of observations in parentheses.
[e] N.D. = not determined
[f] + = present; – = not found.
[g] Focal: confined to the dorsal nucleus of the raphe.

Source: Prusiner (1992)

DETERMINING THE GENETIC ORIGIN OF PrP

By the early 1980s, all our results pointed toward one startling conclusion: the infectious pathogen in scrapie and related diseases lacked a nucleic acid and consisted mainly, if not exclusively, of protein. We had deduced that DNA and RNA were absent because, like Tikvah Alper, we saw that procedures known to damage nucleic acid did not reduce infectivity. And we knew protein was an essential component because procedures that denature or degrade protein reduced infectivity.

Once the PrP was discovered, the major question became where might the instructions specifying the sequence of amino acids in PrP reside? Were they carried by an undetected piece of DNA that traveled with PrP, or were they, perhaps, contained in a gene housed in the chromosomes of cells? The key to this riddle was the determination of the sequence of 15 amino acids at the N-terminus of PrP 27-30 in a collaborative study with Leroy Hood and Stephen Kent (Prusiner et al 1984).

The partial sequence of PrP was determined using the Edman degradation method in which a series of reagents that cleave off the terminal amino acid of a protein are repeatedly employed, so that the amino acids in the polypeptide chain are liberated one by one. In about half of the cleavage cycles, we detected multiple amino acids, with one strong signal being accompanied by weaker ones. Initially we thought the minor signals indicated the existence of a spectrum of PrP molecules with slightly different amino acid sequences. We subsequently found, however, that the variant sequences differ from the major sequence only in their starting point. One minor sequence could be brought into alignment with the major sequence by moving the minor sequence forward four amino acid positions; in another case, the minor sequence must be moved backward two residues. These variations in sequence were observed because the PrP molecules have "ragged ends," presumably caused by limited digestion with proteases used for purification.

Learning this small portion of the amino acid sequence of PrP opened many new avenues of research. Immediately, a synthetic peptide corresponding to the known PrP sequence was constructed and used as an antigen to produce monospecific antisera (Barry et al 1986). At the same time, a mixture of iso-coding DNAs was synthesized which corresponded to the known portion of the amino acid sequence of PrP. This mixture was then used to probe a library of cDNAs constructed from mRNA isolated from the brain of a Syrian hamster.

MOLECULAR CLONING OF PrP cDNAs

Mixtures of synthetic iso-coding DNAs allowed us and others to identify cognate cDNA clones encoding PrP. The synthetic DNAs were produced by Leroy Hood and colleagues and given to Charles Weissmann who, with the help of Bruno Oesch, constructed a cDNA library from the mRNA of the brain of a Syrian hamster (Oesch et al 1985). At about the same time, Bruce

Chesebro made his own DNA probes based on the published N-terminal sequence of Syrian hamster PrP 27-30 (Prusiner et al 1984). Chesebro and colleagues identified a mouse cDNA that encoded PrP (Chesebro et al 1985). The molecular cloning of PrP cDNAs made it possible to isolate the entire PrP gene from the chromosomes of hamsters, mice, humans, and all other mammals examined and to establish that the PrP gene does not reside in prions.

Since these animals synthesize PrP without getting sick, we questioned how this could possibly be. One interpretation of such findings was that we had made a terrible mistake: PrP had nothing to do with prion diseases. Another possibility was that PrP could be produced in two forms, one that generated disease and one that did not. We soon showed the latter interpretation to be correct. The critical clue was the fact that the PrP found in infected brains resisted digestion by proteases, in striking contrast to most cell proteins, which are readily degraded by proteolytic digestion. I therefore suspected that if a normal, benign form of PrP existed, it too would be susceptible to degradation by proteases and suggested to Ronald Barry that he search for a protease-sensitive form of PrP. Once he found this form of PrP, designated PrP^C, it became clear that the scrapie form of PrP, called PrP^{Sc}, is a variant of PrP^C (Figure 3) (Barry and Prusiner 1986; Meyer et al 1986; Oesch et al 1985).

FIGURE 3. STRUCTURE AND ORGANIZATION of the chromosomal PrP gene. In all mammals examined, the entire open reading frame (ORF) is contained within a single exon. The 5' untranslated region of the PrP mRNA is derived from either one or two additional exons (Basler et al 1986; Puckett et al 1991; Westaway et al 1994a; Westaway et al 1991). Only one PrP mRNA has been detected. PrP^{Sc} is thought to be derived from PrP^C by a post-translational process (Basler et al 1986; Borchelt et al 1990; Borchelt et al 1992; Caughey and Raymond 1991; Taraboulos et al 1992). The amino acid sequence of PrP^{Sc} is identical to that predicted from the translated sequence of the DNA encoding the PrP gene (Basler et al 1986; Stahl et al 1993), and no unique post-translational chemical modifications have been identified that might distinguish PrP^{Sc} from PrP^C. Thus, it seems likely that PrP^C undergoes a conformational change as it is converted to PrP^{Sc}.

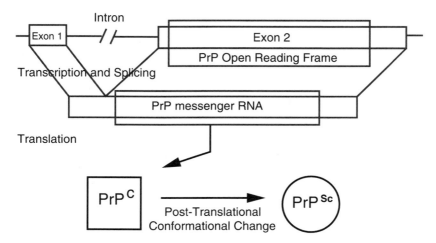

PRION DISEASES CAN BE INHERITED

As early as 1924, physicians began to identify families in whom about half of the members developed Creutzfeldt-Jakob disease (CJD), but these observations provided no clues as to etiology or mechanism of disease pathogenesis (Kirschbaum 1924). Subsequent studies showed that only about 10% of CJD patients had a familial form of the disease; the remaining patients were considered to have a sporadic form, meaning that they were unrelated to other CJD patients (Masters et al 1978).

With the transmission of CJD and later Gerstmann-Sträussler-Scheinker (GSS) disease to apes, the issue of familial forms of these diseases was revisited and the cases well documented (Masters et al 1981). Why CJD and GSS should be both transmissible and inherited posed an interesting conundrum, the significance of which seems to have been unappreciated. Many investigators viewed CJD as being caused by a slow, unusual virus that was vertically transmitted among family members either because of proximity or because they had inherited a particular susceptibility.

In 1988, Karen Hsiao and I took another approach to the familial prion diseases. With blood from a man dying of GSS, we isolated the DNA and produced molecular clones encoding the PrP gene. Then we compared the sequence of his PrP gene with PrP genes obtained from a healthy population and found a point mutation at codon 102 (Hsiao et al 1989a).

With the help of Tim Crow and Jurg Ott as well as their colleagues, we discovered the same mutation in genes from a large number of patients with GSS disease, and we showed that the high incidence in the affected families was statistically significant (Hsiao et al 1989a). In other words, we established a genetic linkage between the mutation and the disease—suggesting that the mutation is the cause. Over the past few years, work by many investigators has uncovered 19 mutations in families with inherited prion diseases; for five of these mutations, enough cases have now been collected to demonstrate significant genetic linkage (Figure 4) (Dlouhy et al 1992; Gabizon et al 1993; Hsiao et al 1989a; Petersen et al 1992; Poulter et al 1992).

The discovery of mutations in the PrP gene gave us a new strategy by which to identify an essential nucleic acid, if it exists, that is bound to PrP^{Sc} and directs the multiplication of infectious prions. To test this hypothesis, we generated transgenic mice carrying a mutated PrP gene. If the presence of the altered gene in these animals led by itself to scrapie, and if the brain extracts prepared from these mice then caused scrapie in inoculated healthy recipients, we would have evidence that the mutant PrP is solely responsible for the transfer of disease (Hsiao et al 1990). Extracts prepared from the brains of the transgenic mice that developed CNS degeneration spontaneously were found to transmit disease to inoculated recipients (Hsiao et al 1994). In fact, the prions that originated in the transgenic mice expressing mutant PrP genes could be serially passaged through inoculated, recipient mice. From these

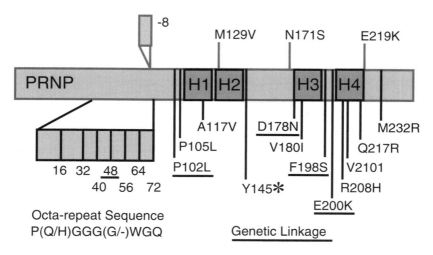

FIGURE 4. HUMAN PRION PROTEIN GENE (PRNP). The open reading frame (ORF) is denoted by the large gray rectangle. Human PRNP wild-type coding polymorphisms are shown above the rectangle, while mutations that segregate with the inherited prion diseases are depicted below. The wild-type human PrP gene contains five octa-repeats [P(Q/H)GGG(G/-)WGQ] from codons 51 to 91 (Kretzschmar et al 1986). Deletion of a single octa-repeat at codon 81 or 82 is not associated with prion disease (Laplanche et al 1990; Puckett et al 1991; Vnencak-Jones and Phillips 1992); whether this deletion alters the phenotypic characteristics of a prion disease is unknown. There are common polymorphisms at codons 117 (Ala→Ala) and 129 (Met→Val); homozygosity for Met or Val at codon 129 appears to increase susceptibility to sporadic CJD (Palmer et al 1991). Additional polymorphisms have been found at codons 171 (Asn→Ser) and 219 (Glu→Lys) (Fink et al 1994; Kitamoto and Tateishi 1994). Octa-repeat inserts of 16, 32, 40, 48, 56, 64, and 72 amino acids at codons 67, 75 or 83 are designated by the small rectangle below the ORF. These inserts segregate with familial Creutzfeldt-Jakob disease (CJD), and genetic linkage has been demonstrated where sufficient specimens from family members are available (Collinge et al 1989, 1990; Crow et al 1990; Goldfarb et al 1990c, 1991a; Owen et al 1989, 1990; Palmer et al 1993). Point mutations are designated by the wild-type amino acid preceding the codon number and the mutant residue follows, e.g., P102L. These point mutations segregate with the inherited prion diseases, and significant genetic linkage (underlined mutations) has been demonstrated where sufficient specimens from family members are available. Mutations at codons 102 (Pro→Leu), 117 (Ala→Val), 198 (Phe→Ser) and 217 (Gln→Arg) are found in patients with GSS (Doh-ura et al 1989; Goldfarb et al 1990a, 1990c, 1990d; Goldgaber et al 1989; Hsiao et al 1989a, 1989b, 1991b; Hsiao and Prusiner 1990; Tateishi et al 1990). Point mutations at codons 178 (Asp→Asn), 200 (Glu→Lys), 208 (Arg→His) and 210 (Val→Ile) are found in patients with familial CJD (Gabizon et al 1991; Goldfarb et al 1990b, 1991b; Hsiao et al 1991a; Mastrianni et al 1995; Ripoll et al 1993). Point mutations at codons 198 (Phe→Ser) and 217 (Gln→Arg) are found in patients with GSS who have PrP amyloid plaques and neurofibrillary tangles (Dlouhy et al 1992; Hsiao et al 1992). Additional point mutations at codons 145 (Tyr→Stop), 105 (Pro→Leu), 180 (Val→Ile) and 232 (Met→Arg) have been recently reported (Kitamoto et al 1993a, 1993b). Single letter code for amino acids is as follows: A, Ala; D, Asp; E, Glu; F, Phe; I, Ile; K, Lys; L, Leu; M, Met; N, Asn; P, Pro; Q, Gln; R, Arg; S, Ser; T, Thr; and V, Val; Y, Tyr.

results, we concluded that scrapie can be generated spontaneously in transgenic mice expressing mutant PrP and that these genetically induced prions can be transmitted to inoculated recipients.

These results in mice resemble those where brain extracts prepared from patients who had died of familial prion diseases were transmitted to monkeys and mice (Masters et al 1981; Tateishi and Kitamoto 1995). Together, the collected transmission studies persuasively argue that prions do, after all, represent an unprecedented class of infectious agents, composed solely of a modified mammalian protein.

Scientists who continue to favor the virus theory might say that we still have not proved our case (Chesebro and Caughey 1993; Manuelidis et al 1995; Ozel and Diringer 1994). If the PrP gene encoded a protein that, when mutated, facilitated infection by a ubiquitous virus, the mutation would lead to viral infection of the brain. Then injection of brain extracts from the animal expressing the mutant PrP gene would spread the infection to another host. Yet in the absence of any evidence of a virus, this hypothesis remains untenable.

STRUCTURES OF PrP^C AND PrP^{Sc}

In addition to showing that a protein can multiply and cause disease without help from nucleic acids, we have gained insight into how PrP^{Sc} forms in cells. Many details remain to be worked out, but one aspect appears quite clear: the main difference between PrP^C and PrP^{Sc} is conformational (Pan et al 1993). Evidently, the PrP^{Sc} propagates itself by binding to PrP^C molecules and somehow causing PrP^C to unfold and flip from their usual conformation to the PrP^{Sc} shape. This change initiates a cascade in which newly converted PrP^{Sc} molecules change the shape of other PrP^C molecules, and so on. These events apparently occur on a membrane within the interior or near the exterior surface of the cell.

We began to think that the differences between cellular and scrapie isoforms of PrP must be conformational after other possibilities began to seem unlikely. Earlier studies established that PrP^C often has the same amino acid sequence as PrP^{Sc}. Of course, molecules that start off being identical can later be chemically modified in ways that alter their activity. But intensive investigations by Neil Stahl and Michael Baldwin have turned up no differences of this kind (Stahl et al 1993).

ONE PRION PROTEIN BUT TWO CONFORMATIONS

How exactly do the structures of cellular and scrapie isoforms of PrP differ? Studies by Keh-Ming Pan indicate that PrP^C consists primarily of α-helices, regions in which the protein backbone twists into a specific kind of spiral; the scrapie form, however, contains β-strands, regions in which the backbone is fully extended. Collections of these strands form β-sheets (Pan et al 1993; Safar et al 1993). Fred Cohen used molecular modeling to try to predict the tertiary structure of PrP^C based on its amino acid sequence. His calculations imply that the protein probably folds into a compact structure with four helices in its core (Huang et al 1994). Less is known about the structures adopted by PrP^{Sc}.

The evidence supporting the proposition that PrP^{Sc} can induce an α-helical PrP^C molecule to adopt a β-sheet conformation comes from several studies. Maria Gasset learned that synthetic peptides consisting of strings of ~15 amino acids each and corresponding to three of the four putative α-helical regions of PrP can fold into β-sheets in aqueous buffers (Gasset et al 1992). A similar behavior was demonstrated for a synthetic peptide of 56 amino acids extending from residue 90 to 145 that contained both the first and second putative α-helices (Zhang et al 1995). Nguyen et al (1995) showed that in their β-sheet conformation, the shorter peptides can impose a β-sheet structure on α-helical PrP peptides. Extending those studies by mixing the long peptide of 56 amino acids with radiolabeled PrP^C, Kaneko et al (1995) found that the PrP^C acquires all the physical properties of PrP^{Sc}: high β-sheet content, insolubility, and protease resistance. Unexpectedly, the synthetic peptide must be in a random-coil conformation before mixing with PrP^C for the conversion to occur. If the peptide has acquired a β-sheet conformation, then the conversion of PrP^C to a PrP^{Sc}-like molecule does not occur. Whether these PrP^{Sc}-like molecules possess infectivity remains to be established.

Another approach was taken by Byron Caughey, Peter Lansbury, and their colleagues who reported that radiolabeled PrP^C mixed with a 50-fold excess of partially denatured PrP^{Sc} could be converted into labeled PrP^{Sc} in a test tube (Kocisko et al 1994; Kocisko et al 1995). Although those investigators reported that PrP^{Sc} could be renatured after removal from 3 M guanidinium HCl, we were unable to verify this finding (Kaneko et al 1995). The interpretation of these experiments awaits a direct demonstration that the isolated radiolabeled PrP has acquired the characteristics of PrP^{Sc}. All of our attempts to separate the radiolabeled PrP after mixing with an excess of PrP^{Sc} have been unsuccessful to date, and thus limit any conclusions from these experiments (Kaneko et al 1995).

PrP molecules arising from mutated genes probably do not adopt the PrP^{Sc} conformation as soon as they are synthesized. Otherwise, people carrying mutant genes would become sick in early childhood. We suspect that mutations in the PrP gene render the resulting proteins susceptible to flipping from an α-helical to a β-sheet shape. Presumably, it takes time until one of the molecules spontaneously flips over and still more time for PrP^{Sc} to accumulate and damage the brain enough to cause symptoms.

Fred Cohen and I think we might be able to explain why the various mutations that have been noted in PrP genes could facilitate folding into the β-sheet isoform, PrP^{Sc} (Cohen et al 1994). Many of the human mutations give rise to the substitution of one amino acid for another within the four putative α-helices or at their borders. Insertion of incorrect amino acids at those positions might destabilize an α-helix, thus increasing the likelihood that the affected α-helix and its neighbors will refold into a β-sheet conformation. Conversely, Hermann Schätzl found that the harmless differences distinguishing the PrP gene of humans from those of apes, monkeys, and many other animals affect amino acids lying outside of the proposed α-helical domains—where the divergent amino acids probably would not profoundly influence the stability of the α-helical regions (Schätzl et al 1995).

POSSIBLE THERAPEUTIC APPROACHES TO PRION DISEASES

No one knows exactly how the accumulation of PrP^{Sc} damages cells. In cell cultures, the conversion of PrP^C into PrP^{Sc} occurs either inside neuronal cells or near the external surface, after which PrP^{Sc} accumulates within intracellular vesicles that appear to be lysosomes (McKinley et al 1991b). In the brain, lysosomes filled with PrP^{Sc} could conceivably burst and damage cells (Arnold et al 1995). As the diseased cells die, creating holes in the brain, their prions would be released to attack other cells.

We do know with certainty that cleavage of PrP^{Sc} is what produces PrP fragments that accumulate as plaques in the brains of some patients. Those aggregates resemble plaques seen in Alzheimer's disease, although the Alzheimer's clumps consist of a different protein, Aβ (Glenner and Wong 1984; Masters et al 1985). The PrP plaques are a useful sign of prion infection, but they seem not to be a major cause of impairment. In many people and animals with prion disease, the plaques do not arise at all (Prusiner et al 1990; Roberts et al 1988).

Even though we do not yet know much about how PrP^{Sc} harms brain tissue, we can foresee that an understanding of the three-dimensional structure of the PrP protein might lead to therapies. If, for example, the four-helix-bundle model of PrP is correct, it might be possible to design a compound that binds to a central pocket formed by the four helices (Cohen et al 1994). So bound, the drug would stabilize these helices and prevent their conversion into β–sheets.

Another idea for therapy is inspired by research in which gene-targeting was used to create mice in which the PrP gene was disrupted so that they could not synthesize PrP^{Sc} (Büeler et al 1992). By "knocking out" a gene and noting the consequences of its loss, one can often deduce the normal function of the gene's protein product. In this case, however, the animals deficient for PrP displayed no detectable abnormalities. If PrP is truly inessential, then physicians might one day consider delivering so-called antisense or antigene therapies to the brains of patients with prion diseases. Such therapies aim to block genes from giving rise to unwanted proteins and could potentially shut down production of cellular PrP. They would thereby block PrP from propagating itself.

The "knockout" mice provided a welcomed opportunity to challenge the prion hypothesis. If the animals became ill after inoculation with prions, their sickness would have indicated that prions could multiply even in the absence of a preexisting pool of PrP molecules. However, inoculation with prions did not produce scrapie, and no evidence of prion replication was detected (Büeler et al 1993; Manson et al 1994; Prusiner et al 1993).

PRION STRAINS

The enigma of how PrP^{Sc} multiplies and causes disease is not the only puzzle beginning to be solved. Another long-standing question—how prions consisting of a single kind of protein can vary markedly in their effects—is beginning to be answered as well. Iain Pattison initially called attention to this phenomenon. Thirty-five years ago, he obtained prions from two separate sets of goats and noted that one isolate made inoculated animals drowsy, whereas the second made them hyperactive (Pattison and Millson 1961). Similarly, it is now evident that some prions cause disease quickly, whereas others do so slowly.

Alan Dickinson, Hugh Fraser, and Moira Bruce examined the differential effects of various isolates in mice and concluded that the existence of multiple prion strains demands that prions contain a nucleic acid, since only pathogens containing nucleic acids are known to occur in multiple strains (Bruce and Dickinson 1987). Hence, they and others assert, the existence of prion "strains" indicates that the prion hypothesis is incorrect and that viruses must be the cause of scrapie and other prion diseases. Yet efforts to find viral nucleic acids have been unrewarding, suggesting that an explanation for the differences among prion isolates must lie elsewhere.

One possibility is that PrP^{Sc} molecules can adopt multiple conformations (Prusiner 1991). Folded in one way, a prion might convert PrP^{C} to the scrapie isoform highly efficiently, giving rise to short incubation times. Folded another way, it might work less efficiently. Similarly, one "conformer" might be attracted to neuronal populations in one part of the brain, whereas another might be attracted to neurons elsewhere, thus producing different symptoms (Bessen et al 1995; Bessen and Marsh 1992; Marsh and Bessen 1994). Considering that PrP can fold in at least two ways, it would not be surprising to find it can collapse into other structures as well.

BREAKING THE SPECIES BARRIER

Since the mid-1980s, we have also sought insight into the phenomenon of species barrier, which refers to the fact that something makes it difficult for prions made by one species to induce disease in animals of another species. The cause of this difficulty is of considerable interest today because of the epidemic of mad cow disease in Britain. We and others have been trying to find out whether the species barrier is strong enough to prevent the spread of prion disease from cows to humans.

The barrier was discovered by Pattison, who in the 1960s found it difficult to transmit scrapie between sheep and rodents (Pattison 1965). To determine the cause of this trouble, Michael Scott and I generated transgenic mice expressing the PrP gene of the Syrian hamster—that is, making the hamster PrP protein (Scott et al 1989). The mouse gene differs from that of the hamster gene at 16 codons out of 254. Normal mice inoculated with hamster prions rarely acquire scrapie, but the transgenic mice became ill within about two months.

We thus concluded that we had broken the species barrier by inserting the hamster genes into the mice. Moreover, this and other experiments suggested that the barrier resides in the amino acid sequence of PrP: the more closely the sequence of scrapie PrP molecule resembles the PrP sequence of its host, the more likely it is that the host will acquire prion disease. In one of those experiments, we examined transgenic mice carrying the Syrian hamster PrP gene in addition to their own mouse gene. Those mice make normal forms of both hamster and mouse PrP. When we inoculated the animals with mouse prions, they made more mouse prions. When we inoculated them with hamster prions, they made hamster prions. From this behavior, we learned that PrP^{Sc} preferentially interacts with homologous PrP^C (Prusiner et al 1990).

The attraction of PrP^{Sc} for PrP^C with the same amino acid sequence probably explains why scrapie of sheep spread to cows in England from a food supplement that included meat and bone meal produced from sheep offal. While sheep and bovine PrP differ only at seven amino acid positions, the sequences of human and bovine PrP molecules diverge at more than 30 positions. Because the variance between human and bovine PrP is substantial, the likelihood of transmission from cows to people is likely to be low. Consistent with this assessment are epidemiological studies by W. Bryan Matthews and colleagues. Matthews found no link between scrapie in sheep and the occurrence of CJD (Cousens et al 1990; Harries-Jones et al 1988). Current deaths from CJD may have nothing to do with the bovine epidemic, but the situation bears watching. Recent studies show that the central region of the PrP molecule is particularly important with respect to the interaction of PrP^{Sc} with PrP^C; presumably the same domain is crucial in breaking the species barrier (Telling et al 1995). If that is the case, then the similarity between cow and human PrP in this critical region may signal a greater danger than would be surmised by a simple comparison of the complete amino acid sequences.

CHIMERIC PrP TRANSGENES AND THE DISCOVERY OF PROTEIN X

We began to consider the possibility that some parts of the PrP molecule might be particularly important to the species barrier after a study related to this blockade took an odd turn. Glenn Telling had created transgenic mice carrying a hybrid PrP gene that consisted of human sequences in the central domain and mouse encoded residues flanking both sides (Telling et al 1994; Telling et al 1995). Then he introduced extracts of brain tissue from patients who had died of CJD or GSS into the transgenic animals. Oddly enough, the animals became ill much more frequently and quickly than did mice carrying a full-length human PrP gene, which diverges from mouse PrP at 28 positions. This outcome implied that similarity in the central domain of the PrP molecule may be more critical than it is in the other segments.

These studies produced another unexpected result when transgenic mice expressing human PrP were crossed with the null ($Prnp^{0/0}$) mice. By this series of genetic crosses, the Tg(HuPrP) mice were rendered susceptible to human prions (Telling et al 1995). From the studies, we concluded that a mouse macromolecule involved in prion replication interacts with PrP^C but not PrP^{Sc} and we provisionally designated this molecule "protein X." We believe that mouse protein X interacts preferentially with mouse PrP^C, but in the absence of mouse PrP^C, it can bind to human PrP^C and mediate the conversion of human PrP^C into PrP^{Sc}. A comparison of the sequences of mouse, human and Syrian hamster PrP suggests that C-terminal residues between 215 and 230 bind to protein X. Earlier studies showed that N-terminally truncated PrP could be converted into PrP^{Sc}, arguing that the binding of PrP^C to protein X occurs through the C-terminus of PrP^C (Rogers et al 1993). Although not established, it seems likely that protein X will turn out to be a molecular chaperone that facilitates the unfolding of PrP^C and its refolding into PrP^{Sc}.

A WIDER VIEW OF PRION DISEASES

An unforeseen story has recently emerged from studies of transgenic mice that produce unusually high amounts of normal PrP proteins. Stephen DeArmond, David Westaway, and George Carlson became perplexed when they noted that some older transgenic mice developed an illness characterized by rigidity and diminished grooming. When we pursued the cause, we found that making excessive amounts of PrP can eventually lead to neurodegeneration and, surprisingly, to destruction of both muscles and peripheral nerves (Westaway et al 1994b). These discoveries widen the spectrum of prion diseases and are prompting a search for human prion diseases that affect the peripheral nervous system and muscles.

Investigations of animals that overproduce PrP have yielded another benefit as well. They offer a clue as to how the sporadic form of CJD might arise. For a time I suspected that sporadic disease might begin when the wear and tear of living led to a mutation of the PrP gene in at least one cell in the body. Eventually, the mutated protein might switch to the scrapie form and gradually propagate itself, until the buildup of PrP^{Sc} crossed the threshold to overt disease. The mouse studies suggest that at some point in the lives of the one in a million individuals who acquire sporadic CJD, PrP^C may spontaneously convert into PrP^{Sc}. The experiments also raise the possibility that people who become afflicted with sporadic CJD overproduce PrP^C, but we do not yet know if, in fact, they do.

All the known prion diseases in humans have now been modeled in mice. With our most recent work, we have inadvertently developed an animal model for sporadic prion disease. Mice inoculated with brain extracts from scrapie-infected animals and from humans afflicted with CJD have long provided a model for the infectious forms of prion disorders. And the inherited prion diseases have been modeled in transgenic mice carrying mutant PrP genes. These murine representations of the human prion afflictions should not only extend understanding of how prions cause brain degeneration, but they should also create opportunities to evaluate therapies for these devastating maladies.

Are the more common CNS degenerative diseases also disorders of protein conformation? Ongoing research may also help to determine whether prions consisting of other proteins play a part in more common neurodegenerative conditions, including Alzheimer's disease, Parkinson's disease, and amyotrophic lateral sclerosis. There are some marked similarities in all these disorders. As is true of the known prion diseases, the more widespread ills mostly occur sporadically but sometimes seem to be familial. Also, all are usually diseases of middle to later life and are marked by similar pathology: neurons degenerate, protein deposits can accumulate as plaques, and glial cells (which support and nourish nerve cells) grow larger in reaction to neuronal damage. Strikingly, in none of these disorders do white blood cells infiltrate the brain. If a virus were involved in these illnesses, white cells would be expected to appear.

Recent findings in yeast encourage speculation that prions unrelated in amino acid sequence to the PrP protein could exist. Reed B. Wickner of the National Institutes of Health reports that a protein called Ure2p might sometimes change its conformation, thereby affecting its activity in the cell (Wickner 1994). In one shape, the protein is active; in the other, it is silent (Chernoff et al 1995).

The collected studies described here argue persuasively that the prion is an entirely new class of infectious pathogen and that prion diseases result from aberrations of protein conformation. Whether changes in protein shape are responsible for common neurodegenerative diseases, such as Alzheimer's, remains unknown, but it is a possibility that should not be ignored.

Acknowledgments

This article was adapted from two articles that were published in *Scientific American* (Prusiner 1984a; Prusiner 1995). The work was supported by grants from the National Institutes of Health and the American Health Assistance Foundation as well as by gifts from the Sherman Fairchild Foundation.

REFERENCES

Alper T, Cramp WA, Haig DA, Clarke MC (1967). Does the agent of scrapie replicate without nucleic acid? Nature 214: 764-766.

Alper T, Haig DA, Clarke MC (1966). The exceptionally small size of the scrapie agent. Biochem Biophys Res Commun 22: 278-284.

Arnold JE, Tipler C, Laszlo L, Hope J, Landon M, Mayer RJ (1995). The abnormal isoform of the prion protein accumulates in late-endosome-like organelles in scrapie-infected mouse brain. J Pathol 176: 403-411.

Barry RA, Kent SBH, McKinley MP, Meyer RK, DeArmond SJ, Hood LE, Prusiner SB (1986). Scrapie and cellular prion proteins share polypeptide epitopes J Infect Dis 153: 848-854.

Barry RA, Prusiner SB (1986). Monoclonal antibodies to the cellular and scrapie prion proteins. J Infect Dis 154: 518-521.

Basler K, Oesch B, Scott M, Westaway D, Wälchli M, Groth DF, McKinley MP, Prusiner SB, Weissmann C (1986). Scrapie and cellular PrP isoforms are encoded by the same chromosomal gene. Cell 46: 417-428

Bellinger-Kawahara C, Cleaver JE, Diener TO, Prusiner SB (1987a). Purified scrapie prions resist inactivation by UV irradiation. J Virol 61: 159-166.

Bellinger-Kawahara C, Diener TO, McKinley MP, Groth DF, Smith DR, Prusiner SB (1987b). Purified scrapie prions resist inactivation by procedures that hydrolyze, modify, or shear nucleic acids. Virology 160: 271-274.

Bellinger-Kawahara CG, Kempner E, Groth DF, Gabizon R, Prusiner SB (1988). Scrapie prion liposomes and rods exhibit target sizes of 55,000 Da. Virology 164: 537-541.

Bendheim PE, Barry RA, DeArmond SJ, Stites DP, Prusiner SB (1984). Antibodies to a scrapie prion protein. Nature 310: 418-421.

Bessen RA, Kocisko DA, Raymond GJ, Nandan S, Lansbury PT, Caughey B (1995). Non-genetic propagation of strain-specific properties of scrapie prion protein. Nature 375: 698-700.

Bessen RA, Marsh RF (1992). Biochemical and physical properties of the prion protein from two strains of the transmissible mink encephalopathy agent. J Virol 66: 2096-2101

Bolton DC, McKinley MP, Prusiner SB (1982). Identification of a protein that purifies with the scrapie prion. Science 218: 1309-1311.

Borchelt DR, Scott M, Taraboulos A, Stahl N, Prusiner SB (1990). Scrapie and cellular prion proteins differ in their kinetics of synthesis and topology in cultured cells. J Cell Biol 110: 743-752.

Borchelt DR, Taraboulos A, and Prusiner SB (1992). Evidence for synthesis of scrapie prion proteins in the endocytic pathway. J Biol Chem 267: 16188-16199

Brown P, Cervenáková L, Goldfarb LG, McCombie WR, Rubenstein R, Will RG, Pocchiari M, Martinez-Lage JF, Scalici C, Masullo C, Graupera G, Ligan J, Gajdusek DC (1994). Iatrogenic Creutzfeldt-Jakob disease: an example of the interplay between ancient genes and modern medicine. Neurology 44: 291-293.

Bruce ME, Dickinson AG (1987). Biological evidence that the scrapie agent has an independent genome. J Gen Virol 68: 79-89.

Büeler H, Aguzzi A, Sailer A, Greiner R-A, Autenried P, Aguet M, Weissmann C (1993). Mice devoid of PrP are resistant to scrapie. Cell 73: 1339-1347.

Büeler H, Fischer M, Lang Y, Bluethmann H, Lipp H-P, DeArmond SJ, Prusiner SB, Aguet M, Weissmann C (1992). Normal development and behaviour of mice lacking the neuronal cell-surface PrP protein. Nature 356: 577-582.

Caughey B, Raymond GJ (1991). The scrapie-associated form of PrP is made from a cell surface precursor that is both protease- and phospholipase-sensitive. J Biol Chem 266: 18217-18223.

Chandler RL (1961). Encephalopathy in mice produced by inoculation with scrapie brain material. Lancet 1: 1378-1379.

Chernoff YO, Lindquist SL, Ono B, Inge-Vechtomov SG, Liebman SW (1995). Role of the chaperone protein Hsp104 in propagation of the yeast prion-like factor [*psi*⁺]. Science 268: 880-884.

Chesebro B, Caughey B (1993). Scrapie agent replication without the prion protein? Curr Biol 3: 696-698.

Chesebro B, Race R, Wehrly K, Nishio J, Bloom M, Lechner D, Bergstrom S, Robbins K, Mayer L, Keith JM, Garon C, Haase A (1985). Identification of scrapie prion protein-specific mRNA in scrapie-infected and uninfected brain. Nature 315: 331-333.

Cohen AS, Shirahama T, Skinner M (1982). Electron microscopy of amyloid. In: Harris JR, ed. Electron microscopy of proteins. Vol 3. New York: Academic Press, 165-206.

Cohen FE, Pan K-M, Huang Z, Baldwin M, Fletterick RJ, Prusiner SB (1994). Structural clues to prion replication. Science 264: 530-531.

Collinge J, Harding AE, Owen F, Poulter M, Lofthouse R, Boughey AM, Shah T, Crow TJ (1989). Diagnosis of Gerstmann-Sträussler syndrome in familial dementia with prion protein gene analysis. Lancet 2: 15-17.

Collinge J, Owen F, Poulter H, Leach M, Crow T, Rosser M, Hardy J, Mullan H, Janota I, Lantos P (1990). Prion dementia without characteristic pathology. Lancet 336: 7-9.

Cousens SN, Harries-Jones R, Knight R, Will RG, Smith PG, Matthews WB (1990). Geographical distribution of cases of Creutzfeldt-Jakob disease in England and Wales 1970-84. J Neurol Neurosurg Psychiatry 53: 459-465.

Crow TJ, Collinge J, Ridley RM, Baker HF, Lofthouse R, Owen F, Harding AE (1990). Mutations in the prion gene in human transmissible dementia. Seminar on Molecular Approaches to Research in Spongiform Encephalopathies in Man. Medical Research Council, London, Dec. 14, 1990 [Abstract]

Cuillé J, Chelle PL (1939). Experimental transmission of trembling to the goat. CR Acad Sci 208: 1058-1060.

DeArmond SJ, McKinley MP, Barry RA, Braunfeld MB, McColloch JR, Prusiner SB (1985). Identification of prion amyloid filaments in scrapie-infected brain. Cell 41: 221-235.

Diener TO, McKinley MP, Prusiner SB (1982). Viroids and prions. Proc Natl Acad Sci USA 79: 5220-5224.

Diringer H, Gelderblom H, Hilmert H, Ozel M, Edelbluth C, Kimberlin RH (1983). Scrapie infectivity, fibrils and low molecular weight protein. Nature 306: 476-478.

Dlouhy SR, Hsiao K, Farlow MR, Foroud T, Conneally PM, Johnson P, Prusiner SB, Hodes ME, Ghetti B (1992). Linkage of the Indiana kindred of Gerstmann-Sträussler-Scheinker disease to the prion protein gene. Nat Genet 1: 64-67.

Doh-ura K, Tateishi J, Sasaki H, Kitamoto T, Sakaki Y (1989). Pro→Leu change at position 102 of prion protein is the most common but not the sole mutation related to Gerstmann-Sträussler syndrome. Biochem Biophys Res Commun 163: 974-979.

Fink JK, Peacock ML, Warren JT, Roses AD, Prusiner SB (1994). Detecting prion protein gene-mutations by denaturing gradient gel-electrophoresis. Hum Mutat 4: 42-50.

Fradkin JE, Schonberger LB, Mills JL, Gunn WJ, Piper JM, Wysowski DK, Thomson R, Durako S, Brown P (1991). Creutzfeldt-Jakob disease in pituitary growth hormone recipients in the United States. JAMA 265: 880-884.

Gabizon R, McKinley MP, Groth DF, Prusiner SB (1988). Immunoaffinity purification and neutralization of scrapie prion infectivity. Proc Natl Acad Sci USA 85: 6617-6621.

Gabizon R, Meiner Z, Cass C, Kahana E, Kahana I, Avrahami D, Abramsky O, Scarlato G, Prusiner SB, Hsiao KK (1991). Prion protein gene mutation in Libyan Jews with Creutzfeldt-Jakob disease. Neurology 41: 160.

Gabizon R, Prusiner SB (1990). Prion liposomes. Biochem J 266: 1-14

Gabizon R, Rosenmann H, Meiner Z, Kahana I, Kahana E, Shugart Y, Ott J, Prusiner SB (1993). Mutation and polymorphism of the prion protein gene in Libyan Jews with Creutzfeldt-Jakob disease. Am J Hum Genet 33: 828-835.

Gajdusek DC (1977). Unconventional viruses and the origin and disappearance of kuru. Science 197: 943-960.

Gajdusek DC (1987). The transmissible dementias and other brain disorders caused by unconventional viruses—Relationship of transmissible to nontransmissible amyloidosis of the brain. In: Kurstak E, Lipowski ZJ, Morozov PV, eds. Viruses, immunity and mental disorders. New York: Plenum Publishing Company, 3-22.

Gajdusek DC (1994). Spontaneous generation of infectious nucleating amyloids in the transmissible and nontransmissible cerebral amyloidoses. Mol Neurobiol 8: 1-13.

Gajdusek DC, Gibbs CJ Jr, Alpers M (1966). Experimental transmission of a kuru-like syndrome to chimpanzees. Nature 209: 794-796.

Gajdusek DC, Zigas V (1957). Degenerative disease of the central nervous system in New Guinea—The endemic occurrence of "kuru" in the native population. N Engl J Med 257: 974-978.

Gasset M, Baldwin MA, Lloyd D, Gabriel J-M, Holtzman DM, Cohen F, Fletterick R, Prusiner SB (1992). Predicted α-helical regions of the prion protein when synthesized as peptides form amyloid. Proc Natl Acad Sci USA 89: 10940-10944.

Gibbs CJ Jr, Gajdusek DC, Asher DM, Alpers MP, Beck E, Daniel PM, Matthews WB (1968). Creutzfeldt-Jakob disease (spongiform encephalopathy): transmission to the chimpanzee. Science 161: 388-389.

Glenner GG, Eanes ED, and Page DL, (1972). The relation of the properties of Congo red-stained amyloid fibrils to the beta-conformation. J Histochem Cytochem 20: 821-826.

Glenner GG, Wong CW (1984). Alzheimer's disease: initial report of the purification and characterization of a novel cerebrovascular amyloid protein. Biochem Biophys Res Commun 120: 885-890.

Goldfarb L, Brown P, Goldgaber D, Garruto R, Yanaghiara R, Asher D, Gajdusek DC (1990a). Identical mutation in unrelated patients with Creutzfeldt-Jakob disease. Lancet 336: 174-175.

Goldfarb L, Korczyn A, Brown P, Chapman J, Gajdusek DC (1990b). Mutation in codon 200 of scrapie amyloid precursor gene linked to Creutzfeldt-Jakob disease in Sephardic Jews of Libyan and non-Libyan origin. Lancet 336: 637-638.

Goldfarb LG, Brown P, Goldgaber D, Asher DM, Rubenstein R, Brown WT, Piccardo P, Kascsak RJ, Boellaard JW, Gajdusek DC (1990c). Creutzfeldt-Jakob disease and kuru patients lack a mutation consistently found in the Gerstmann-Sträussler-Scheinker syndrome. Exp Neurol 108: 247-250.

Goldfarb LG, Mitrova E, Brown P, Toh BH, Gajdusek DC (1990d). Mutation in codon 200 of scrapie amyloid protein gene in two clusters of Creutzfeldt-Jakob disease in Slovakia. Lancet 336: 514-515.

Goldfarb LG, Brown P, McCombie WR, Goldgaber D, Swergold GD, Wills PR, Cervenakova L, Baron H, Gibbs CJJ, Gajdusek DC (1991a). Transmissible familial Creutzfeldt-Jakob disease associated with five, seven, and eight extra octapeptide coding repeats in the *PRNP* gene. Proc Natl Acad Sci USA 88: 10926-10930.

Goldfarb LG, Haltia M, Brown P, Nieto A, Kovanen J, McCombie WR, Trapp S, Gajdusek DC (1991b). New mutation in scrapie amyloid precursor gene (at codon 178) in Finnish Creutzfeldt-Jakob kindred. Lancet 337: 425.

Goldfarb LG, Petersen RB, Tabaton M, Brown P, LeBlanc AC, Montagna P, Cortelli P, Julien J, Vital C, Pendelbury WW, Haltia M, Wills PR, Hauw JJ, McKeever PE, Monari L, Schrank B, Swergold GD, Autilio-Gambetti L, Gajdusek DC, Lugaresi E, Gambetti P (1992). Fatal familial insomnia and familial Creutzfeldt-Jakob disease: disease phenotype determined by a DNA polymorphism. Science 258: 806-808.

Goldgaber D, Goldfarb LG, Brown P, Asher DM, Brown WT, Lin S, Teener JW, Feinstone SM, Rubenstein R, Kascsak RJ, Boellaard JW, Gajdusek DC (1989). Mutations in familial Creutzfeldt-Jakob disease and Gerstmann-Sträussler-Scheinker's syndrome. Exp Neurol 106: 204-206.

Hadlow WJ (1959). Scrapie and kuru. Lancet 2: 289-290.

Harries-Jones R, Knight R, Will RG, Cousens S, Smith PG, Matthews WB (1988). Creutzfeldt-Jakob disease in England and Wales 1980–1984: a case-control study of potential risk factors. J Neurol Neurosurg Psychiatry 51: 1113-1119.

Hsiao K, Baker HF, Crow TJ, Poulter M, Owen F, Terwilliger JD, Westaway D, Ott J, Prusiner SB (1989a). Linkage of a prion protein missense variant to Gerstmann-Sträussler syndrome. Nature 338: 342-345.

Hsiao K, Dlouhy S, Farlow MR, Cass C, Da Costa M, Conneally M, Hodes ME, Ghetti B, Prusiner SB (1992). Mutant prion proteins in Gerstmann-Sträussler-Scheinker disease with neurofibrillary tangles. Nat Genet 1: 68-71.

Hsiao K, Meiner Z, Kahana E, Cass C, Kahana I, Avrahami D, Scarlato G, Abramsky O, Prusiner SB, Gabizon R (1991a). Mutation of the prion protein in Libyan Jews with Creutzfeldt-Jakob disease. N Engl J Med 324: 1091-1097.

Hsiao K, Prusiner SB (1990). Inherited human prion diseases. Neurology 40: 1820-1827.

Hsiao KK, Cass C, Schellenberg GD, Bird T, Devine-Gage E, Wisniewski H, Prusiner SB (1991b). A prion protein variant in a family with the telencephalic form of Gerstmann-Sträussler-Scheinker syndrome. Neurology 41: 681-684.

Hsiao KK, Doh-ura K, Kitamoto T, Tateishi J, Prusiner SB (1989b). A prion protein amino acid substitution in ataxic Gerstmann-Sträussler syndrome. Ann Neurol 26: 137.

Hsiao KK, Groth D, Scott M, Yang S-L, Serban H, Rapp D, Foster D, Torchia M, DeArmond SJ, Prusiner SB (1994). Serial transmission in rodents of neurodegeneration from transgenic mice expressing mutant prion protein. Proc Natl Acad Sci USA 91: 9126-9130.

Hsiao KK, Scott M, Foster D, Groth DF, DeArmond SJ, Prusiner SB (1990). Spontaneous neurodegeneration in transgenic mice with mutant prion protein. Science 250: 1587-1590.

Huang Z, Gabriel J-M, Baldwin MA, Fletterick RJ, Prusiner SB, Cohen FE (1994). Proposed three-dimensional structure for the cellular prion protein. Proc Natl Acad Sci USA 91: 7139-7143.

Jarrett JT, Lansbury PT Jr (1993). Seeding "one-dimensional crystallization" of amyloid: a pathogenic mechanism in Alzheimer's disease and scrapie? Cell 73: 1055-1058.

Kaneko K, Peretz D, Pan K-M, Blochberger T, Gabizon R, Griffith OH, Cohen FE, Baldwin MA, Prusiner SB (1995). Prion protein (PrP) synthetic peptides induce cellular PrP to acquire properties of the scrapie isoform. Proc Natl Acad Sci USA 92: 11160-11164.

Kellings K, Meyer N, Mirenda C, Prusiner SB, Riesner D (1992). Further analysis of nucleic acids in purified scrapie prion preparations by improved return refocussing gel electrophoresis (RRGE). J Gen Virol 73: 1025-1029.

Kirschbaum WR (1924). Zwei eigenartige Erkrankungen des Zentralnervensystems nach Art der spastischen Pseudosklerose (Jakob). Z Ges Neurol Psychiatr 92: 175-220.

Kitamoto T, Iizuka R, Tateishi J (1993a). An amber mutation of prion protein in Gerstmann-Sträussler syndrome with mutant PrP plaques. Biochem Biophys Res Commun 192: 525-531.

Kitamoto T, Ohta M, Doh-ura K, Hitoshi S, Terao Y, Tateishi J (1993b). Novel missense variants of prion protein in Creutzfeldt-Jakob disease or Gerstmann-Sträussler syndrome. Biochem Biophys Res Commun 191: 709-714.

Kitamoto T, Tateishi J (1994). Human prion diseases with variant prion protein. Philos Trans R Soc Lond B 343: 391-398.

Kitamoto T, Tateishi J, Tashima I, Takeshita I, Barry RA, DeArmond SJ, Prusiner SB (1986). Amyloid plaques in Creutzfeldt-Jakob disease stain with prion protein antibodies. Ann Neurol 20: 204-208.

Kocisko DA, Come JH, Priola SA, Chesebro B, Raymond GJ, Lansbury PT Jr, Caughey B (1994). Cell-free formation of protease-resistant prion protein. Nature 370: 471-474.

Kocisko DA, Priola SA, Raymond GJ, Chesebro B, Lansbury PT Jr, Caughey B (1995). Species specificity in the cell-free conversion of prion protein to protease-resistant forms: a model for the scrapie species barrier. Proc Natl Acad Sci USA 92: 3923-3927.

Kretzschmar HA, Stowring LE, Westaway D, Stubblebine WH, Prusiner SB, DeArmond SJ (1986). Molecular cloning of a human prion protein cDNA. DNA 5: 315-324.

Laplanche J-L, Chatelain J, Launay J-M, Gazengel C, Vidaud M (1990). Deletion in prion protein gene in a Moroccan family. Nucleic Acids Res 18: 6745.

Manson JC, Clarke AR, McBride PA, McConnell I, Hope J (1994). PrP gene dosage determines the timing but not the final intensity or distribution of lesions in scrapie pathology. Neurodegeneration 3: 331-340.

Manuelidis L, Sklaviadis T, Akowitz A, Fritch W (1995). Viral particles are required for infection in neurodegenerative Creutzfeldt-Jakob disease. Proc Natl Acad Sci USA 92: 5124-5128.

Marsh RF, Bessen RA (1994). Physicochemical and biological characterizations of distinct strains of the transmissible mink encephalopathy agent. Philos Trans R Soc Lond B 343: 413-414.

Marsh RF, Kimberlin RH (1975). Comparison of scrapie and transmissible mink encephalopathy in hamsters. II. Clinical signs, pathology and pathogenesis, J Infect Dis 131: 104-110.

Masters CL, Gajdusek DC, Gibbs CJ Jr (1981). Creutzfeldt-Jakob disease virus isolations from the Gerstmann-Sträussler syndrome. Brain 104: 559-588.

Masters CL, Harris JO, Gajdusek DC, Gibbs CJ Jr, Bernouilli C, Asher DM (1978). Creutzfeldt-Jakob disease: patterns of worldwide occurrence and the significance of familial and sporadic clustering. Ann Neurol 5: 177-188.

Masters CL, Simms G, Weinman NA, Multhaup G, McDonald BL, Beyreuther K (1985). Amyloid plaque core protein in Alzheimer disease and Down syndrome. Proc Natl Acad Sci USA 82: 4245-4249.

Mastrianni JA, Iannicola C, Myers R, Prusiner SB (1995). Identification of a new mutation of the prion protein gene at codon 208 in a patient with Creutzfeldt-Jakob disease (Abstr). Neurology 45 [Suppl]: 201.

McKinley MP, Bolton DC, Prusiner SB (1983a). A protease-resistant protein is a structural component of the scrapie prion. Cell 35: 57-62.

McKinley MP, Masiarz FR, Isaacs ST, Hearst JE, Prusiner SB (1983b). Resistance of the scrapie agent to inactivation by psoralens. Photochem Photobiol 37: 539-545.

McKinley MP, Masiarz FR, Prusiner SB (1981). Reversible chemical modification of the scrapie agent. Science 214: 1259-1261.

McKinley MP, Meyer R, Kenaga L, Rahbar F, Cotter R, Serban A, Prusiner SB (1991a). Scrapie prion rod formation in vitro requires both detergent extraction and limited proteolysis. J Virol 65: 1440-1449.

McKinley MP, Taraboulos A, Kenaga L, Serban D, Stieber A, DeArmond SJ, Prusiner SB, Gonatas N (1991b). Ultrastructural localization of scrapie prion proteins in cytoplasmic vesicles of infected cultured cells. Lab Invest 65: 622-630.

Medori R, Tritschler H-J, LeBlanc A, Villare F, Manetto V, Chen HY, Xue R, Leal S, Montagna P, Cortelli P, Tinuper P, Avoni P, Mochi M, Baruzzi A, Hauw JJ, Ott J, Lugaresi E, Autilio-Gambetti L, Gambetti P (1992). Fatal familial insomnia, a prion disease with a mutation at codon 178 of the prion protein gene. N Engl J Med 326: 444-449.

Merz PA, Rohwer RG, Kascsak R, Wisniewski HM, Somerville RA, Gibbs CJ Jr, Gajdusek DC (1984). Infection-specific particle from the unconventional slow virus diseases. Science 225: 437-440.

Merz PA, Somerville RA, Wisniewski HM, Iqbal K (1981). Abnormal fibrils from scrapie-infected brain. Acta Neuropathol (Berl). 54: 63-74.

Merz PA, Wisniewski HM, Somerville RA, Bobin SA, Masters CL, Iqbal K (1983). Ultrastructural morphology of amyloid fibrils from neuritic and amyloid plaques. Acta Neuropathol (Berl). 60: 113-124.

Meyer RK, McKinley MP, Bowman KA, Braunfeld MB, Barry RA, Prusiner SB (1986). Separation and properties of cellular and scrapie prion proteins. Proc Natl Acad Sci USA 83: 2310-2314.

Nguyen J, Baldwin MA, Cohen FE, Prusiner SB (1995). Prion protein peptides induce α-helix to β-sheet conformational transitions. Biochemistry 34: 4186-4192.

Oesch B, Westaway D, Wälchli M, McKinley MP, Kent SBH, Aebersold R, Barry RA, Tempst P, Teplow DB, Hood LE, Prusiner SB, Weissmann C (1985). A cellular gene encodes scrapie PrP 27-30 protein. Cell 40: 735-746.

Owen F, Poulter M, Lofthouse R, Collinge J, Crow TJ, Risby D, Baker HF, Ridley RM, Hsiao K, Prusiner SB (1989). Insertion in prion protein gene in familial Creutzfeldt-Jakob disease. Lancet 1: 51-52.

Owen F, Poulter M, Shah T, Collinge J, Lofthouse R, Baker H, Ridley R, McVey J, Crow T (1990). An in-frame insertion in the prion protein gene in familial Creutzfeldt- Jakob disease. Mol Brain Res 7: 273-276.

Ozel M, Diringer H (1994). Small virus-like structure in fraction from scrapie hamster brain. Lancet 343: 894-895.

Palmer MS. Dryden AJ, Hughes JT, Collinge J (1991). Homozygous prion protein genotype predisposes to sporadic Creutzfeldt-Jakob disease. Nature 352: 340-342.

Palmer MS, Mahal SP, Campbell TA, Hill AF, Sidle KCL, Laplanche J-L, Collinge J (1993). Deletions in the prion protein gene are not associated with CJD. Hum Molec Genet 2: 541-544.

Pan K-M, Baldwin M, Nguyen J, Gasset M, Serban A, Groth D, Mehlhorn I, Huang Z, Fletterick RJ, Cohen FE, Prusiner SB (1993). Conversion of α-helices into β-sheets features in the formation of the scrapie prion proteins. Proc Natl Acad Sci USA 90: 10962-10966.

Pattison IH (1965). Experiments with scrapie with special reference to the nature of the agent and the pathology of the disease. In: Gajdusek DC, Gibbs Jr CJ, Alpers MP, eds. Slow, latent and temperate virus infections, NINDB Monograph 2. Washington DC: US Government Printing, 249-257.

Pattison IH (1988). Fifty years with scrapie: a personal reminiscence. Vet Rec 123: 661-666.

Pattison IH, Millson GC (1961). Scrapie produced experimentally in goats with special reference to the clinical syndrome. J Comp Pathol 71: 101-108.

Petersen RB, Tabaton M, Berg L, Schrank B, Torack RM, Leal S, Julien J, Vital C, Deleplanque B, Pendlebury WW, Drachman D, Smith TW, Martin JJ, Oda M, Montagna P, Ott J, Autilio-Gambetti L, Lugaresi E, Gambetti P (1992). Analysis of the prion protein gene in thalamic dementia. Neurology 42: 1859-1863.

Poulter M, Baker HF, Frith CD, Leach M, Lofthouse R, Ridley RM, Shah T, Owen F, Collinge J, Brown G, Hardy J, Mullan MJ, Harding AE, Bennett C, Doshi R, Crow T J (1992). Inherited prion disease with 144 base pair gene insertion. 1. Genealogical and molecular studies. Brain 115: 675-685.

Prusiner SB (1982). Novel proteinaceous infectious particles cause scrapie. Science 216: 136-144.

Prusiner SB (1984a). Prions. Sci Am 251: 50-59.

Prusiner SB (1984b). Some speculations about prions amyloid and Alzheimer's disease. N Engl J Med 310: 661-663.

Prusiner SB (1991). Molecular biology of prion diseases. Science 252: 1515-1522.

Prusiner SB (1992). Molecular biology and genetics of neurodegenerative diseases caused by prions. Adv Virus Res 41: 241-280.

Prusiner SB (1995). The prion diseases. Sci Am 272: 48-57.

Prusiner SB, Bolton DC, Groth DF, Bowman KA, Cochran SP, McKinley MP (1982). Further purification and characterization of scrapie prions. Biochemistry 21: 6942-6950.

Prusiner SB, Groth D, Serban A, Koehler R, Foster D, Torchia M, Burton D, Yang S-L, DeArmond SJ (1993). Ablation of the prion protein (PrP) gene in mice prevents scrapie and facilitates production of anti-PrP antibodies. Proc Natl Acad Sci USA 90: 10608-10612.

Prusiner SB, Groth DF, Bolton DC, Kent SB, Hood LE (1984). Purification and structural studies of a major scrapie prion protein. Cell 38: 127-134.

Prusiner SB, Groth DF, Cochran SP, Masiarz FR, McKinley MP, Martinez HM (1980). Molecular properties, partial purification, and assay by incubation period measurements of the hamster scrapie agent. Biochemistry 19: 4883-4891.

Prusiner SB, Hadlow WJ, Eklund CM, Race RE (1977). Sedimentation properties of the scrapie agent. Proc Natl Acad Sci USA 74: 4656-4660.

Prusiner SB, Hadlow WJ, Eklund CM, Race RE, Cochran SP (1978a). Sedimentation characteristics of the scrapie agent from murine spleen and brain. Biochemistry 17: 4987-4992.

Prusiner SB, Hadlow WJ, Garfin DE, Cochran SP, Baringer JR, Race RE, Eklund CM (1978b). Partial purification and evidence for multiple molecular forms of the scrapie agent. Biochemistry 17: 4993-4997.

Prusiner SB, McKinley MP, Bowman KA, Bolton DC, Bendheim PE, Groth DF, Glenner GG (1983). Scrapie prions aggregate to form amyloid-like birefringent rods. Cell 35: 349-358.

Prusiner SB, McKinley MP, Groth DF, Bowman KA, Mock NI, Cochran SP, Masiarz FR (1981). Scrapie agent contains a hydrophobic protein. Proc Natl Acad Sci USA 78: 6675-6679.

Prusiner SB, Scott M, Foster D, Pan K-M, Groth D, Mirenda C, Torchia M, Yang S-L, Serban D, Carlson GA, Hoppe PC, Westaway D, DeArmond SJ (1990). Transgenetic studies implicate interactions between homologous PrP isoforms in scrapie prion replication. Cell 63: 673-686.

Puckett C, Concannon P, Casey C, Hood L (1991). Genomic structure of the human prion protein gene. Am J Hum Genet 49: 320-329.

Ripoll L, Laplanche J-L, Salzmann M, Jouvet A, Planques B, Dussaucy M, Chatelain J, Beaudry P, Launay J-M (1993). A new point mutation in the prion protein gene at codon 210 in Creutzfeldt-Jakob disease. Neurology 43: 1934-1938.

Roberts GW, Lofthouse R, Allsop D, Landon M, Kidd M, Prusiner SB, Crow T J (1988). CNS amyloid proteins in neurodegenerative diseases. Neurology 38: 1534-1540.

Rogers M, Yehiely F, Scott M, Prusiner SB (1993). Conversion of truncated and elongated prion proteins into the scrapie isoform in cultured cells. Proc Natl Acad Sci USA 90: 3182-3186.

Safar J, Roller PP, Gajdusek DC, Gibbs CJ Jr (1993). Conformational transitions, dissociation, and unfolding of scrapie amyloid (prion) protein. J Biol Chem 268: 20276-20284.

Schätzl HM, Da Costa M, Taylor L, Cohen FE, Prusiner SB (1995). Prion protein gene variation among primates. J Mol Biol 245: 362-374.

Scott M, Foster D, Mirenda C, Serban D, Coufal F, Wälchli M, Torchia M, Groth D, Carlson G, DeArmond SJ, Westaway D, Prusiner SB (1989). Transgenic mice expressing hamster prion protein produce species-specific scrapie infectivity and amyloid plaques. Cell 59: 847-857.

Sigurdsson B (1954). Rida, a chronic encephalitis of sheep with general remarks on infections which develop slowly and some of their special characteristics Br Vet J 110: 341 -354.

Stahl N, Baldwin MA, Teplow DB, Hood L, Gibson BW, Burlingame AL, Prusiner SB (1993). Structural analysis of the scrapie prion protein using mass spectrometry and amino acid sequencing. Biochemistry 32: 1991-2002.

Tagliavini F, Prelli F, Ghisto J, Bugiani O, Serban D, Prusiner SB, Farlow MR, Ghetti B, Frangione B (1991). Amyloid protein of Gerstmann-Sträussler-Scheinker disease (Indiana kindred) is an 11-kd fragment of prion protein with an N-terminal glycine at codon 58. EMBO J 10: 513-519.

Taraboulos A, Raeber AJ, Borchelt DR, Serban D, Prusiner SB (1992). Synthesis and trafficking of prion proteins in cultured cells. Mol Biol Cell 3: 851-863.

Tateishi J, Kitamoto T (1995). Inherited prion diseases and transmission to rodents. Brain Pathol 5: 53-59.

Tateishi J, Kitamoto T, Doh-ura K, Sakaki Y, Steinmetz G, Trenchant C, Warter JM, Heldt N (1990). Immunochemical, molecular genetic, and transmission studies on a case of Gerstmann-Sträussler-Scheinker syndrome. Neurology 40: 1578-1581.

Telling GC, Scott M, Hsiao KK, Foster D, Yang S-L, Torchia M, Sidle KCL, Collinge J, DeArmond SJ, Prusiner SB (1994). Transmission of Creutzfeldt-Jakob disease from humans to transgenic mice expressing chimeric human-mouse prion protein. Proc Natl Acad Sci USA 91: 9936-9940.

Telling GC, Scott M, Mastrianni J, Gabizon R, Torchia M, Cohen FE, DeArmond SJ, Prusiner SB (1995). Prion propagation in mice expressing human and chimeric PrP transgenes implicates the interaction of cellular PrP with another protein. Cell 83: 79-90.

Telling GC, Haga T, Torchia M, Tremblay P, DeArmond SJ, Prusiner SB (1996). Interactions between wild-type and mutant prion proteins modulate neurodegeneration in transgenic mice. Gene Dev (in press).

Vnencak-Jones CL, Phillips JA (1992). Identification of heterogeneous PrP gene deletions in controls by detection of allele-specific heteroduplexes (DASH). Am J Hum Genet 50: 871-872.

Wells GAH, Wilesmith JW (1995). The neuropathology and epidemiology of bovine spongiform encephalopathy. Brain Pathol 5: 91-103.

Westaway D, Cooper C, Turner S, Da Costa M, Carlson GA, Prusiner SB (1994a). Structure and polymorphism of the mouse prion protein gene. Proc Natl Acad Sci USA 91: 6418-6422.

Westaway D, DeArmond SJ, Cayetano-Canlas J, Groth D, Foster D, Yang S-L, Torchia M, Carlson GA, Prusiner SB (1994b). Degeneration of skeletal muscle peripheral nerves, and the central nervous system in transgenic mice overexpressing wild-type prion proteins. Cell 76: 117-129.

Westaway D, Mirenda CA, Foster D, Zebarjadian Y, Scott M, Torchia M, Yang S-L, Serban H, DeArmond SJ, Ebeling C, Prusiner SB, Carlson GA (1991). Paradoxical shortening of scrapie incubation times by expression of prion protein transgenes derived from long incubation period mice. Neuron 7: 59-68.

Wickner RB (1994). [URE3] as an altered URE2 protein: evidence for a prion analog in Saccharomyces cerevisiae. Science 264: 566-569.

Zhang H, Kaneko K, Nguyen JT, Livshits TL, Baldwin MA, Cohen FE, James TL, Prusiner SB (1995). Conformational transitions in peptides containing two putative α-helices of the prion protein. J Mol Biol 250: 514-526.

Epilogue

*"...Do not let yourselves be tainted by a deprecating and barren skepticism,
do not let yourselves be discouraged....Live in the serene peace
of laboratories and libraries....until the time comes when you may have
the immense happiness of thinking that you have contributed
in some way to the progress and good of humanity..."*
J. Lister, in a speech to students of the Sorbonne

PAUL DE KRUIF—*Microbe Hunters*

OVERVIEW AND OUTLOOK FOR THE FUTURE

It has become clear that we deal with well-known viruses and bacteria on the one hand, while on the other, we are faced with new, previously unknown infectious agents. Many infectious diseases have persisted and display a remarkable ability to re-emerge after lengthy periods of stability, demonstrating the cyclical nature of disease trends (Satcher 1995). Demographic factors together with the ongoing evolution of viral and microbial variants and selection for drug resistance ensure that infectious diseases will continue to exist and/or to emerge and possibly even increase in the immediate future. Increased contacts with animals in areas so far hardly explored, such as Amazonia, may contribute to the situation. On rare occasions, a new variant may evolve and cause a new disease, as was the case with HIV or a new morbillivirus resulting in a fatal disease in horses and humans (Murray et al 1995). Whether or not new and quite unusual infectious agents such as prions will be found in the future remains a matter of speculation. Table 1 lists some of the etiologic agents and infectious diseases identified since 1973. At any rate, new infectious diseases will continue to be discovered and new agents will be identified, particularly with the technique of nucleic acid amplification. All this is rather predictable and microbe hunters will be kept as busy in the future as they were in the past. The new nucleic acid techniques have enabled modern microbe hunters not only to identify new agents, but also to look for the genes that render an agent pathogenic. Unfortunately, it turned out that, with very few exceptions, the pathogenicity of bacteria or viruses is not caused by single genes but by a concerted action of many genes. In Yersinia, for example, a single plasmid encodes 11 different proteins, all of which are important pathogenic factors, and *V. cholerae* produces not only one but several toxins (Heesemann 1995). So far, the hunt for pathogenic factors of a given agent has not been as successful in providing new means of prevention or therapy as one might have hoped some years ago.

TABLE 1. MAJOR ETIOLOGIC AGENTS of infectious diseases identified since 1973*

Year	Agent	Disease
1973	Rotavirus	Major cause of infantile diarrhea worldwide
1975	Parvovirus B19	Fifth disease; aplastic crisis in chronic hemolytic anemia
1976	*Cryptosporidium parvum*	Acute enterocolitis
1977	Ebola virus	Ebola hemorrhagic fever
1977	*Legionella pneumophila*	Legionnaires' disease
1977	Hantaan virus	Hemorrhagic fever with renal syndrome (HFRS)
1977	*Campylobacter* sp.	Enteric pathogens distributed globally
1980	Human T-cell lympho-tropic virus I (HTLV I)	T-cell lymphoma/leukemia
1981	*Staphylococcus* toxin	Toxic shock syndrome assoc. with tampon use
1982	*Escherichia coli* O157:H7	Hemorrhagic colitis; hemolytic uremic syndrome
1982	HTLV II	Hairy cell leukemia
1982	*Borrelia burgdorferi*	Lyme disease
1983	Human immunodeficiency virus (HIV)	Acquired immunodeficiency syndrome (AIDS)
1983	*Helicobacter pylori*	Gastric ulcers
1988	Human herpesvirus-6 (HHV-6)	Roseola subitum
1989	*Ehrlichia chaffeensis*	Human ehrlichiosis
1989	Hepatitis C	Parenterally transmitted non-A, non-B hepatitis
1991	Guanarito virus	Venezuelan hemorrhagic fever
1992	*Vibrio cholerae* O139	New strain associated with epidemic cholera
1992	*Bartonella (Rochalimaea) henselae*	Cat-scratch disease; bacillary angiomatosis
1993	Hantavirus isolates	Hantavirus pulmonary syndrome
1994	Sabiá virus	Brazilian hemorrhagic fever

*Compiled by CDC staff. Dates of discovery are assigned on the basis of the year in which the isolation or identification of etiologic agents was reported. Source: D. Satcher (1995)

However, the outlook for the future yields a more rewarding picture if we consider several new and potentially very important aspects emerging from the search for pathogenicity factors, both with respect to viruses and bacteria. Recent investigations revealed that some open reading frames of viruses, such as of Epstein-Barr virus (EBV) and myxoma viruses, encode proteins with homology to known immunoregulatory molecules or to their receptors. An EBV gene gives rise to the production of IL-10, a cytokine which is usually produced by T-helper cells. IL-10 inhibits the synthesis of interferon-γ and IL-2, and the viral equivalent thus helps the virus to escape the antiviral effects of interferon and therefore must be regarded as a pathogenicity factor. Similarly, the abundant production of a viral equivalent of interferon receptor molecules released by myxoma virus-infected cells is likely to capture interferon, again preventing its antiviral effects. There are additional and different virus-encoded molecules which are to be regarded as immune modulators (Spriggs 1994).

The search for pathogenicity factors of intracellular bacteria has taught us a great deal about the biology of the host cells. Also, the identification of genes of pathogenic factors of extracellular bacteria, and possibly of viruses, will in the not too distant future provide us with quite novel means of prophylactic vaccination against bacteria and viruses.

Of course, immune prophylaxis has already turned out to be the only really successful method in controlling virus diseases since chemotherapy was, with very few exceptions, rather a disappointment. This is due not only to the intimate relationship of virus and host cell during virus replication, which makes a decent therapeutic index so difficult, but more importantly, due to the differential sensitivity of viruses to different chemicals, demanding a great number of different drugs; in all cases, drug-resistant variants are selected within a relatively short time. Thus I personally do not believe in the chemotherapy of virus diseases as a meaningful tool of the future.

As an immunologist, I even believe that, as was the case with viruses, prophylactic immunization will also turn out to be one of the most powerful methods of the future in coping with bacterial infections. It is clear that we can no longer regard both viruses and bacteria simply as pathogens that cause diseases but must view them increasingly as novel instruments in the control of infectious as well as of other diseases. This potential role has already been experimentally investigated in the case of viruses. Viruses suitably altered by gene technology can be used as specific vehicles to transport new genes into specific cells. One example, recently reported by Jeffrey Leiden of Chicago, is the use of adenoviruses containing the suppressor *Rb* gene of reticuloblastoma viruses during balloon dilatation of coronary vessels. One of the complications following dilatation is the growth of smooth muscle cells which obstruct the arteries again in about 40% of patients treated. So far only in animal experiments, the recombinant adenoviruses have been successfully applied locally to infect smooth muscle cells in the immediate environment of dilatation. There they produce, under the control of smooth muscle cell specific promoters, the Rb gene product which arrests the muscle cells in Go phase, thus preventing their outgrowth. It is quite obvious that recombinant viruses are the most promising and specific vectors for gene therapy in general and it can be expected that viruses in addition to adeno- and herpesviruses, for example, will turn out to be suitable candidates for just this purpose.

How about recombinant bacteria as novel tools in modern medicine? Some bacteria, such as salmonellae and Listeriae, grow intracellularly. Bacterial vaccines can be visualized containing the genes of distinct pathogenicity factors such as exotoxins, lacking the active center. Bacteria can also be altered in such a way as to home to and remain in specific tissues such as Peyer's patches. Attenuated salmonellae can be tailored to express a recombinant gene either intraplasmatically or in the cell membrane, or to secrete the gene product (Goebel 1995). These attenuated bacteria no longer replicate in the cell, but they persist at least for several days and produce the recombinant gene products. Any bacterial protein, provided it is an important pathogenic factor and a strong antigen, can thus be expressed by host cells in a suitably immunogenic way and can give rise to either a humoral or a cellular immune response. Obviously, antibodies produced by such a prophylactic immunization will be able to cope with infections by extracellular bacteria and their exotoxins, whereas intracellular bacteria require the induction of specific cytotoxic T cells. In view of the constant emergence of bacterial variants resistant to ever new antibiotics, the possibility of prophylactic vaccinations using these modern techniques is indeed likely to develop into a potentially very powerful instrument in the hands of the microbe hunters of the future. This method may even be used for immunization against virus antigens and one can visualize that such manipulated salmonellae can be made to express not only one but several antigens at the same time. Another great advantage of these antigen-producing bacteria would be their use as oral vaccines that can be easily and cheaply applied. However, as is the case with most recombinant viruses for gene therapy, the immunization experiments with recombinant salmonellae have so far been conducted only in animal models. One reason for this limitation is the hesitation to spread recombinant microbes in the environment and the corresponding legal obstacles which, particularly in Germany, are quite forbidding.

Undoubtedly, there are problems that must be addressed before viruses and bacteria are tamed to become drugs of the future. However, I believe that this goal will be reached in the near future and that we will come to regard these microbes as friends, not only as foes.

Eberhard Wecker
Würzberg, Germany

REFERENCES

Goebel W (1995). Personal communication.

Heesemann J (1995). Personal communication.

Murray K, Rogers R, Selvey L, Selleck P, Hyatt A, Gould A, Gleeson L, Hooper P, Westbury H (1995). A novel morbillivirus pneumonia of horses and its transmission to humans. Emerging Infect Dis 1: 31-33.

Satcher D (1995). Emerging infections: Getting ahead of the curve. Emerging Infect Dis 1: 1-6.

Spriggs MK (1994). Current Opinion Immunol. Alt F and Marrack Ph, eds. Vol 6, 526-529.

Index